ARCHITECTURE CREATES CITIES. CITIES CREATE ARCHITECTURE.

:城市创造

SHENZHEN BIENNALE
OF URBANISM\ARCHITECTURE
URBANIZING\COMMITTEE
深圳市\建筑双年展组委会 编

图书在版编目（CIP）数据

城市创造 "2011深圳·香港城市\建筑双城双年展" / 深圳城市\建筑双年展组织委员会编.—北京：中国建筑工业出版社，2014.1
ISBN 978-7-112-16068-6

Ⅰ.①城… Ⅱ.①深… Ⅲ.①城市建筑—建筑设计—作品集—中国 Ⅳ.①TU984

中国版本图书馆CIP数据核字（2013）第261369号

责任编辑：吴　绫　李东禧
责任校对：姜小莲　赵　颖
装帧设计：唐天辰

城市创造 "2011深圳·香港城市\建筑双城双年展"
深圳城市\建筑双年展组织委员会　编

中国建筑工业出版社出版、发行（北京西郊百万庄）
各地新华书店、建筑书店经销
精一印刷（深圳）有限公司印刷

开本：889×1194毫米　1/16　印张：34¼　字数：907千字
2014年2月第一版　2014年2月第一次印刷
定价：388.00元
ISBN 978-7-112-16068-6
　　　　（24808）
版权所有　翻印必究
如有印装质量问题，可寄本社退换
（邮政编码　100037）

OPENING PROGRAM: DEC. 7TH- 11TH, 2011 开幕活动：2011年12月7日 – 11日
OPENING CEREMONY: DEC. 8TH, 2011 开幕典礼：2011年12月8日
DURATION: DEC. 7TH, 2011 – FEB. 19TH, 2012 展期：2011年12月7日 – 2012年2月19日
VENUE: SHENZHEN CIVIC SQUARE, OCT-LOFT 展场：深圳市民广场、华侨城创意文化园
CHIEF CURATOR: TERENCE RILEY 总策展人：泰伦斯·瑞莱

GREETING, SHENZHEN BIENNALE OF URBANISM\ARCHITECTURE ORGANIZING COMMITTEE

深圳城市\建筑双年展组织委员会致辞

GREETING, SHENZHEN BIENNALE OF URBANISM\ARCHITECTURE ORGANIZING COMMITTEE

Architecture creates cities. Cities create architecture.

The Shenzhen & Hong Kong Bi-City Biennale of Urbanism/Architecture (SZHKB) is the only biennale whose fixed theme for years has been "Urban and Urbanization", putting it in a league of its own and bringing it to the forefront among the large number of biennales. The 4th SZHKB presents the theme "Architecture creates cities. Cities create architecture," and furthers the discussion about how to create cities. This discussion is closely interrelated with China, especially with such characteristics of the country as the high-speed urbanization taking place within the Pearl River Delta region, and it provides us with a platform for re-thinking and re-examining these busy and noisy cities.

This SZHKB focuses on the following goals: to give attention to several specific fields such as urban planning, urban design, architecture, urban public space, and urban public art; to discover, observe, discuss, and study by means of visual culture those urban issues found in today's society; to gather city-related ideas and wisdom that will contribute creative proposals to solving urban problems. At the same time, the Biennale acts as an urban cultural event driven by public participation and professional communication.

Against the background of globalization, SZHKB is concerned with common as well as specific issues in the development of international cities. The 2011 SZHKB coincided with the 30th Anniversary of the Establishment of the Shenzhen Special Zone, a historical moment with special significance, a fresh starting point for the urban development of Shenzhen over the following thirty years. Within the context of new concepts in urban development from "Shenzhen Speed" and "Shenzhen Manufacture" to "Shenzhen Quality" and "Shenzhen Innovation", the 2011 SZHKB placed more focus on local problems in Shenzhen's urban development that fit within a larger international dialogue. A goal of the Biennale is to absorb more creative inspiration and obtain new energy for the innovative development of this city. This Biennale will explore several urban issues in Shenzhen, and even greater China, within the context of globalization. It simultaneously introduces global experiences to China and compares them to the situation there. This will bring about a very creative and provocative discussion.

While considering localized issues, the Bi-City Biennale is also endowed with an international vision and curatorial team. Unlike with previous editions of the SZHKB, for the 2011 Biennale, Terence Riley was invited to act as chief curator. Riley is an architect, critic, museum expert, and curator for world-renowned architecture exhibitions. He proposed "Architecture creates cities. Cities create architecture" as a theme that would expound upon the infinite relationships that exist between architecture and cities, and he inspired provocative discussions about sustainable development and the vitality of the city. In an era of innovative culture and a knowledge-based economy, how can a city draw the world's attention? The answer is through its creativity and culture. It is not only the culture of the city's past that is important, but also its most current creative culture, and only ceaseless creation of culture can ensure a future for the city.

Based on the success of the last three biennales in 2005, 2007, and 2009, this year, in service to the goal of "two cities, one theme, one exhibition", we collaborated closely with Hong Kong, focused on common urban problems, and curated a series of academic activities and exhibitions so that the Bi-City Biennale could become one of the year's most prominent cultural events in both Shenzhen and Hong Kong.

This year, about 60 works will be exhibited in the main venue (including invitational exhibitions by five countries), a special topic exhibition, and the fifteen satellite exhibitions also presented to the public. We believe the 2011 SZHKB not only brought to Shenzhen a brand-new international exchange and cultural dissemination, but also broadened the horizon of the Biennale itself, helping us to achieve our goal of becoming a world-renowned biennale.

Shenzhen Biennale of Urbanism\Architecture Organizing Committee
December 2011

深圳城市\建筑双年展组织委员会致辞

Member of the Standing Committee of Guangdong Province, secretary of municipal Party committee of Shenzhen, Mr. Wang Rong, to declare the opening of this biennale
广东省委常委、深圳市委书记王荣宣布本届双年展开幕

Mr. Lv Ruifeng; Mr. LI Huanan; Ms. Yan Xiaopei and Mr. Huang Zhongwei came to the stage and present the souvenirs special made for these four Curators.
开幕式上,深圳市常务副市长吕锐锋(左一)、市委秘书长李华楠(右四)、市人大常委会副主任闫小培(左三)、市政协副主席黄中伟(右三)、向四位总策展人及代表张永和(左四)、马清运(代表,右一)、欧宁(右二)、泰伦斯·瑞莱(左二)赠送纪念品。

城市创造

深圳·香港城市\建筑双城双年展是唯一一个以"城市或城市化"作为固定主题的双年展,这使其在双年展的庞大家族中别具一格,脱颖而出。迄今已经第四届的深圳双年展,提出"城市创造"主题,继续着如何创造城市的讨论。这一讨论与中国,尤其是高速城市化的珠三角地域特征密切相关,为永远处于忙碌、急躁的城市提供了再思和反思的平台。

展览关注城市规划与城市设计、建筑、公共空间与公共艺术等各个具体领域,注重用视觉文化的方式,观察、发现、探讨当下的城市问题,积聚与城市相关的思想和智慧,对解决城市问题贡献各种富有创造性的方案。展览同时也成为公众参与、专业交流的城市文化盛事。

双城双年展不仅关注全球化背景中国际城市的共同性问题,它还十分关注国际城市发展中具有地域性的特殊问题。本届深圳·香港城市\建筑双城双年展恰逢"深圳特区成立30年"这一有特殊意义的历史节点,这是深圳未来30年城市发展的新的起点。展览把深圳放在全球6个不到60岁的新城市中,进行对比展示,在从"深圳速度"、"深圳制造"向"深圳质量"、"深圳创造"的城市转型、发展的语境下,将以更加注重深圳本土色彩的城市化问题参与国际对话,希望在全球性和地域性的互动中,获得更多的创新性的灵感,为城市的创新发展获取新的能量。双年展研讨深圳甚至大中华地区的城市案例,并将之扩大到全球化语境中,在引进全球经验的同时,又与"全球经验"相比较。这将是一个极富创意性的深度论题。

与双城双年展的本土化诉求相映成趣的,是它更加国际化的视野和更加国际化的策展团队。与往届双城双年展不同的是,2011深圳·香港双城双年展邀请国际著名建筑主题展策展人、建筑师、博物馆专家及评论家泰伦斯·瑞莱(Terence Riley)先生担纲策展工作,他提出了"城市创造"的策展主题,这一主题阐释了建筑与城市之间无止境的互动关系,并激起对城市可持续发展和城市生命力的深度讨论。在创意文化和知识经济时代,一个城市怎么最能吸引全球的目光?是它的文化和创造力。这种文化不仅是它过去的文化,更是它当下能与时俱进的创新性文化,唯有永不止息的文化创造,才有一个城市的未来。

继2005、2007、2009三届展览成功地举办的基础上,本届展览继续与香港紧密合作,双方朝着"两个城市、一个主题,一个展览"的目标,关注共同的城市问题,携手策划系列学术活动与精彩展览,让双城双年展成为深港两地的年度性文化事件。

本届双年展将有约60件作品参加主展场展出(包括5个国家邀请展),除此以外,还有1个深圳大运会专题展特别项目、15个外围展,我们相信本届双年展将给深圳带来一场全新的国际交流活动和文化品牌传播的同时,也将进一步开阔深港双城双年展展览本身的国际化视角,使双城双年展逐步跻身于全球知名双年展行列。

深圳城市\建筑双年展组织委员会
2011年12月

CONTENTS / 目录

	GREETING, SHENZHEN BIENNALE OF URBANISM\ARCHITECTURE ORGANIZING COMMITTEE 深圳城市\建筑双年展组织委员会致辞	7
	INTRODUCTION / 绪言	15
SHENZHEN **深圳**	10,000 FLOWER MAZE / 万花阵	39
	ULTRA-LIGHT VILLAGE / 超轻村	47
	SHENZHEN BUILDS / 深圳建造	87
	BOOM! SHENZHEN / 轰隆！深圳	121
SHENZHEN AND HONG KONG **深圳和香港**	COUNTERPART CITIES / 对应双城	133
	SHENZHEN: BECOMING A CITY ... FUTURE METROPOLIS OF THE WORLD [Forum Transcript] 深圳：成为一座城……世界未来都市（论坛讲稿整理）	170
SHENZHEN AND CHINA **深圳和中国**	8 URBAN PLANS FOR CHINA / 八个城市项目	187
	CHINESE CITIES IN TWO VIEWS / 双城记	227
	URBAN CHINA / INFORMAL CHINA / 城市中国 / 自发中国	233
	10 MILLION UNITS: HOUSING AN AFFORDABLE CITY / 广厦千万·居者之城	257
	2011 REBIRTH BRICK DEVELOPMENT / 再生砖进展 2011	271

GLOBAL PERSPECTIVES
全球视野

6 UNDER 60 / 6 小于 60	281
AND THEN, IT BECAME A CITY: SIX CITIES UNDER 60 然后，它成了一座城：未满 60 岁的 6 座城市	311
ROUNDTABLE: CITIES UNDER 60 [Forum Transcript] / 圆桌会议：6 小于 60（论坛讲稿整理）	332
GO WEST PROJECT: ALLMETRO POLIS / 西游项目：所有都市	341
THE STREET / 街道	363
"STRADA NOVISSIMA" AND "THE STREET" / "新趋势"和"街道"	418
STREET THEATRE / 街道剧场	424
THE PRESENCE OF THE PAST REVISITED / 再造访"过往的呈现"	427
FAVELA PAINTING / 彩绘都市	439
GHANA THINKTANK: DEVELOPING THE FIRST WORLD 加纳智库：发展第一世界	445

INTERNATIONAL PARTICIPATION
国际参与

HOUSING IN VIENNA / 维也纳住房	461
RECLAIM, THE KINGDOM OF BAHRAIN / 再生 – 巴林王国	465
GIMME SHELTER! CHILEIN / 给我避难所！智利馆	471
NEWLY DRAWN - EMERGING FINNISH ARCHITECTS 新绘图——芬兰新生代建筑师	477
SOLUTION FINLAND: THE WELFARE GAME / 芬兰方案：福利博弈	483
HOUSING WITH A MISSION, DUTCH AND CHINESE ARCHITECTS' DESIGNS FOR THE ANTS TRIBE, THE NETHERLANDS 住宅的使命，荷兰与中国建筑师为蚁族而设计，荷兰	487

A CATALYST REACTION OF OUR CITY: "SHENZHEN AND UNIVERSIDE" SPECIAL EXHIBITION 城市触媒：大运与深圳专题展	501
EXHIBITIONS OF PUBLIC WELFARE OF LEADING SPONSORS / 赞助企业公益文化展	507
SATELLTE EXHIBITIONS AND EVENTS / 外围展及活动	518
EPILOGUE / 结语	534
AWARDS LIST / 获奖名单	538

INTRODUCTION
绪言

by Terence Riley

INTRODUCTION

The Theme

The theme for the 2011-2012 edition of the Shenzhen & Hong Kong Bi-City Biennale of Urbanism\ Architecture is rooted in the very conception of the event, which first appeared in 2005. Uniquely, this biennial exhibition features not only architecture but urbanism, which is rarely addressed in such events. The combination is immensely appropriate, as these two human activities are inseparable even if they are often treated as separate disciplines. Throughout history, they have been totally interdependent in the transformation and growth of human settlements. Hence, the theme:

Architecture creates cities. Cities create architecture.

These simple, axiomatic statements can be made specific without losing their meaning: The Louvre creates Paris. Paris creates the Louvre. However, it is not only monumental architecture that creates cities. In the same way, the brownstone creates Manhattan and the machiya creates Kyoto, and vice versa.

The theme represents a relationship between cities and architecture that can be seen as global, especially in terms of contemporary culture. However, the SZHKB looks at cities and architecture in the context of time as well as place. Looking at history, it is clear that every city does not continue to create architecture indefinitely. The reasons for the decline of any one city or cities and their corresponding culture—architectural and otherwise—are usually complex.

However, Jared Diamond—a scientist whose studies have included many different fields from ornithology to geography—has demonstrated that the decline of a great many cultures was preceded by a period of systemic disregard for their environment. As such, the Biennale's theme could be modified: Architecture creates cities and cities create architecture as long as they understand and respect the environmental conditions that support them.

Sustainability + Vitality

The meaning of the word "sustainable" has become distorted in recent years by its usage as a label for selling so-called green products, whether toothpaste, automobiles or office buildings. The ability for our global commercial culture to trivialize such matters may well be a reoccurrence of the attitudes of the disappeared cultures studied by Jared Diamond - the Norse and Inuit of Greenland, the Maya, the Anasazi, the indigenous people of Easter Island, and others.

Developing sustainable practices with regards to energy usage, waste management, and air and water quality are, without a doubt, the greatest challenges to cities and architecture today. Yet, for architects and planners, efforts to make cities and architecture more sustainable must be matched by efforts to make them more vital culturally.

Just as the word "green" has become almost banal through misuse, the phrase "quality of life" has become a pale reflection of what might be called true urban vitality. That vitality is impossible to quantify and only slightly less difficult to define positively. Suffice it to say that, in the context of this Biennale, it does not refer to the size of apartments, the number of cars, or the scale of shopping malls. If we don't readily understand the idea of urban vitality, perhaps that is part of a larger common problem.

As such, the full significance of the Shenzhen Biennale's theme can only be understood in its graphic representation. While the theme itself is rather simply stated, the intended implication of "Architecture creates cities—Cities create architecture" is more complex. Expressed graphically by the Guangzhou design firm wx-design, the two sentences are strung together like a Möbius strip, without a beginning or an end. The implied endlessness intentionally raises the issue of sustainability. The DNA-like appearance suggests the vital qualities of cities and architecture. In this way, it takes two simple sentences and makes them an aspiration rather than an observation.

Shenzhen

Appropriately, there is a definite focus in the Shenzhen Biennale on the city of Shenzhen itself, one of the world's transformative cities. Recalling a series of shows in the 1950's and 1960's at the Museum of Modern Art in New York, an exhibition titled Shenzhen Builds focuses on five major architectural works with significant urban implications. Works by Atelier FCJZ (Shenzhen TV Tower), Coop Himmelb(l)au (The Museum of Contemporary Art and Planning Exhibition), Massimiliano and Doriana Fuksas (Terminal 3, Shenzhen Bao'an International Airport), OMA/Rem Koolhaas (the Shenzhen Stock Exchange), and Urbanus (Shenzhen Bay Metro Plaza) are all presented in detail, highlighting their urban genesis and development. Architecture creates Shenzhen. Shenzhen creates architecture. While the goals of Shenzhen Builds had been presented to the architects, how they each decided to present their work was, in a sense, yet another design problem.

As each architect focused on their work and how it creates the city, another exhibit provides a historical context, addressing broader aspects of Shenzhen's urban culture over the past decades. Dr. Mary Ann O'Donnell, a research associate at the College of Architecture, Shenzhen University, curated an installation entitled Boom! Shenzhen, which gives playful visual expression to the boring statistics that describe Shenzhen's explosive growth over the last decades in a graphic and three-

INTRODUCTION

dimensional manner. Topics addressed include the city's Gross Domestic Product, the cost of housing, the intensity of investment, the geological effects of urbanization, and the city's illusive and disappearing history.

Shenzhen's Civic Center, with its over-scaled and underused plaza, is a principal focus of the Shenzhen Biennale. The plaza is the site of a transformative installation by New York architects John Bennett and Gustavo Bonevardi. Their installation was a maze-like assemblage of thousands of orange traffic cones, the ubiquitous emblem of new building or road construction worldwide. Titled 10,000 Flower Maze, the installation was inspired by the labyrinthine garden of the same name that was designed by the Italian architect Giuseppe Castiglione for the Old Summer Palace just outside the Forbidden City in 1756.

During the opening ceremonies of the Biennale, the maze was taken over by dozens of Shenzhen teenagers on roller skates. In an informal, free form choreographic performance, the volunteer performers skated between, over and around the cones. The largest and most ceremonial of public spaces was thus transformed by a small common object and a typical urban activity that can be found on any street.

The upper plaza of the Civic Center is also a large space, in essence one of the stages upon which Shenzhen's urban theater unfolds. Another project—Ultra—Light-Village—consists of six structures constructed along the axis that connects the main plaza and the Lianhua Mountain Park to the north, passing through the Government Center. These structures were built not on terra firma but on what is essentially the top surface of the structure below. Hence they necessarily needed to be lightweight, which in the context of this Shenzhen Biennale is a virtue, not a limitation. The renowned engineer and philosopher, R. Buckminster Fuller, was an early proponent of the thrifty use of resources in building construction and would provocatively ask his colleagues, "How much does your building weigh?"

The architects chosen to design structures for the Ultra—Light-Village have all demonstrated a clear attitude toward the relationship between conceptualization and building, and how full-scale construction can be used to convey theoretical concepts in a more expressive manner. They are Amateur Architecture Studio (Hangzhou, China), Clavel Arquitectos (Murcia, Spain), MOS (New York, USA), OBRA (New York, USA), Studio UP (Zagreb, Croatia), and Regional Construction Studio led by Wei Chunyu (Changsha, China).

Shenzhen and Hong Kong

While the Bi-City Biennale consists of two coordinated but separated events, it would be a mistake to pretend that the cities of Shenzhen and Hong Kong can be thought of distinct autonomous entities. The two cities do have different histories, governance and currencies. However, they are part of the same ecosphere, one that is dominated by the outflows of the Pearl River into the South China Sea. A key component of the SZHKB is the exhibition Counterpart Cities, which takes as its starting point the fact that water respects no political boundaries.

Based on an exhibition concept developed by Barry Bergdoll, Philip Johnson chief curator for Architecture and Design at the Museum of Modern Art in New York, the project focuses on three sites that all face environmental challenges over the next decades due to both rising sea levels as well as water table depletion. Jonathan Solomon and Dorothy Tang, both professors at Hong Kong University, have adapted the MoMA model to address specific issues relevant to the region and took responsibility for the overall organizational tasks related to the project. Six teams—three from Shenzhen and three from Hong Kong—were assembled to research and provide design solutions with technical support provided by Ricky Tsui and Iris Hwang, of ARUP.

The three Hong Kong team leaders that were selected are Stefan Al, Tom Verebes, and Vincci Mak, and three team leaders from Shenzhen include Doreen Liu, Feng Guochuan, and Zhu Xiongyi.

Shenzhen and China

In addition to the projects focused on Shenzhen and Hong Kong, three projects address critical issues in urbanism in China. It is commonplace now to refer to China as the urban laboratory of the world with its 12 cities of over 5 million population. Curated by Jeffrey Johnson, director of China Lab at Columbia University, and Li Xiangning, associate professor at Tongji University, the exhibition 8 Urban Projects presents a critical view of contemporary urban projects for Shenzhen and from across China. With designs by both domestic and international studios, the exhibition illustrates how China's unique conditions and challenges have generated new urban modalities. Each project has been selected based on a set of critical issues or themes that define the urban project in China today, such as community, lifestyle, identity, harmony, ecology, economy, temporality, and historic etc.. preservation. Additionally, the projects selected are based on their location within a specific spatial condition that characterizes the unique contemporary urban terrain that has emerged over the past thirty years, including the peri-urban, suburbia, satellite cities, the infrastructural and networked, and the existing urban core.

The featured projects and their lead designers are Qianhai (Shenzhen), James Corner, Field Operations; Jiading Ad Base (Shanghai) Yung Ho Chang/Atelier FCJZ; Rockbund (Shanghai), David Chipperfield Architects; Hangzhou Shan-Shui (Hangzhou), Steven Holl Architects; Shenzhen Eye (Shenzhen), Urbanus and OMA; Ordos20+10 (Ordos), Fang Zhenning and thirty architectural design

INTRODUCTION

firms; XiXi Wetlands (Hangzhou), WSP+ 10 architects and Woods Bagot Asia; and Zhongshan New Information Industry District (Zhongshan), Wu Zhiqiang/Shanghai Tongji Urban Planning & Design Institute.

As with Shenzhen Builds, contemporary urban planning efforts require substantial contextual treatment. While many biennial exhibitions are often characterized by a kind of historical amnesia, 8 Urban Projects is complimented by another installation, Chinese Cities in Two Views. In this visual essay, Dr. Tang Keyang, deputy director at the Center for Visual Studies, Peking University, presents the history of Chinese urbanism in two correlated approaches. In one view, a city might be the sum of all its historical fragments and is often represented by its culminating stage. Observers of such cities usually turn to general typological principles that generate "identikit" of them. In another view, Chinese cities are constantly changing entities with specific causes for their transformation. For such cities the purpose of our show is not only to provide established perspectives of a city but also to examine how it was transformed through time.

The exhibition Urban China / Informal China,[1] curated by Jiang Jun, former editor of Urban China Magazine and founder of Underline Office, and Su Yunsheng, urban planner at Tongji University, presents another overview of Chinese urbanism. Conceived of as a magazine "remixed" in the form of wallpaper, the 35-meters long installation offers a concise narrative of the history of China's urbanization, presented as an ongoing struggle between systems of control and laissez-faire, literally "let it be." The text is primarily two colors—with red representing formal, ordered, or planned governmental decisions, and blue representing informal, organic, or ad-hoc reactions to policies or events. In addition to color distinctions, the wallpaper is organized in three sections from top to bottom, with the top emphasizing national beliefs and policies, the middle highlighting how that impacts city architecture, and the bottom showing the effect on the families and its objects. The numbers correspond to key historical events or ideas that have and continue to shape policies and planning; economic growth; architecture; and unregulated, informal transformation. Informal systems—spatial, economic, and utilitarian—showed their abilities in subverting the highly structured nature of planned Chinese.

China's rapid economic and urban growth is mirrored in the rapid increase in the need for low cost housing as millions of people immigrate to the nation's cities from the countryside. To address the inadequacies of affordable housing provisions in the current market-based real estate development, China is launching a large-scale social housing initiative the central government has completed a planning directive for 10 million units of affordable housing to be distributed throughout the country within the year 2011. Curated by Juan Du, assistant professor and director of the Masters of Architecture program at the University of Hong Kong, 10 Million Units: Housing an Affordable City, brings together government, enterprises, scholars, architects, planners, engineers, developers, and the public to examine the challenges and opportunities of providing low and mid-income housing in the dense environment of the contemporary city. The exhibition presented winning proposals of Shenzhen's "1 Unit -100 Families-10000 Residents Affordable Housing Design Competition". The exhibition also showcased innovative designs and research into the current issues of affordable housing, spanning different stages and scales of intervention ranging from construction details to national policy.

The exhibit is divided in various sections, which were curated by Du Juan and presented with her collaborators:PEOPLE (Bai Xiaoci), IDEAS (Winners of the 1.100.10000 Competition), DETAIL (MIT 10K House Studio), UNIT (Urbanus), BUILDING (Standard architecture), CITY (HKU Urban Ecologies Studio), COUNTRY (Urban China Research Center), CONSTRUCTION (Zhuoyue Group + Xiepeng Design) and PREFABRICATION (M3house + UAO Creations).

The Rebirth Brick program was started by Chengdu architect Liu Jiakun in July 2008, following the major earthquake that destroyed much of the city of Wenchuan, in Sichuan province. The project was originally initiated to help local people conduct self-help production and reconstruction work. The basic idea of the rebirth brick is: taking the fractured ruins materials as aggregate, blending the straw as fiber, adding cement, etc., to make lightweight bricks at local factories for the purpose of reconstruction work in the affected area. The Rebirth Brick represents not only the "regeneration" of waste materials, but also the mental and emotional "regeneration" of post-disaster reconstruction.

To highlight both the physical as well as the poetic properties of the Rebirth Brick, the architect Liu Jiakun designed an installation that abstractly represented a rural Chinese house using the Rebirth Brick. The installation reminds us that post-disaster reconstruction involves not only rebuilding structures, but also rebuilding lives.

Global Perspectives

When we say "the world is getting smaller", it usually means that communication and transportation have shrunk what used to be considered great distances. However, if you consider sustainability, the world might be thought of as even smaller still. All of human settlements past and present lie within what might be called the "urbanosphere", the thin layer of the earth's habitable environment. With the towns near the Dead Sea being the earth's lowest settlements (425 meters below sea level) and La Rinoconado, Peru, being the earth's highest (5,100 meters above sea level), the urbanosphere can be thought of as an incredibly thin veneer of habitability covering the planet's surface.

INTRODUCTION

While Shenzhen shares an environmental link with Hong Kong, it is also part of a unique subset of cities around the globe, commonly referred to as "new cities"[2]. The creation of ex novo cities is without a doubt one of the most intensive of human efforts in terms of planning, cost, and physical effort. While new cities have been planned and built since antiquity, the scale of such projects in the second half of the 20th century is unprecedented. Given the equally enormous human investment in these projects, an important part of the SZHKB includes an exhibition and research project that looks retrospectively at the successes and failures of new cities around the world. The project's original working title—Cities < 60—underscores the fact that in all of these new cities, the city itself was younger than its older inhabitants.

With the participation of David van der Leer, assistant curator of Architecture and Urban Studies at the Guggenheim Museum, and Rochelle Steiner, dean of the Roski School of Fine Arts at the University of Southern California, the project evolved into two coordinated efforts, both focused on the cities of Shenzhen, Almere, Gaborone, Las Vegas, Brasilia, and Chandigarh. The first of these efforts is a research and exhibition project—6 Under 60—led by a multi-disciplinary team at the University of Southern California involving not only the School of Fine Arts, but School of Architecture (Ma Qingyun, dean, and Stefano di Martino Director of M. Arch Program) and School of Cinematic Arts (Scott Fisher, professor and chair, and Jennifer Stein, research associate). Using traditional research strategies as well as crowd-sourcing and other digital techniques for acquiring data and images, the research project traces each city's development from conception to realization through transformation, with a critical evaluation of both the intentions and the results. At the Biennale, this research is presented through an innovative installation of digital technologies, which allow for easy access through touch screens and group participation through large projections. Rather than a static presentation, the project continues to grow and develop through an online presence.

The second part of the project, titled "...and then it became a city: 6 Cities Under 60" is a more decidedly visual and sensory approach, which reflects the desire to understand the cities from a contemporary vantage point: the cities as they are, rather than as they were planned. Six videographers were selected by David van der Leer for their expertise and accomplishment as well as their proximity to the place and the culture of the cities in question. The participating videographers are: Surabhi Sharma (Chandigarh), Cao Guimarães (Brasilia), Sam Green (Las Vegas), Miki Redelinghuys (Gaborone), Astrid Bussink (Almere), and Wang Gongxin (Shenzhen). The videos are displayed alongside the USC data presentation, as well as on-line. They have also been presented in a large-scale outdoors format for the vernissage and on board an innovative education bus that travels through the city as a mobile classroom and urban laboratory.

The similarities and differences between the Chinese city of Shenzhen and the Dutch city of Almere, have been studied by others. As the installation Allmetropolis makes clear, both areas started their urban development in the late 1970s and both are located in a river delta, adjacent to major international metropolises; yet they have developed, in the view of the curators, in completely different ways. The exhibition was organized by the Go West Project (Michiel Hulshof and Daan Roggeveen) to put into perspective the economic, social, spatial and ecological developments of Shenzhen and Almere as a starting point of a research that shows the possibilities and opportunities of using the Shenzhen model in a European context.

As was the original Venice Biennale of 1895, the SZHKB is essentially an international cultural marketplace of ideas, a concept probably most clearly evident in the exhibition The Street. Arrayed in rows of six facing each other, the exhibition consists of 12 installations of work by architects or architectural teams from around the world, in the manner of the landmark exhibition of the 1980 Venice Biennale of Architecture, La Strada Novissima. While diverse in their approaches to architecture and urbanism, all of the participating architects are of the generation that came of age under the growing awareness that contemporary practices of both architecture and urbanism required radical rethinking in light of unprecedented technological opportunity and equally unprecedented environmental challenges.

Working with identical volumes of space, the twelve architects were asked to present work that most addresses the themes of the exhibition in a manner that best expresses their own design philosophies. Each architect had the opportunity to also design a "facade" that identifies their installation and offers the possibility of further describing their work in full-scale. Together, the twelve facades will create a "street", literally reflecting the theme of the Shenzhen Biennale. The twelve participating architects and/or teams of architects are Atelier Deshaus (Shanghai, China); Alejandro Aravena, Arquitecto (Santiago, Chile); Fake Industries Architectural Agonism (New York, USA and Barcelona, Spain); spbr (Sao Paolo, Brazil); SO-IL (New York, USA); J. Mayer H. (Berlin, Germany); Johnston MarkLee (Los Angeles, USA); OPEN Architecture (Beijing, China); Aranda Lasch (New York, USA); MAD Architecture (Beijing, China); Mass Studies (Seoul, Korea); and Hashim Sarkis Studios (Beirut, Lebanon and Cambridge, USA).

The Strada Novissima exhibition featured 20 architects, many of them in their 30's and 40's. Since then, a number have become the leaders in the field of architecture today, including Rem Koolhaas, Frank Gehry and Arata Isozaki, amongst others. To bring the discussion full circle, Aaron Betsky, director of the Cincinnati Museum of Art and architectural critic and curator, interviewed the key figures from that exhibition 30 years ago. Their interviews are presented on monitors adjacent to the contemporary installations, providing a historic dimension. The Presence of the Past Revisited includes participating interviewees by Denise Scott Brown, Frank O. Gehry, Michael Graves, Allen

INTRODUCTION

Greenberg, Leon Krier, Thomas Gordon Smith, Robert A. M. Stern, and Stanley Tigermann.

While urbanism is a profession that is defined by a high degree of technical skill and usually undertaken by large corporate and/or governmental entities, it would be a mistake to think that technocrats are the only ones with important ideas about cities, or that the only way to affect change is through corporate or bureaucratic mechanisms. Inasmuch, various alterative perspectives have been included in the constellation of exhibitions that makes up the 2011 SZHKB.

As people migrate to urban areas in masses, many cities are growing at an unprecedented rate. A majority of these new city dwellers live in informal additions to the urban landscape, often unwanted, neglected or simply forgotten. The Favela Painting Project, organized by Jeroen Koolhaas and Dre Urhahn, seeks to integrate these areas and their inhabitants into society by using art as a catalyst for social change. Their installation documents their success in transforming entire neighborhoods with little other than cans of paint, including Vila Cruzeiro and Santa Marta in Rio de Janeiro.

Urbanism as a technical discipline is not only associated with large scale corporate and government entities, it is also packaged as an export industry of the wealthier "developed" countries with the poorer countries being the consumers. Ghana ThinkTank: Developing the First World is a project that stands this relationship on its head, rejecting the idea that the only people with solutions to urban problems are the ones with the capital to profit from them. The Ghana ThinkTank is a network of Third World think tanks in Ghana, Cuba, El Salvador, Gaza Strip, Iran, Serbia, Mexico, and in the USA devising solutions for this problems.

Problems are collected in cities in the developed world, then sent to the think tanks to analyze. They devise solutions, which are implemented back in the community where the problems originated. The results are then delivered back to the think tanks, for post completion analysis. The Shenzhen installation, which was organized by John Ewing, Carmen Montoya and Christopher Robbins, consists of stylized displays constructed of recycled materials and demonstrating each stage of the process. The audience is invited to submit their problems and to help implement the solutions.

International Participation

One of the strengths of the Venice Biennale is the manner in which it is structured. While the principal curator or curators have great discretion over the central exhibits, that voice is variously complimented, contradicted or ignored by an array of exhibits independently curated and installed in the national pavilions or in off-site venues.

In an effort to not only internationalize the Shenzhen Biennale, but to also expand the number of voices involved, various nations and/or prominent national institutions were invited to participate for the first time. Representations from Austria, Bahrain, Chile, Finland, and the Netherlands, accepted the invitation and agreed to present exhibitions.

Architekturzentrum Wien (Vienna, Austria) presented an exhibition entitled Housing in Vienna, Innovative, Social and Ecological, which documents Vienna's history as a leader in the area of social housing. The exhibition not only highlights important architectural works by such significant contemporary architects as Coop Himmelb(l)au and Jean Nouvel, it also describes the role of housing as an essential component of that city's urban planning over more than a century. Curated by Wolfgang Förster, Gabriele Kaiser, Dietmar Steiner, and Alexandra Viehhauser, the exhibition also included projects by Manfred Wehdorn, Wilhelm Holzbauer, BKK-2, BKK-3, Franziska Ullmann, Lieselotte Peretti, Gisela Podrekka, Elsa Prochazka, and Elke Delugan-Meiss and Roman Delugan.

The installation presented by the Ministry of Culture, Kingdom of Bahrain, appeared in the 2010 Venice Biennale and won the award for Best National Participation. Reclaim consists of three fishing platforms—the informal waterfront structures that used to line the sea and served as lively social spaces before the real estate boom of recent decades reconfigured the city's waterfront. Interspersed within and among the structures are video screens with interviews with Bahrainis recalling the fishing platforms as a part of the cities social fabric as well as images of the waterfront under redevelopment. The exhibit, which was commissioned by Shaikha Mai Al Khalifa, was curated by Noura Al Sayeh and Fuad Al Ansari and was designed by Harry Gugger and Leopold Banchini.

The Consejo Nacional de la Cultura y las Artes (Chile) presented Gimme Shelter!—an exhibition with a broad view of the Chilean experience of reconstruction and urban regeneration after the 2010 earthquake and tsunami, and the recent innovation processes in social housing. The exhibition was coordinated by Cristobal Molina Baeza, the Consejo's coordinator for architecture, and designed and curated by Hugo Mondragon and Sebastian Irarrazaval. The innovative installation was constructed with materials associated with emergency situations: mattresses, water bottles, strings of lights, etc.. The curatorial content, both texts and images of recent Chilean housing projects, were delivered in the form of projections and on digital monitors.

The exhibition representing Finland was organized by Martta Louekari, producer for World Design Capital Helsinki 2012, and included two components. The first is entitled NEWLY DRAWN—Emerging Finnish Architects and served to introduce a younger generation of Finnish architects, their latest projects, visions and ways of working. The work of Hollmén Reuter Sandman Architects, Verstas Architects, NOW, Anttinen Oiva Architects, Lassila Hirvilammi Architects, Avanto Architects, ALA,

INTRODUCTION

AFKS Architects and K2S Architects is presented. The second component, entitled Solution Finland: The Welfare Game, is based on a project developed by a Finnish publisher that challenges authors to address important social problems with original solutions. The Welfare Game is authored by architect Martti Kalliala with writer and curator Jenna Sutela and architect Tuomas Toivonen and proposes solutions to their native country's quandaries, ranging from the practical (rescuing ailing public space through climatization), to the absurd (dividing the country into two interlocking sub-nations: City and Wilderness) and the earnest, if far-reaching (the repurposing of the country to host the world's nuclear waste).

The Nederlands Architectuur Institut (the Netherlands), under the direction of Ole Bouman, Director of the NAi with the assistance of Jorn Konijn, organized an exhibition titled Housing with a Mission, which consists of housing projects specifically designed for the millions of young graduates of China's education system that are just starting out their careers. These graduates are called "ant tribes". Five architects from China and five from the Netherlands worked together to develop new forms of housing that fit the minimum existing housing requirements in China and the Netherlands and developed a series of hybrid guidelines, all of which were made evident in the exhibition in the form of a three full scale $8m^2$, $28m^2$, and $14m^2$, model apartments. As the centerpiece, the exhibition featured large scale housing projects by the individual architects, based on the hybrid requirements.

The participating architects form the Netherlands are NL Architects, Arons & Gelauff Architects, NEXT Architects, Barcode Architects and KCAP and their counterparts from China are URBANUS, Standard architects, NODE, O-Office and CAFA University.

Conclusion

Since the opening of the first Venice Biennale in 1895, most subsequent biennial exhibitions around the world have followed the model established in Venice: one part national representation and one part international representation. In addition, Venice also established the precedent of assuring that the process of selecting the participating artists was not too narrowly constructed. All three of these positions are intentionally incorporated into the 2011-2012 SZHKB.

In formulating the exhibition program for the 2011 SZHKB, the intention was to fully explore contemporary urbanism and architecture in the overlapping contexts of Shenzhen, Shenzhen-Hong Kong, China and the world. Moreover, the goal was to create a city-wide event, not just a convention for architects, engineers and other specialists. At the same time, the 2011 Biennale was intended to be a truly international encounter, building on the groundwork established in the previous three editions.

In researching possible participants for the 2011-2012 SZHKB, an informal group of advisors assisted in ensuring that the research was undertaken in as wide a field as possible. This group, which included Ma Qingyun (China/USA), Barry Bergdoll (USA), Andres Lepik (Germany), Aric Chen (China/USA), Kenneth Frampton (USA), Luis Fernandez-Galliano (Spain), Erwin Viray (Singapore/Japan), Raymund Ryan (USA), Mohsen Mostafavi (Harvard, USA), Stan Allen (Princeton, USA), and Nader Tehrani, (MIT, USA) recommended dozens of potential participants, many of them little-known up until now.

While a well-conceived program is not necessarily a guarantee of critical success, the program that has been described herein is an exercise in urbanism and architecture in and of itself. The theme—Cities create architecture. Architecture creates cities. —can now be seen as it was intended: not a prescription for a specific manner of critical thinking but as a theoretical fabric woven of disparate strands of architectural and urban ideas.

1 "Urban China / Informal China" appeared at the Museum of Contemporary Art in Chicago (October 16, 2010-April 3, 2011).
2 "New cities" would include not only cities that were built tabula rasa, such as Brasilia, but those with an earlier history. For example, Shenzhen actually had a pre-history as a Pearl River fishing town, a history that has been erased by the superimposition of the new vision of a Special Economic Zone.

主题

2011-2012年深圳双年展的主题植根于始自2005年以来的活动构想。这项双年展独具特色，展现的不只是建筑，还有城市——这在同类活动中是罕见的。城市与建筑的组合极其恰当——即使常被看作不同的学科，但这两项人类活动密不可分。历史上，在人居变迁与发展中，它们全然彼此依存，由此，本届双年展的主题为：

城市创造
(Architecture creates cities.
Cities create architecture.)

这一简单、自明的陈述可以说得更具体而不失其意：卢浮宫创造巴黎，巴黎创造卢浮宫。然而，不仅是大型、纪念性的建筑创造了城市，同样地，褐砂石建筑创造了曼哈顿，町屋[1]创造了京都，反之亦然。

尤其就当代文化而言，这一主题所表征的城市与建筑之间的关系可视为全球性的。而本届深圳双年展既在地点脉络中，也在时间脉络中看待城市与建筑。回视历史，显然每座城市并非持续无止境地创造建筑。任何一座城市、一群城市，及其相应的文化——建筑及其他——其衰落原因，则常是复杂的。

Jared Diamond，一位其研究囊括从鸟类学到地理学众多不同领域的科学家，已证实：许多伟大文化衰落前，都有一段完全忽视其环境的时期。依此，本届双年展主题或可修改为：建筑与城市理解并尊重其环境支持条件，那么，建筑创造城市，城市创造建筑。

可持续 + 活力

近些年来，"可持续（sustainable）"，作为售卖所谓"绿色产品"（不管是牙刷、汽车或是办公楼）的标签，其词义已然扭曲。我们全球化的商业文化轻视此事的态度，俨然是Jared Diamond研究过的那些消逝的文化——格林兰的古挪威与因纽特、玛雅、阿那萨齐[2]、复活节岛上的原住民和其他——的种种态度之复现。

毫无疑问，就能源使用、废物管理、空气与水质量发展可持续实践，是今日城市与建筑最大的挑战。但是，对建筑师和规划师来说，让城市与建筑更加可持续的努力，必须和文化上更具活力悉心配合。

一如"绿色"这个词因误用而变得几近平庸乏味，"生活质量"这个词组苍白地反映着或能称为"真实的城市活力"的东西。活力不可量化，要正面定义，它亦颇为勉强。就本届双年展而言，可以这么说，活力不是指公寓大小、汽车数量或购物中心规模。如果我们不能轻松地理解"城市活力"这个概念，这可能会是一个更大的共同问题中的一部分。

在严格意义上，本届双年展主题的全部重要性只能在其平面视觉再现中领略。虽然主题自身表述得相当简单，"城市创造"[3]的预期含义却较为复杂。广州的王序设计用平面视觉的方式表达了主题，将"建筑创造城市"、

"城市创造建筑"两句话串在一起，仿佛莫比乌斯环，无始无终。暗含的无尽意在提出"可持续"这一话题；近乎 DNA 的外表暗示着城市与建筑的活力，由此，两个简单句成为一种渴望，而非观察。

深圳

深圳是世界上最具变革性的城市之一。深圳双年展明确关注这座城市本身，恰如其分。呼应 20 世纪五六十年代在纽约现代美术馆举行的一系列展览，名为"深圳建造"的展览聚焦五项有重要城市含义的主要建筑作品。非常建筑（深圳电视塔）、蓝天组（深圳当地艺术馆与规划展览馆）、马西米利亚诺与多莉阿娜·福克萨斯（深圳宝安国际机场 3 号航站楼）、大都会建筑事务所／雷姆·库哈斯（深圳证券交易所）与都市实践（深圳市地铁前海湾车辆段上盖物业）都详细呈现了其作品，彰显城市生长与发展。建筑创造深圳，深圳创造建筑。虽然"深圳建造"的目标已展现给各位建筑师，他们各自如何呈现其作品，在某种意义上，是另一个设计问题。

当各位建筑师专注于其作品以及作品如何成就城市，另一项展事则提供了历史性的文脉，诉说着过往几十年中深圳城市文化的更广阔面向。马立安博士（Dr. Mary Ann O'Donnell），深圳大学建筑学院的一位研究工作者，策划了名为"轰隆！深圳"的装置展，将描述深圳过往岁月爆炸性发展的枯燥数据以平面与三维方式化为欢快的视觉表达，议题包括这座城市的国内生产总值、房价、投资强度、城市化的地理影响以及虚幻的、消逝中的历史。

深圳市民中心及其超尺度的、使用不足的广场，是本届双年展的主要焦点。广场是纽约建筑师 John Bennett 与 Gustavo Bonevardi 创作的可变装置的所在地。装置是千万个橙色交通锥——这种无所不在的全球新建筑或道路施工的标志——的迷宫状组合。名为"万花阵"的这一作品，其灵感来自 1756 年意大利建筑师郎世宁为紫禁城外旧日夏宫[4] 所设计的同名花园迷宫。

在本届双年展的开幕式中，这一迷宫由几十位滑旱冰的深圳少年控制。这些志愿表演者以一种轻松的、自由的编排，滑行在交通锥之间，越过交通锥之上，穿行在交通锥周围，于是，最大而最具仪式性的公共空间，由一种小小的普通物件和一种街上随处可见的典型城市行为，得到了改变。

市民中心的上层广场也是一个巨大的空间，本质上是深圳的城市剧场展开的舞台之一。另一项目"超轻村"，由 6 个建筑组成，建造在经过市民中心的连接主广场至北向莲花山公园的轴线上。这些建筑，并非建于陆地之上，其实是建于其下建筑的顶层表面，于是，这些建筑得是轻质的，这在本届深圳双年展的文脉中亦为长处，而非限制。著名工程师、哲学家富勒（R. Buckminster Fuller）是建筑施工领域资源节俭使用的早期倡导者，他或许会问其同事："你的房子有多重？"

获选为"超轻村"设计构筑物的建筑师们都展现了对概念与建造间关系的清晰态度，以及足尺构筑如何能以更具表现力的方式传达理论性的观念。他们是业余建筑工作室（中国杭州）、Clavel Arquitectos（西班牙穆尔西亚）、MOS（美国纽约）、OBRA（美国纽约）、Studio Up（克罗地亚萨格勒布）和魏春雨领导的地方营造工作室（中国长沙）。

深圳与香港

尽管双城双年展由两项相关但分离的公众活动所组成，如果人们认为深圳与香港两座城是不同的自治实体就是错的。这两座城确有不同的历史、管治与货币，然而，它们是珠江入南海占首要地位的同一生物圈的一部分。深圳双年展的一个重要组成部分是"对应双城"展，展览的起点即是水不遵从政治世界的事实。

"对应双城"展基于由位于纽约的现代美术馆菲利普·约翰逊建筑与设计总策展人 Barry Bergdoll 所发展的一项展览概念。它聚焦在未来岁月中因海平面上升与地下水耗竭而面临环境挑战的三块场地。Jonathan Solomon 与邓信惠同为香港大学教授，他们调整了现代美术馆的模式以表达与这个区域相关的特别议题，并承担了这一项目的全部组织工作。六个团队——深圳三个、香港三个——集合起来研究，并在奥雅纳的 Ricky Tsui 与 Iris Hwang 的技术支持下，提出设计对策。

入选的三位香港团队负责人为 Stefan Al、Tom Verbes 与 Vincci Mak，三位深圳团队负责人则包括刘珩、冯果川与朱雄毅。

深圳与中国

在关注深圳与香港的项目之外，另有三个项目表达中国城市建筑、城市规划的重要议题。中国有 12 个城市人口超过 500 万，将中国称为世界城市实验室已是寻常之论。哥伦比亚大学中国实验室主任 Jeffrey Johnson 与同济大学副教授李翔宁策划的"八个城市项目"展现了深圳以及中国各地的当代城市项目的一种批判性视野。设计者既有国内事务所，也有国际工作室，这个展览勾勒出中国特有的境况和挑战是如何产生新城市形态的。每一个项目的入选，基于与当今中国城市项目的定义有关的一系列重要问题及主题，譬如社区、生活方式、身份、和谐、生态、经济、临时性及历史保护等。此外，项目入选也基于它们具有特定空间状况的地理位置，这些状况决定了过去 30 年间出现的独特当代都市地域，譬如近郊城市、郊区、卫星城、基础设施和网络化相关项目，以及现存核心城区。

入展项目及其主创设计师为：深圳前海规划，James Corner, Field Operations 事务所；上海嘉定中广国际广告创意产业基地，张永和/非常建筑事务所；上海洛克·外滩源，戴卫·奇普菲尔德建筑事务所；山水杭州（杭氧·杭锅地块国际旅游综合体设计），斯蒂文·霍尔建筑事务所；深圳眼，大都会建筑事务所和深圳都市实践；鄂尔多斯 20+10，方振宁与 30 家建筑设计事务所；杭州西溪国家湿地公园，维思平与伍兹贝格亚洲；中山信息产业新区，吴志强/上海同济城市规划设计研究院。

一如"深圳建造"，当代城市规划的努力需要坚实的文脉处理。虽然许多双年展多有历史失忆之嫌，"八个城市项目"却由另一展项"双城记"的补充而相得益彰。北京大学视觉与图像研究中心副主任唐克扬博士，在这一视觉论文中，以两种相关的视角来呈现中国城市的发展。在其中一种视角里，城市是它所有历史碎片的总和，通常呈现为它的极盛状态。依据一般的类型原则，如此打量这座城市的人生造出了一种"特征标准像"。在另一种视角里，中国城市则是变化的个体，盛衰都自有其情境，对于这些城市而言，这一展览不仅仅希望秀出它最为人熟知的一面，也想揭示它历时的变化。

姜珺是《城市中国》杂志前主编、下划线工作室创办人，由其与同济大学的城市规划师苏运升策展的"城市中国/自发中国"[5]，展现了中国城市建筑的另一番概貌。展览构思为以墙纸形式"重新混合"出的一本杂志，这一35米长的装置给出了中国城市化的简明历史叙事，呈现出一种控制系统与放任（也就是说"随它去"）之间正在进行的斗争。文本主要分两色——红色代表正式的、有规则的以及有计划的政府决策；蓝色代表非正式的、有机的、对政策与事件的点对点反应。除了色彩区分，这些墙纸从上到下分为三栏：顶栏强调的是国族信仰与政策，中栏内容突出前者对城市建筑的冲击，底栏则展现对家庭及其物件的影响。数字与关键的历史性事件和观念相对应，后者曾经也仍在塑造着政策与规划、经济发展、建筑以及无限制的、自发的转变。自发系统——空间的、经济的、实用主义的——在已被规划的、具有高度结构化本质的中国展现出了它们的颠覆能力。

中国经济与城市的快速增长，反映在因大量农村人口涌入城市造成的低造价住宅需求的快递增长中。为应对在当前市场主导的房地产发展中保障性住房供应不足，中国正大力发展大规模社会住宅建设举措。中央政府已确定2011年全国保障性住房建设任务是1000万套。由香港大学副教授、建筑学硕士课程主任杜鹃策划的"广厦千万·居者之城"保障性住房设计展，广泛联合政府、企业、学者、规划师、建筑师、工程师与普通公众，探讨保障中低收入者居住问题对当代城市的高密度环境带来的机遇和挑战。展览展示了深圳"一户·百姓·万人家"保障房设计竞赛获奖方案，也展示了从建造细节到国家政策的不同尺度的保障房问题的创新设计及研究。

展览分为不同的多个部分，由策展人杜鹃和其合作者共同呈现：居民（白小刺）、概念（"一户·百姓·万人家"设计竞赛获奖者）、节点（麻省理工学院一万元住宅课题组）、单元（都市实践）、房屋（标准营造）、城市（香港大学城市生态课题组）、国家（城市中国研究中心）、建造（协鹏设计+卓越置业集团）和预制（魔力方新型房屋+UAO Creations）。

"再生砖"计划由成都建筑师刘家琨于2008年7月启动，紧随四川汶川大地震。这个项目原本意在帮助当地民众生产自救、重建家园。其基本原理是：用破碎的废墟材料作为骨料，掺合切断的秸秆作纤维，加入水泥等，由灾区当地原有的制砖厂，制成轻质砌块，用作灾区重建材料。"再生砖"不仅表征着废弃材料的"再生"，更表征着灾后重建在精神和情感方面的"再生"。

"再生砖进展2011"特展中，建筑师刘家琨用再生砖抽象地再现了一座中国乡村住宅，既表达其物质属性，又显露诗意情怀。这个装置提醒我们，灾后重建，不只是重建建筑，更是重建生命。

国际视野

当我们谈到"世界越来越小"时，通常意味着交流与交通缩短了我们曾经以为的遥远距离。然而，当回溯这一历程时，整个世界确实也可以被认为是缩小了。无论是过去抑或当下的人类定居处，都存在于一种可以被称为"都市区域"（urbanosphere）内，这薄薄的一层就是地球上可居住的环境，即死海附近的城镇是地球最低的居住地（海拔负425米以下），而秘鲁的林诺科那多则是最高的（海拔5100米以上）。

绪言

除了深圳与香港的一衣带水,它是全球都市独一无二的类型中的一部分——通常被称为"新城"[6]。毫无疑问,这种完全新建的城市花费了人类大量的各种努力,规划、财力和体力劳动。自古以来,如 20 世纪后半叶那样如此大兴土木的修建新城市,可谓是前所未有。在这些项目中,也同样消耗了大量的投资。作为本届双年展中,重要的部分原定名为"城市小于 60"是主要回顾世界各地新城市的成功与失败的展览。这个最初的题目强调了在所有这些新城里,城市自身可能比它们的年长的居民还要年轻。

这个项目的参与者包括古根海姆博物馆的建筑与城市研究助理策展人 David van der Leer,南加州大学罗斯科艺术学院院长 Rochelle Steiner,他们两位合作努力将关注点放在了深圳、阿勒梅尔、哈博罗内、拉斯维加斯、巴西利亚和昌迪加尔。首先,他们实现了由跨学科团队进行的研究和展览——"6 小于 60"——这个团队不仅包括南加州大学的艺术学院,还包括其建筑学院(院长马清运和建筑研究生项目的主任 Stefano d Martino)以及电影艺术学院(院长及教授 Scott Fisher 与研究助理 Jennifer Stein)的参与。他们运用了传统的研究策略以及群众外包(crowd-sourcing)的策略,以及其他的数码技术,由此获取了各种数据与图像。这一研究项目追溯了每个城市的发展历程,从其概念的发起到转型,对其目的与结果进行了批判性的考察。在这次双年展上,研究通过一种非常有创意的数字技术装置来呈现,并且观众可以进行简单的触屏和大屏幕互动,就能参与项目。相比起静态展示,这一项目通过线上部分不断得到延续与发展。

这一项目的第二部分名为"然后,它就成了一座城:6 小于 60",这一部分更看重视觉和感觉的方式,显示出一种希望通过当代视点来反映理解这些城市的意愿,即城市是它们现在的样子,而不是曾被规划的。David van der Leer 根据他们的技能和成就,以及他们与这些被讨论城市的地域和文化上的相近,挑选了六名导演参与其中。这六名参与者分别是:Surabhi Sharma(昌迪加尔)、Cao Guimarães(巴西利亚)、Sam Green(拉斯维加斯)、Miki Redelinghuys(哈博罗内)、Astrid Bussink(阿尔梅勒)和王功新(深圳)。这些短片与加州大学的数据展示一起,并且加入了线上部分。这六个录像还在展览开幕当天在巴士项目上播放,和观众们一起穿越城市,如同一个移动的课堂与都市实验室。

中国都市深圳与荷兰新城阿尔梅勒之间的相同与差异一直有相应的研究。正如装置"所有都市"(Allmetropolis)让这些相同与不同都清晰地展示出来,两个城市都在 20 世纪 70 年代开始了它们的城市化进程,而且两座城市都坐落在三角洲地区,毗邻主要的国际大都市;在策展人看来,尽管它们获得了发展,但是其发展方式却十分迥异。这个展览由"西游项目"(Michiel Hulshof 和 Daan Roggeveen)进行策划,将经济、社会、空间和生态发展都囊括进考察系统中,以此作为研究的出发点来探讨将深圳模式放在欧洲语境中的可能性与机遇。

如 1895 年的第一届威尼斯双年展一样,深圳双年展也是一个关于想法的国际文化市场,这一观念兴许在展览"街道"中体现得淋漓尽致。12 件来自世界各地的建筑师或建筑团队的装置作品呈两列排开,两两相对,以此向具有里程碑式意义的 1980 年的威尼斯建筑双年展"街道"致敬。一方面,对待建筑与都市主义的方式在不断地多样化,所有的参与建筑师们这代人,以过去所未有的技术机遇和不能预期的环境危机为前提,对当代建筑和建筑实践进行彻底反思显得越来越必要。

这 12 名建筑师在相同大小的空间内，让展览的主题通过他们的设计哲学表现出来。每一名建筑师都有机会设计一个"立面"，并以装置的形式实现出来，这便意味着让建筑师把他们的创想以最大限度的方式展示出来。除此之外，这 12 个立面会一同呈现出一条"街道"，以此呈现出深圳双年展的主题。参与者包括了建筑师/团队有大舍建筑师事务所（中国上海）、Alejandro Aravena, Arquitecto（智利圣迭戈）、Fake Industries Architectural Agonism（美国纽约 & 西班牙巴塞罗那）、spbr（巴西圣保罗）、SO-IL（美国纽约）、J. Mayer H.（德国柏林）、Johnston MarkLee（美国洛杉矶）、开放建筑（中国北京）、Aranda Lasch（美国纽约）、MAD（中国北京）、Mass Studies（韩国首尔）以及 Hashim Sarkis 工作室（黎巴嫩贝鲁特和美国剑桥）。

在当年的威尼斯双年展上，受邀的 20 名艺术家正值壮年，只有 30 到 40 岁。他们中的大部分人都已经成为当今建筑界的领袖，其中就包括库哈斯、盖里和矶崎新等人。为了让这一讨论更加圆满，辛辛那提美术馆的建筑批评家与策展人 Aaron Betsky 采访了 30 年前参与这次展览的关键人物。他们的采访视频与当代的装置作品放在一起，提供了一个历史维度。"再造访'过往的呈现'"中的被采访者包括 Denise Scott Brown、盖里、Michael Graves、Allen Greenberg、Leon Krier、Thomas Gordon Smith、Robert A. M. Stern 以及 Stanley Tigermann。

当都市主义成为由高级技能所定义的领域，并由大公司和政府机构完成，并且把技术专家们当作城市与思想最重要的人士，或者说影响变革的唯一途径是通过机构或者是官僚体制才得以实现，然而，这样的观点是不可取的。正是基于此，各种多样的观念都被放进了展览中，构成了这一届的深圳双年展。

当人们蜂拥入都市空间，城市们都以前所未有的效率在成长。新城市的大部分移民都居住在都市景观的非正式附属地区，这些地区常常是不被需要的、被忽略的或被遗忘的。项目"彩绘都市"由 Jeroen Koolhaas 和 Dre Urhahn 组织，试图把艺术当作一种社会变革的催化剂，促使这些区域与居民融入社会。他们的装置记录了其如何用颜料改变整个邻里街区的成果，其中包括了里约的 Cruzeiro 村和圣玛塔。

都市主义是一套技术化的学科，不仅仅与大公司和政府机构有关，同时还包装成一个更富裕的"发达"国家向作为消费者的贫穷国家的出口产业。"加纳智库: 发展第一世界"则是一个调转这一关系的项目，它拒绝"能解决城市问题的人只是那些资本受益者"的想法。加纳智库是一个第三世界的网络联合，包括加纳、古巴、萨尔瓦多、加沙地带、伊朗、塞尔维亚、墨西哥，也包括美国，来自这些地区的参与者们都在为这一问题出谋划策。

他们在发达国家的城市里收集问题，然后发送到智库进行分析。由他们设计出解决方案，试图注入那个问题发生的社区去。然后结果再反馈到智库，进行事后分析。深圳的展览由 John Ewing、Carmen Montoya 和 Christopher Robbins 一同组织，其中包括了由回收品搭建成的风格化展示并说明过程步骤。观众都被邀请来提出他们的问题，并参与到解决方案的制定当中。

绪言

国际参与

威尼斯双年展所具有的众多优点之一，是其所组建的方式。总策展人与策展人团队可以对主展自行做主，但这些主张可以通过一系列独立策划的国家馆，或非主展场项目来补充，挑战或忽略。

这一做法并非只是为了让深圳双年展更加国际化，但也同时能让各种声音加入，各个国家和/或前沿的国际机构都首次应邀参展：奥地利、巴林、智利、芬兰和荷兰都接受了我们的邀请。

维也纳建筑博物馆以"维也纳住房，革新的，社会的以及生态的"为题，记录了维也纳在社会性住房领域的领先性。这个展览强调的不仅是重要当代建筑师的作品，例如蓝天组和 Jean Nouvel，同时还试图勾勒出社会性住房作为一个世纪以来城市规划的核心部分。这个项目由 Wolfgang Förster、Gabriele Kaiser、Dietmar Steiner 和 Alexandra Viehhauser 策划，其中包括了 Manfred Wehdorn、Wilhelm Holzbauer、BKK-2、BKK-3、Franziska Ullmann、Lieselotte Peretti、Gisela Podrekka、Elsa Prochazka 和 Elke Delugan-Meiss 及 Roman Delugan 的作品。

由巴林王国文化部推出的项目"再生"曾在 2010 年的威尼斯建筑双年展展出，并获得了最佳国家馆奖。这个展览由 3 座渔屋组成，它们曾是海边活跃的社会空间，而近几十年的地产泡沫则重构了城市的海岸线。这些渔屋之间点缀着采访录像，令人们回忆起了当年还是城市社会肌理一部分的渔屋，和正在再发展的前海地区的景象。这个展览由 Shaikha Mai Al Khalifa 委托，Noura Al Sayeh 和 Fuad Al Ansari 策划，展览由 Harry Gugger 和 Leopold Banchini 设计。

由智利国家艺术文化委员会呈现的"给我避难所！"则是一个 2010 年大地震和海啸之后智利进行灾后重建和都市再生的全纪录，以及社会性住房领域的创新过程。这个展览由国家艺术文化委员会的建筑部分协调人 Cristobal Molina Baeza 委托，展览由 Hugo Mondragon 和 Sebastian Irarrazaval 设计与策划。这个具有创新意义的装置由各种紧急情况所需材料组成：床垫、水平、应急灯，等等。在策展文案中，图文描绘了近期的智利住房项目，并在展览中以投影和数字电视的方式展现给观众。

芬兰的展示由世界设计之都赫尔辛基 2012 的制作人 Martta Louekari 担纲，其中包含了两个部分。第一部分名为"新绘图——新兴芬兰建筑师"，介绍了新一代的芬兰建筑师以及他们的作品、视野和工作方式。这次在深圳的展示包括：Hollmén Reuter Sandman Architects、Verstas Architects、NOW、Anttinen Oiva Architects、Lassila Hirvilammi Architects、Avanto Architects、ALA、AFKS Architects 和 K2S Architects 的作品。第二部分名为"芬兰方案：福利博弈"，这是芬兰的出版项目，要求作者为重要的社会问题提出解决方案。"福利博弈"的作者是建筑师 Martti Kalliala 和作家兼策展人 Jenna Sutela，与建筑师 Tuomas Toivonen 协作通力完成，尝试为他们的祖国所面临的疑难提出解决之道，从实践（以顺应气候的方式挽救困窘的公共空间）到荒诞（把整个国家分为两个连锁国家：城市与荒野），再到长远的愿望（重新调整国家来承载全世界的核废料）。

荷兰建筑师协会 NAi 在总监 Ole Bouman 的带领和 Jorn Konijn 的协助下组织了这一场名为"住宅的使命"的展览，关于专门为中国教育体系下刚刚踏入社会的数以百万计的年轻毕业生的住房项目设计。这些毕业生被称为"蚁族"。分别有五个来自中国和荷兰的建筑师们协作，为这一类的住房发展出了一些新的形式，以符合中国和荷兰最基本的住房需求，他们一同发展出了一系列混合指导原则，分别呈现在 8 平方米、28 平方米和 14 平方米的样板房空间中，其核心部分是展示基于混合指导原则的个体建筑师的大规模住房项目。

参与其中的荷兰建筑师包括 NL Architects、Arons & Gelauff Architects、NEXT Architects、Barcode Architects 和 KCAP，以及来自中国的都市实践、标准营造、南沙原创、源计划和中央美术学院。

结论

自 1895 年第一届威尼斯双年展以来，众多国际上前仆后继的双年展都在追随威尼斯模式：一部分是本国展示，另一部分则是国际展览。除此之外，威尼斯双年展还创建了确保选择参展艺术家过程不会太狭窄的先例。所有这些部分都被有意地融合进了本届深圳双年展中。

在形成本届双年展的展项前，当代都市主义和建筑话题在深圳、深圳 – 香港、中国与世界这一相互叠的语境中得到了全方位的研究。本届双年展不仅仅希望成就建筑师、工程师和其他领域专家之间的交流，更要形成一个城市性的活动。与此同时，在前三届的基础上，本届深圳双年展试图做到真正的国际化。

在寻找本届双年展的潜在参与者时，一个非正式的顾问组确保了研究在一个尽可能广泛的领域被实施。这一顾问组包括马清运（中国 / 美国）、Barry Bergdoll（美国）、Andres Lepik（德国）、Aric Chen（中国 / 美国）、Kenneth Frampton（美国）、Luis Fernandez-Galliano（西班牙）、Erwin Viray（新加坡 / 日本）、Raymund Ryan（美国）、Mohsen Mostafavi（美国哈佛）、Stan Allen（美国普林斯顿）和 Nader Tehrani（美国麻省理工），他们都分别推荐了不少可能的参与者，其中不乏此前名不见经传的新秀。

尽管一个周密的策划不一定能确保最后的成功，但是它也能被描述为都市主义和建筑中的一项实践。这一主题——都市创造——可以被看作成最初策划的一样：并不是一个批评性思考的特殊处方，而是用各种建筑和都市思考编织起的理论布料。

1 町屋，machiya，日语中指传统的城市住宅，与农村住宅"民家"（minka）相对。译注。
2 阿那萨齐人（Anasazi）为美国西南部一古代美洲印第安人，公元前 200 年至公元 1500 年为鼎盛期，其最早的文化阶段通称作竹篮人（Basket Maker）文化时期，现在的普韦布洛（Pueblo）文化从该文化的晚期发展而来。译注。
3 本届双年展的汉语主题为"城市创造"，英语主题则表达为"Architecture creates cities—Cities create architecture"，直译即为双年展标志所示"建筑创造城市——城市创造建筑"，故后文称"两句话"。译注。
4 即圆明园。译注。
5 "城市中国 / 自发中国"曾于芝加哥当代美术馆展出（2010 年 10 月 16 日 – 2011 年 4 月 3 日）。编注。
6 "新城"不仅包括平地而起的城市，如巴西利亚，而且包括那些有历史的，例如，深圳实际上有着作为珠江渔村的历史，而这段历史被经济特区的新愿景淡化了。编注。

SHEN
深圳

10,000 FLOWER MAZE
万花阵

Designers/设计师 : John Bennett, Gustavo Bonevardi

Sound / 声音设计: Michael Rosen
Special Thanks to / 鸣谢: Steven Barr, Tom DeKay, Kris Haberman, Paul Martantz, Michael Rogan

| SHENZHEN/ 深圳 | 10,000 FLOWER MAZE/ 万花阵 |

In 1756 at Emperor Qianlong's request, Italian architect Giuseppe Castiglione created European style buildings and a garden for the Old Summer Palace just outside the Forbidden City. The garden took the form of a maze—a motif popular in Europe at the time, with a history stretching back to ancient Greece—and was known as the Wan Hua Zhen (10,000-flower maze). On the occasion of the Mid-Autumn Festival, the Emperor would preside over an evening race through the maze, with participants carrying yellow lanterns.

The maze was tragically destoryed during the Old Summer Palace was burnt down by the Anglo-French Allied Forces, and it was reconstructed in the 1980s. This autumn, in celebration of the 2011 Shenzhen & Hong Kong Bi-City Biennale of Architecture and Urbanism, a temporary and contemporary installation on Citizen's Plaza in Shenzhen—built entirely of orange safety cones—nods to that historic garden while looking to the future.

On roads and at construction sites around the world orange safety cones are so ubiquitous that they seem to proclaim the coming of a new building, highway or other public amenity. Symbols of modernity, they help to usher in the future even as they speak of dangers. Each cone restricts a movement and warns of a local hazard; collectively, they are reminders of ongoing overdevelopment. Brightly colored and banded with reflective tape, these iconic objects demand our attention, heralding change like a kind of alarm.

The orange cones of the new installation are spread across the plaza like bright orange paint across an artist's canvas. Toppled, precariously balanced or decoratively arrayed in floral patterns, the cones are elegantly but irreverently arranged, robbing them of their controlling nature. They become playful and joyful and can be appreciated abstractly as merely brightly colored, almost toy like objects. The largest and most ceremonial of public spaces is transformed by a small common object that can be found on any street.

The transformation is spatial as well as visual. The plaza, at the foot of City Hall, is a vast and normally austere, and somewhat inhospitable expanse. Though it may be appropriate for public ceremonies, it is largely unused by daily passersby. It only comes to life at night, as kids gather to make use of the nearly pristine surface by gracefully skating across it. At night the plaza belongs to them.

Echoing the obstacle courses skaters create out of cones, the installation is an open and welcoming composition (To ensure that it would not interfere with the evening skaters, a team of rollerbladers was consulted in the layout). It can be entered or exited at any spot, and everyone is invited to venture in and explore. As they do, they will be accompanied by the sound of one hundred crickets emanating from one hundred of the installation's thousands and thousands of orange cones. On the evening of the Biennale's opening ceremony rollerbladers skate through the installation in a short exhibition. Dressed in black and, like the cones, trimmed with highly reflective tape, the skaters glow as if carrying lanterns as they stream by.

After the close of the Biennale the installation will be disassembled and nearly all of the materials recycled. In the end the only significant resources consumed will have been the energy used to transport the parts and the labor to install them. Although the thousands of orange cones will have temporarily transformed the plaza, they themselves will be unharmed and subsequently put into service as they were intended. They will serve out the rest of their existence warning both of physical risk and of impending changes in the world.

1756年,在清乾隆皇帝的要求下,意大利建筑师郎世宁为紫禁城外的圆明园设计了一系列欧洲风格的建筑和花园,其中一个花园呈迷宫状,被称为"万花阵"。源自古希腊,迷宫在当时的欧洲正是一种潮流。中秋节夜晚,皇帝便在万花阵举行比赛,参赛者提着黄色灯笼看谁先走出迷宫。

不幸的是,万花阵在当年英法联军火烧圆明园时被毁坏,于20世纪80年代重建。2011年秋天,为了庆祝2011年深圳·香港城市\建筑双城双年展,我们在深圳市民中心广场创作了一个临时性的装置作品,该装置全部由橙色安全锥构成,是对万花阵这座历史花园的致敬,也是对未来的展望。

在世界各地的建筑工地和马路上,橙色安全锥无处不在,这预示着一座新建筑、一条新公路或一个公共设施即将建成。安全锥作为现代化的象征之一,意味着危险,但也召唤着未来。安全锥对人们的行动是一种限制,也是对当地危险的一种警告,整体上看,它们是对正在进行的过度开发的一种提醒。安全锥色彩明亮,缠绕着反光胶带,吸引了人们的注意力,并用警告的形式预示着变化的到来。

这件装置的橙色安全锥就像艺术家画布上流淌的明亮橙色颜料一样,散布在广场四周。安全锥的摆放可以用摇摇欲坠来形容,无法自己保持平衡,被装饰性地排列成花的形状,不再以咄咄逼人而是以优雅的姿态呈现着,其控制的本性被剥夺了。它们变成一种有趣并令人愉悦的东西,可以只被当成一种有明亮色彩的玩具来欣赏。最大而最具仪式的公共空间,由一种小小的、街上随处可见的普通物件,得到了改变。

这种转变是空间上的,也是视觉上的。这个位于市民中心脚下的广场是一个巨大但简陋的空间,也是一个没什么亲切感的场所。虽然它可能适合举行公共仪式,但在日常生活,并不被人们所使用。直到晚上,它才被赋予了生命:晚上,广场是属于孩子们的,他们在光滑的地面优雅地滑着旱冰跑过。

该装置是一个开放而且好客的合成物,为此,安全锥被搭建成一个障碍跑道(为了保证不妨碍人们在夜间溜旱冰,我们在设计布局时咨询了一组滑旱冰专业人士的意见)。人们可以从任何一个点进入或者退出广场,每个人都被邀请到装置里进行探险。当人们在广场上滑行时,伴随他们的是蟋蟀的声音。这些声音从100个安全锥里传出,而这100个安全锥则分布在构成装置的成千安全锥中。在双年展开幕当晚的庆典上,将会有旱冰表演者在装置中进行旱冰表演,他们会身着镶有反光胶带的黑色衣服,滑过广场时会发光,好像手执灯笼一样。

双年展结束后,这个装置将会被拆解,所有拆解下来的材料都会被回收利用。这个过程是绿色节能的,唯一被消耗是在运输这些装置材料时所消耗的能源,以及安装它们所需要的人力。尽管成千上万的橙色安全锥被临时用来组成这个巨大的装置,但是被拆解后都完好无损,可以重新用来发挥其应有的功能:用于警示潜在的危险以及预示世界上即将发生的改变。

| SHENZHEN/ 深圳 | 10,000 FLOWER MAZE/ 万花阵 |

Testing the display styles of the safety cones in New York
设计师在纽约测试安全锥的摆放方式

ULTRA-LIGHT VILLAGE
超轻村

Curator: Terence Riley 　　策展人：泰伦斯·瑞莱

| SHENZHEN/ 深圳 | ULTRA-LIGHT VILLAGE / 超轻村 |

In exhibitions of architecture, it is typical to represent a structure with drawings, models or photographs. Notable full-scale exceptions to this rule have become more and more integral to architectural exhibitions, recalling Aldo Rossi's Teatro del Mundo at the 1980 Venice Biennale and the temporary summer pavilions built at MoMA PS1 in New York and Serpentine Gallery in London, including works by Frank Gehry, Zaha Hadid and Jean Nouvel.

Like those efforts, participation in the Ultralight Village project gave the architects and/or teams of architects the opportunity to explore their ideas in full-scale within an important civic space. Ultimately, six architects were invited to design structures: Amateur Architecture Studio from Hangzhou, China (WANG Shu and LU Wenyu principals); Clavel Arquitectos from Murcia, Spain (Manuel Clavel Rojo, principal); MOS from New York, USA (Michael Meredith and Hilary Sample, principals); OBRA from New York, USA (Pablo Castro and Jennifer Lee, principals); Studio UP from Zagreb, Croatia (Toma Plejic and Lea Pelivan, principals); and Wei Chunyu from Changsha, China. All of the participants were nominated by noted international critics, curators, and writers from the field of architecture and each of them has demonstrated not only an ability to design innovative structures but the ability to communicate ideas through the process of building.

Unlike other similar efforts, the Ultra-Light Village was conceived of as a grouping of structures with the possibility that the sum of the parts would offer yet another spatial and physical perspective on contemporary architecture. Furthermore, the Ultra-Light Village was constructed on the upper plaza of the Shenzhen Civic Center, giving the architects a specific kind of public space with which to react.

Unlike permanent architecture, these structures have dealt with several unique challenges and opportunities. As the plaza is the uppermost surface of a multi-level structure, no foundations could be excavated and the structures could not be anchored directly to the ground. Hence, the designers were challenged to conceive of structures using the lightest weight materials and to use them in the most efficient way. Such an attitude recalls the spirit of the legendary engineer and theoretician Buckminster Fuller, who challenged his colleagues with the famous polemic question: "How much does your building weigh?" As the biennale is addressing themes of sustainability, Fuller's query regarding the inefficiencies of traditional construction is even more relevant.

The six structures erected on the plaza respond to the issues at hand in unique and innovative ways. With their installation, Amateur Architecture Studio converted the plaza from a strictly ceremonial space to a potentially functional one. Using a design they developed from their research on temporary emergency housing, the architects constructed a 14-meter long tube of space supported by a lightweight wood structure and sheathed in polycarbonate sheets for protection. Intended to provide shelter for refugees, the installation is not only visually interesting, but also very inviting. Throughout the biennale, workers used it for their rest breaks, adding a touch of realism to the experimental prototype.

Clavel Arquitectos' Centrifugal Village extended the notion of "lightness" in various ways. Three spinning parasols, the underside of which was a reflective gold fabric, add a certain levity or lightness of mood to the installation. Under each canopy, this mood is also evident in the design of the seats, which in turn revolve on their central support. The introduction of motion emphasizes the relationship between dynamism and lightness and provides a welcome note of informality on the otherwise austere government plaza.

MOS' contribution was constructed of lightweight, machine-cut metal components with a painted pinkish color. From above, the structure could be seen as a series of interlinked, discrete volumes of spaces, each of which was roughly large enough to contain a single person. As such, the installation can be seen as describing a social structure of interaction as much as a physical structure. In contrast to the monumental scale of the government center, MOS's contribution emphasizes the critical scale of the individual over the collective.

OBRA's Oxymoron Pavilion addressed multiple themes seamlessly.

Covered in translucent textile mesh the structure is light-filled by day and glows like a lantern by night. It addresses the fundamentals of shelter as well as the importance of light in the transformation of architecture. The installation is also reminiscent of Tatlin's Monument to the Third International, whose spiraling form is often interpreted as symbolic of Communism's constant emphasis on progress and the future. With Deng Xiaoping's statue behind it in the distance, OBRA's design reminded us that architectural forms possess potentially strong narrative ability.

Studio UP's installation is a 14-meter long extrusion of space, expanding from 2.4m high at the entrance to 6.7m high at the far end. In section, the structure took the shape of a house, at least the kind of a house a child might draw. All of the surfaces are covered with acrylic sheets—a material that has a subtle reflective quality despite its dark appearance. Oriented towards the city, the interior walls reflect the urban scene like a kaleidoscope, especially so at night. In this instance, however, the viewer could enter the kaleidoscope and their image became intertwined with the reflections of the urban landscape, producing a sense of wonder and visual delight.

Wei Chunyu also took the opportunity to expand on his design work in that his installation of three grouped structures is essentially a very large model of the project he is developing with the Regional Construction Studio in Guzhang County in Western Hunan province. The polycarbonate constructions, the tallest of which is 4 meters, represent a polemical position that calls for rural homes to be aggregated into larger community structures. Thus, the architect again transforms the civic center from a ceremonial use to a didactic one.

During the opening ceremonies, the Ultralight Village became animated with hundreds of visitors. Temporary video screens added another dimension of lightness: the flickering, transience of digital images appearing amongst the six structures.

| SHENZHEN/ 深圳 | ULTRA-LIGHT VILLAGE / 超轻村 |

在建筑展中，使用图纸、模型和照片来表现结构，已经是参展的典型做法。对这一规则做出全方位的并引人关注的改动，日益成为建筑展一直尝试的策展策略，其中就有1980年威尼斯双年展上阿尔多·罗西（Aldo Rossi）的世界剧院（Teatro del Mundo），纽约MoMA PS1的短期夏季展馆和伦敦的蛇形画廊，包括弗兰克·盖里（Frank Gehry）、扎哈·哈迪德（Zaha Hadid）和让·努弗尔（Jean Nouvel）的作品都位列其中。

"超轻村"为建筑师/建筑师小组提供了这样的契机，他们可以在重要的城市空间全方位地探索他们的创见。最终，6名建筑师被邀请参与做结构设计：来自中国杭州的业余建筑事务所（王澍与陆文宇），来自西班牙穆尔西亚的Clavel Arquitectos（Manuel Clavel Rojo），来自美国纽约的MOS（Michael Meredith和Hilary Sample），来自美国纽约的OBRA（Pablo Castro和Jennifer Lee），来自克罗地亚萨格勒布的UP（Toma Plejic和Lea Pelivan），以及来自中国长沙的魏春雨。这几名参与者都是由著名的国际批评家、策展人及建筑领域的写作者提名，他们中的每一个人都不只是在设计创新建造方面显得才华出众，在建造过程中，他们也擅长沟通自己的创见。

与其他项目不同的是，"超轻村"被设想为一组结构，同时提出了当代建筑中有关空间与物理新观念的可能性。此外，"超轻村"被安排在深圳市民中心上层广场——在这一特定空间中，建筑师的创建也能对空间有所呼应。

与永久性建筑不同，这些建造物必须面对几个独特的机遇和挑战。因为指定的广场位于一座多层建筑的最顶层，建筑师既不能挖掘地基，而建造物也不能直接接地。所以，设计师要面对的首要挑战是他们必须使用最轻材质，并要发挥出其最大效用。由此，这些建造物依循于传奇工程师、理论家巴克明斯特·富勒（Buckminster Fuller）的精神，众所周知，他有一个颇能引发辩论的提问："你的建筑有多重？"——这直指传统建筑的低效用。因为这一届双年展提到关于持续性的论题，富勒的质询与这个话题的关系也更为密切。

竖立在广场上的六个建造物用独特和创新的方式回应了在即问题。业余建筑事务所通过装置将广场从严格意义上的纪念性空间转换为有潜力的功能空间。他们的作品运用了其从临时应急住房的研究中生发出的设计，建筑师建造出一条14米长由轻量木结构和聚碳酸酯板材支撑保护的管状空间。这一建造旨在为难民提供庇护场所，它不仅在视觉上充满趣味性,而且对人们来说颇具吸引力:双年展期间,工人把它用于休息,在这一实验性模型中加入了一丝现实主义。

Clavel Arquitectos 的离心村从各个维度延伸了"轻盈"的概念。三个底面是可反光的金色的纺纱阳伞成为富有变化或轻盈感的装置。在伞下绕转中央支撑物的座位设计中,这种动态体现得十分明显。运动状态的引入强调了动态和轻盈之间的关系,为严肃的政府广场提供了非正式的致意。

MOS 呈现的是轻巧的由机器切割的粉色涂绘金属部件。从上向下俯视,建造物看上去是相互连接而又相互离散的空间体块,而每个空间大约可容纳一人。因此,此件装置被视为社会交互与物理结构的结合。相对于政府中心的巨大规模,MOS 的贡献在于强调了个人尺度优于整体。

OBRA 的矛盾馆将多个主题连接于一体。建造物覆盖于半透明的织网之中,充满了来自白昼的光亮,夜晚又如同灯笼一般闪烁。它阐明了庇护的基本要素,以及光在建筑转变中的重要性。这件装置也让人联想到塔特林第三国际塔,其不断上升的形式往往被视为象征着共产主义的进步和未来。位于 OBRA 的矛盾馆后方的邓小平雕像,也提醒着我们建筑形式所隐含的强烈的叙事能力。

Studio UP 建筑事务所的装置是一个 14 米长的扩展空间,其高度由入口处的 2.4 米不断扩展为最远端的 6.7 米。空间中的一段采用了房子——毋宁说是孩子绘制出的房子——形状。建造物的表面包裹着亚克力板——这是一种有着暗色外表却有微妙反光的材质。面向城市的内墙如万花筒般映射出都市镜像,到了夜晚景色更甚美妙。正是在这种环境中,观众得以进入这个万花筒,他们自己的形象与都市风景的影像交织在一起,能够产生出惊奇的视觉愉悦感。

魏春雨也用参展方案来继续延展自己位于湖南省古丈县和地方营造一起设计的三组建筑物的模型。这组聚碳酸酯构筑物中,最高的为 4 米,代表了需要农村家庭聚集成更大群落结构的讨论立场。因此,建筑师再次将市民中心从纪念性使用转型为教化功能。

开幕式期间,因数百名游客的参与,"超轻村"显得生气勃勃。临时电视屏幕增加了轻盈的另一个维度:闪烁着的临时的数字图像出现在六座构筑物之间。

| SHENZHEN/ 深圳 | ULTRA-LIGHT VILLAGE / 超轻村 |

Amateur Architecture Studio / 业余建筑工作室

Hangzhou, China / 中国杭州

1\ The design proposal was given for the invitation of the project "Home-for-All" after the earthquake in Japan. The project aims to solve problem of house shortage after the earthquake. It calls for light and fast-in-construction design proposals. Therefore, we conceived a construction activity that every disaster victim could participate. To keep the lowest level of neighborhood relationship, the residential units, which are light structures, for a temporary village. It consists of a serial specific ideas and procedures.

2\ Assume a temporary building with its material easy to get, with the least amount and smallest scale, easy to move and simple to construct.

3\ The kind of building those residents, who needs temporary house, could join the construction. Because the method is simple, even those who don't understand construction are able to learn how to construct.

4\ Building with the most basic neighborhood; every unit with two apartments and share one entrance and a communication platform.

5\ Elevate the ground floor in order to isolate moisture and rain; the double layer roof forms thermal insulation system and ventilation system.

6\ The mixture of bedroom and restroom provides flexibility in function. Every unit is able to accommodate four single beds, where 800mmx2000mm tatami is able to lay down. Under the bed is a storage cabinet.

7\ There is a half out-space tea terrace in every apartment. Though it is a temporary building, but the residents live with dignity and enjoy the comfort.

8\ The apartment does not have bathroom and kitchen. It is planned to combine a group of buildings to form a community unit and furnish kitchen and bathroom in one standard building unit.

9\ We believe that the light anti-disaster village is well-matched with the theme of biennale 2011. It provides a good opportunity to test the feasibility of the construction experiment. But because of the restriction of budget provided by the Biennale, half standard unit of the proposal could be built, that is, a basic form of one temporary residential apartment; besides, the original proposal is simplified.

1. 这一方案是应日本震后项目"人类家园"（Home-for-All）之邀的设计，以解决震后房屋短缺为目标。它要求轻质以及速成的设计方案。因此，我们构想了一个受灾人员能够迅速开展起来的建筑工程。为了保证邻里的基本关系，所有轻质结构的居住单元形成了一个临时的村落。该项目包含了一系列具体概念和步骤。

2. 构想出一种材料易获得、规模和用量极小，易于移动和搭建方式极简单的临时建筑。

3. 使用这类建筑的居民，是那些需要临时房屋的人群，他们都能够加入这个建造过程中。正是因为方法简易，即便是那些不真正懂得建筑的人，也可以习得搭建的方式。

4. 利用最基本的邻里关系进行建造；每一个单元都有共用同一个入口及交流平台的两间公寓。

5. 从地面抬高是为了隔离湿气和雨水；双层屋顶形成了隔热系统和通风系统。

6. 卧室与卫生间的结合形成了一种功能上的灵活度。每一个单元都放得下4张单人床，放得下800毫米x2000毫米的榻榻米。床下则是储物箱。

7. 在每间公寓中，都有一个半户外的茶室。尽管建筑都是临时搭建的，但居民们依然可以保有尊严，享受闲适。

8. 在设计中，公寓并没有浴室和公寓。原意是希望把一组建筑组合在一起形成一个社区团体，并在一套标准公寓中设计其厨房和浴室。

9. 我们相信，这个轻质的抗灾村庄在主题上与2011年的双年展主题十分契合。它提供了一个良好的契机，用以测试这一建筑实验的可行性。但是正是因为预算的限制，能建成的标准公寓——临时公寓的基本形式——仅为一半；除此之外，原本的方案进行了简化。

SHENZHEN / 深圳　　ULTRA-LIGHT VILLAGE / 超轻村

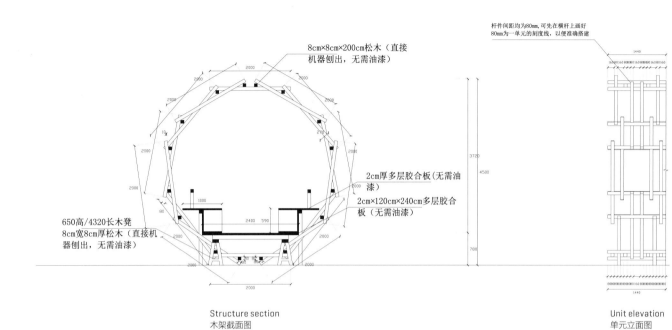

Structure section
木架截面图

Unit elevation
单元立面图

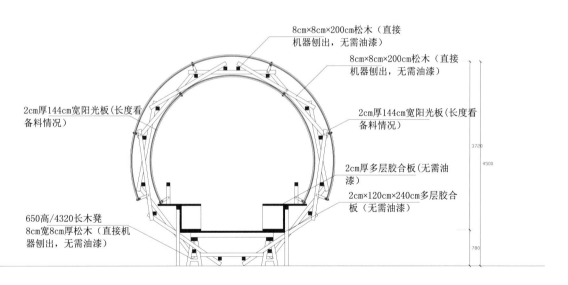

Lateral section
短向剖面图

Amateur Architecture Studio / 业余建筑工作室

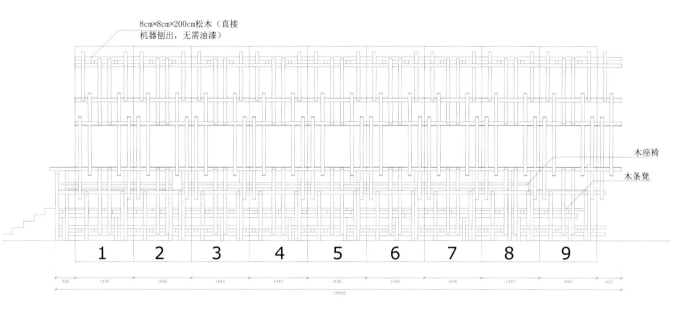

Structure elevation
木架立面图

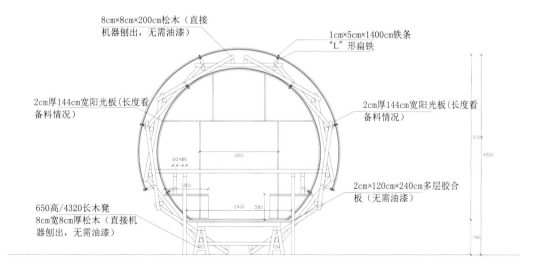

Lateral elevation
短向立面图

| SHENZHEN/ 深圳 | ULTRA-LIGHT VILLAGE / 超轻村 |

Lateral section diagram
横向截面示意图

Step 1: Preparation
第一步：准备

Step 2: Sorting
第2步：选料

Step 3: Building base
第3步：搭建底座

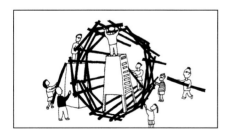

Step 6: Installation of top beams
第6步：安装顶端横梁

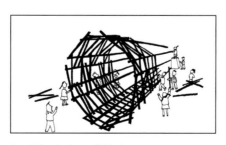

Step 7: Continuing to finish the rest
第7步：依此步骤继续安装余下部分

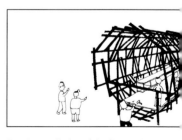

Step 8: Installation of the floor and seat
第8步：安装地板和座椅

Amateur Architecture Studio / 业余建筑工作室

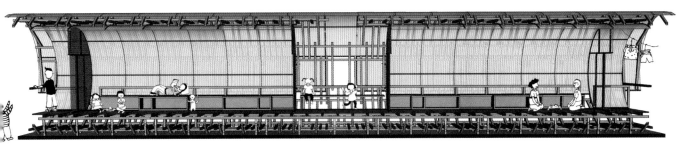

Longitudinal section diagram
纵向截面示意图

Step 4: Installation of lateral beams and vertical poles
第4步：安装侧面横梁和纵向杆件

Step 5: Building the rest of vertical poles
第5步：搭建其余横纵向构件

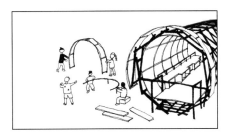

Step 9: Installation of sun shade panels
第9步：安装阳光板

Step 10: Completion
第10步：完成

| SHENZHEN/ 深圳 | ULTRA-LIGHT VILLAGE / 超轻村 |

Clavel Arquitectos

Murcia, Spain

Centrifugal Village

"How much does your building weigh?" Asked Buckminster Fuller and now again Terence Riley asks the six teams were invited to the Ultra Lightweight Village project.

Apart from the obvious answer we could conclude that any structure avoiding the effects of the gravity would weight nothing. So can we make gravity disappear? No, but we could compensate for it.

Centrifugal Village is based on that concept. Three circles of waterproof light fabric of 7.8, 6.4 and 5.4 meter of diameter spin around their axis at only 1.5 turns per second. In the biggest one we can reach a cantilever of nearly 4 meters with an only 2 mm roof thickness. It is interesting to check how similar the movement with the aquatic animal one is. Thanks to the rotation the gravity apparently disappears and only aerodynamical forces shape the fabric. The soft waves produced the surface create a smooth breeze that improves the thermal conditions behind during the hot and sunny days.

In a world where we charge our mobile phones wirelessly and we send information through the air, why should we not dream about a not so far future in which our buildings are sustained by other forces, leaving gravity in a second line.

PUBLIC SPACE: The pavilion creates a specific place for children in the huge square. Children can experiment with the centrifugal forces and generate electricity to open the structures and activate the lighting. The faster the children's chairs rotate, the more light the structure receives.

WHAT HAPPENS LATER: Children's chairs and the surrounding soft floor can be installed in other parts of the city, extending the life of the plays after the biennale. Electric materials can be used for industrial purposes. The red and gold fabric can be recycled in bags.

The renowned German engineer, Werner Sobke, experimented with rotating umbrellas as early as the 1990's. For more information, visit www.wernersobek.com or see the book *Werner Sobek: Art of Engineering = Ingenieur-Kunst* by Werner Blaser (Birkhauser, 2000).

The umbrella spins when children play with the chair underneath.
当孩子们玩下面的椅子时，伞也旋转。

| SHENZHEN/ 深圳 | ULTRA-LIGHT VILLAGE / 超轻村 |

ONE DAY OTHER FORCES WILL SUPPORT OUR VILLAGES

Clavel Arquitectos

西班牙穆尔西亚

离心村

你的建筑有多重？巴克明斯特·富勒（Buckminster Fuller）曾经问过，而现在泰伦斯·瑞莱对受邀参加超轻型村庄项目的六个团队提出了同一个问题。

除了显而易见的答案，我们还可以推断说任何避免重力影响的建构都会没有重量。因此，我们能使重力消失吗？不能，但我们可以作出补偿。

离心村正是基于这样的理念来实现的。直径分别为 7.8 米、6.4 米和 5.4 米的三圈防水轻纤维以每秒一圈半的速度绕轴旋转。体积最大的那一个圈，可以做出近 4 米高的悬臂，而顶板厚度只有 2 毫米。仔细观察，你会发现它与水生动物的形态与动作如此相仿，确实十分有趣。因为旋转的缘故，重力明显地消失了，只有空气动力在形塑着纤维。柔软的波动在表面造出"O"形，制造出轻柔的微风，改善了热天和暴晒的日子里的酷热。

在这样一个我们为自己的移动电话进行无线充电，并由空气传递信息的世界，为什么我们不能梦想在一个不那么遥远的未来，我们的建筑物也会由外部其他的力量来维持，而将重力放在第二位。

公共空间：展馆在巨大的广场上专为孩子们设立了一个空间。孩子们可以体验到用离心力发电来开启结构物，并启动照明设施。儿童椅旋转得越快，照明设施接收到的能量就越多。

之后会发生什么：在双年展结束之后，儿童椅和围绕着它们的软地板会被安装到城市的其他地方，这些游乐设施的生命会得以延续。电气材料会被用于工业。红色和金色的纤维材料可以被回收。

享有盛誉的德国工程师 Werner Sobke 早在 20 世纪 90 年代就以滑翔伞进行了实验。获取更多的信息，请登陆 www.wernersobek.com，或参看《维尔纳·索贝克：工程的艺术 = 工程师 – 艺术家》，维尔纳·布拉瑟著。

The profotype was tested at the workshop in Spain.
在西班牙的工作坊里测试原型。

SHENZHEN/ 深圳 | ULTRA-LIGHT VILLAGE / 超轻村

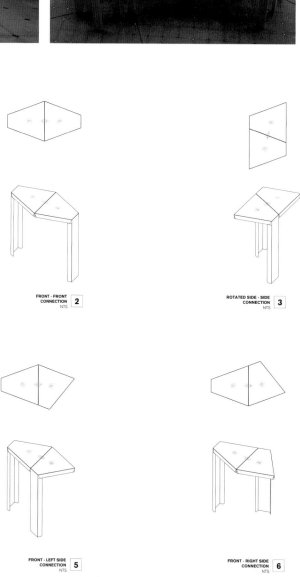

1 BACK - BACK CONNECTION — NTS
2 FRONT - FRONT CONNECTION — NTS
3 ROTATED SIDE - SIDE CONNECTION — NTS
4 MIRRORED SIDE - SIDE CONNECTION — NTS
5 FRONT - LEFT SIDE CONNECTION — NTS
6 FRONT - RIGHT SIDE CONNECTION — NTS

Connection options of adjacent units
相邻单元的连接方式

MOS

New York, USA / 美国纽约

This pavilion is repetitive, it's pink, and it was done quickly on a shoestring budget. We came up with a system, produced tons of variations and then we chose one—variation #18. When you tell people that, inevitably they say "Well how could you could choose this variation over that variation?" That is one of our least favorite questions—it assumes tautological arguments are the penultimate architectural goal. Ugh. We can't think of anything worse than a perfectly closed system—just because you use a positivist process to design (or not design) doesn't mean you should use that process to evaluate it.

该馆是一个粉色的重复性场馆，只花费了很短的时间以及十分有限的预算。我们构思了一个系统，并生产了大量的变体，我们从中选择了第 18 号。当与别人说起这个话题时，不可避免地会被别人问道："你是如何从这么多变体中做出选择的呢？"这是我们最不爱回答的问题，它假设了重复的争论是排在倒数第二位的建筑目标。啊！我们很难想到还有比一个完全封闭的系统更加糟糕的事物——因为你引入的是一个实证主义的设计（或不设计）过程，但这并不意味着你应该用这一过程去评价设计本身。

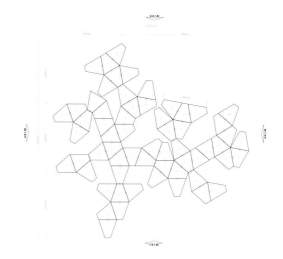

Option A plan
组合方式 A 平面图

| SHENZHEN/ 深圳 | ULTRA-LIGHT VILLAGE / 超轻村 |

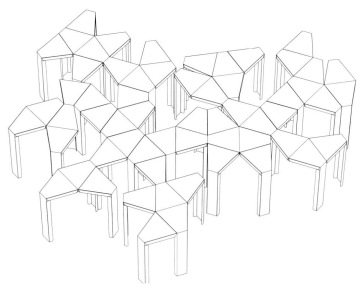

Option A axonometric / unit count: 60
组合方式 A 轴测 / 单元数量：60

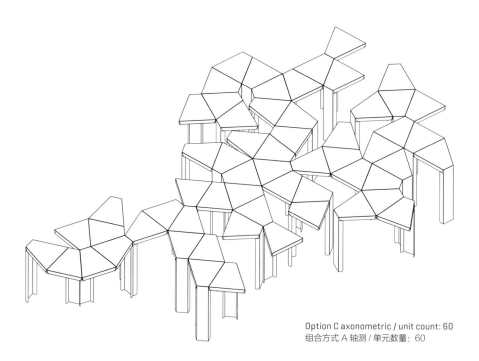

Option C axonometric / unit count: 60
组合方式 A 轴测 / 单元数量：60

MOS

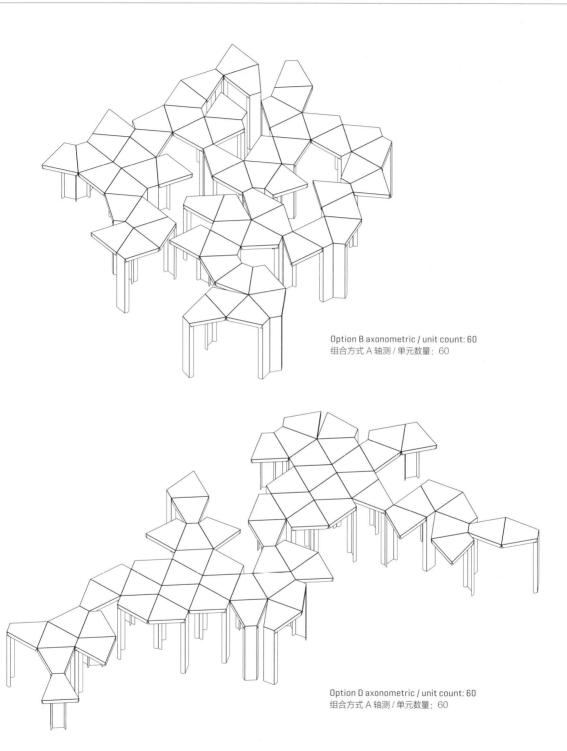

Option B axonometric / unit count: 60
组合方式 A 轴测 / 单元数量：60

Option D axonometric / unit count: 60
组合方式 A 轴测 / 单元数量：60

SHENZHEN/ 深圳 | ULTRA-LIGHT VILLAGE / 超轻村

OBRA
New York, USA

Oxymoron Pavilion

Will the current ubiquitous environmentalism save our cities and the world? On Mondays, Wednesdays and Fridays. it seems like it could. On Tuesdays, Thursdays and Saturdays. it appears rather unlikely. The development of the project for an Ultralight Village in Shenzhen presents to us an opportunity to suspend relative disbelief and attempt an exemplary contribution.

Beyond the quantifiable benefits of "lightness" in architecture as a strategy to preservation of scarce resources, the idea of the Ultralight Village appears as a providentially appropriate way to consider the nuances of an interdependence between architecture, city and environment.

A work of architecture in the city is traditionally regarded as belonging to one of two types, either a monument or a part of the fabric. The monument of course celebrates momentous events, perpetuates foundational myths or aggrandizes the humanities of celebrated individuals; the fabric is everything else and is supposed to create the context in which the monuments can exist. The monuments are symbols, "exemplary" and artificial moments in the city. Precisely because their role is to perpetuate certain conditions, they are deprived of the possibility of changing – ergo all that stone and bronze. And unchanged, they are confident and certain, the architectural equivalent of an unshakable conviction, or of a faith, they are lifeless. Life is happening elsewhere in the doubt, confusion and chaos that typically be found in the city's fabric.

The Ultralight Village of the SZHKB is an oxymoron, a beautiful self-contradiction both monumental—because anything worth building in a biennale shares a measure of the monumental—and ephemeral. This paradox becomes our guiding principle in the development of this modest work of architecture. This little building can become a salute to the millions of souls living in Shenzhen. Like most other people that live in other gigantic metropolises of the earth, they share a life unfolding amongst the fragments of a larger reality no one understands, and their lives may be sometimes overwhelmed by difference and hesitation. This pavilion can be anti-monument to them, an ephemeral monument for those who do not want to impose their opinion or perpetuate their situation. A pavilion as an ephemeral "monument", an Oxymoron Pavilion.

| SHENZHEN/ 深圳 | ULTRA-LIGHT VILLAGE / 超轻村 |

Jennifer Lee & Pablo Castro, Partners of OBRA
OBRA 事务所合伙人 Jennifer Lee & Pablo Castro

OBRA

美国纽约

矛盾馆

当下十分流行的环境保护论可以拯救我们的城市和这个世界吗？在星期一、星期三和星期五时觉得似乎有这个可能，星期二、星期四和星期六时，则又觉得不具备这样的可能性。在深圳，项目超轻村的发展似乎提供给我们一次机会，暂止疑虑，直接进行一次有益尝试。

除了把在建筑中的可计量的"轻质"作为保存稀有资源的策略之外，超轻村的概念则显然是一种考量建筑、城市和环境彼此间相互依靠的微妙关系的有益方式。

从传统上来说，城市中的一项建筑工程只会被看作是两种形式之一：要么是纪念碑式的，要么是肌理的一部分。纪念碑式的建筑物当然是用于纪念式的庆典，让人不要忘怀奠基性的传说，或者用于吹捧名人的人文精神；而肌理是所有其他的，并被认为是创造纪念碑可以存在的语境。纪念碑是城市的标志、"典范的"以及人造的片刻。确切地说，正是因为它们的角色是铭记某些特定的情形，所以它们缺乏改变的可能性——于是决定采用石头和铜。正是因为稳定性，它们是自在及肯定的，与不可撼动的坚信或者信仰的建筑等量，可以说是无生命的。生命，它在别处有着怀疑、迷茫和混乱，不出所料地会发生在城市的肌理中。

此次深圳双年展中的"超轻村"好比一个逆喻，一个美丽的自我矛盾，但同时却具有纪念碑性——因为双年展中的任何值得建构的事物都分享了纪念碑性这一特点——以及短暂性。这一自相矛盾成为了这次建筑设计的纲领。这座小小的建筑物可以成为对居住在深圳的百万灵魂的致敬。正如大多数居住在地球上其他超大城市的人们一样，他们共享着没人可以理解的、掩藏在更大现实的碎片中的生命，他们的生活有时会被差异和犹豫所淹没。这一项目可以说是"反纪念碑"的，它是一座为那些不想强加他们的观点或铭刻他们的境况而建的短暂的纪念碑。矛盾馆，是一座短暂的"纪念碑"。

SHENZHEN / 深圳   ULTRA-LIGHT VILLAGE / 超轻村

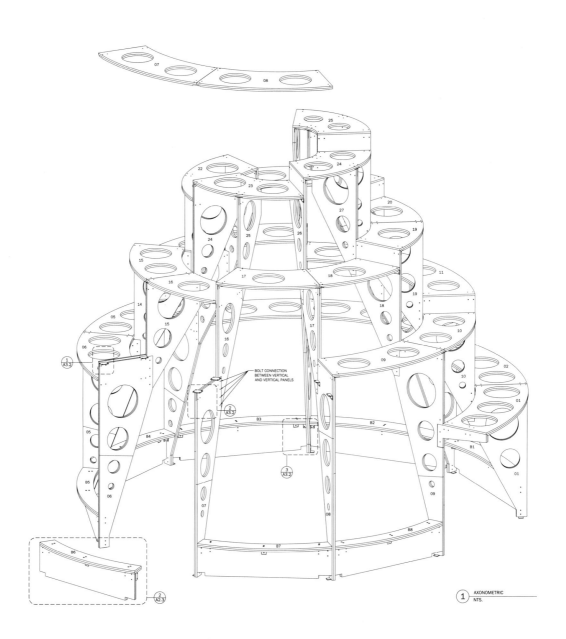

Axonometric
轴测图

Working model
工作模型

SHENZHEN / 深圳 | ULTRA-LIGHT VILLAGE / 超轻村

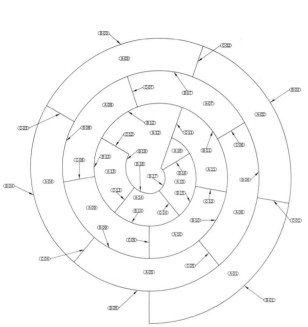

Surface fabric plan
表面布料平面

Surface fabric axonometric
表面布料轴测面

OBRA

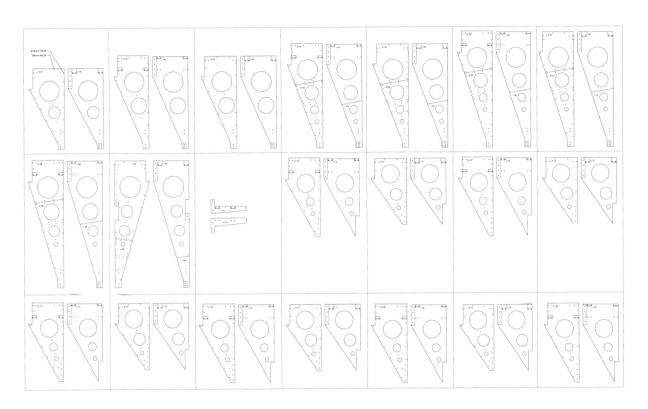

Vertical panel layout
垂直面板

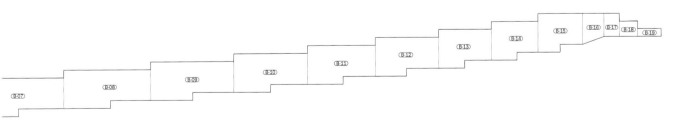

Unrolled vertical surface fabric
垂直面布料展开图

SHENZHEN/ 深圳 | ULTRA-LIGHT VILLAGE / 超轻村

Studio Up

Zagreb, Croatia / 克罗地亚萨格勒布

Gizmo / 小发明

As part of the Ultra-Light Village, STUDIO UP's GIZMO questions the concept of lightness not as a physical condition, but as an experience. It serves as a device, challenging the ideas of interior and exterior, public and private space, nature and manufactured landscape.

It confronts people and space around them in a paradoxical way—one is more aware of the outside world when is inside. Such situation initiates one's senses and thoughts, provoking a certain reaction to the exaggerated world they see.

The two opposing entrances relativized the border between collective and individual scale, creating yet another paradoxical scene. At singular entrance, the exterior is hypertrophied; at mass entrance, the focus is the individual.

The reflective interior generates the nomadic aspect. Depending on location and position, Gizmo is each time a different space: the reflections of the exterior change the space within. Radically.

作为"超轻村"的一部分，Studio Up 的作品《小发明》质疑了"轻质"的概念不作为一种物理情形，而被当作一种体验。它作为一种装置，挑战了内部和外部的观念、公共和私人的空间、自然和制造的景观。

它以自相矛盾的方式同时面对人和其周围的空间——在内部，人们就会更加关心外部的世界。这一情形激发了人的感觉和思考，也唤起了人们对所见的夸张世界的某些反应。

两个对着的入口将集体领域和个人领域的边界对立起来，由此创造了另一种模棱两可的场景。若从小的入口进入，观看到的是一个多重的外部世界；若是从大的入口进入，其焦点是单一的对象。

这一反射性质的内部则产生了游牧性的一面。《小发明》根据不同的地点和位置，在不同的时间都呈现了不同的空间：外部的各种反射改变了内部空间，彻底地将其改变。

 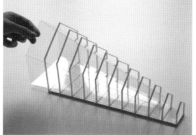

Working model
方案工作模型

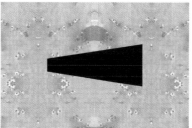 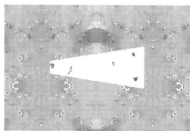

Top view
顶视图

Plan
平面图

SHENZHEN/ 深圳　　ULTRA-LIGHT VILLAGE / 超轻村

Regional Construction Studio [College of Architecture Hunan University] / 地方营造建筑室（湖南大学建筑学院）

Changsha, China / 中国长沙

YanPaiXi Village / 岩排溪村

YanPaiXi Village is located in Guzhang County, western Hunan. The houses in the village surrounded by terraces were built along the mountain. The terraces are divided into a number of large and small blocks by channels which that are generated by nature. The society is steady and the residents are sincere and honest. It is a picture of Shangri-La.

Inspired by this village, this installation attempts to explore a intergrowth prototype, implanting this "light" model in urban and rural space.

Light:
YanPaiXi Village developed a sustainable way of traditional self-sufficing, and formed a intensive land-use mode which integrated farming into daily life. That is the real "Light", the people obtained most suitable living space with minimal consumption of resources. This is our understanding of "light" in "ultra-light village".

Intergrowth:
The plan uses a free enclosed settlement mode, as little occupation of the land as possible, to ingress the rural life into the modern space. The design was derived from imagination about the kind of "arable" urban space which was three-dimensional and intensive residential space. It intends to create a green circulation system which includes manufacturing, leisure, and entertainment etc.

岩排溪村位于湖南省古丈县，村落依山而上，周边梯田环绕。梯田、渠道、浅丘层层叠叠，宛若天成。村民安居乐业，民风淳朴，一派桃源景象。

受此启迪，本次装置试图探索一种共生原型，将一种"轻"的模式植入城乡。

"轻"：
岩排溪村紧凑节地的聚落形态及生产与生活一体化的生存状态是一种传统的自给自足的可持续方式，是真正的"轻"：以最少的资源消耗获得最适宜的生存空间。这是我们理解的"超轻村"的"轻"的本质内涵。

"共生"：
设计以围合聚落形态自由组合，尽可能少地占用土地，将田园生活导入现代空间，畅想一种"可耕作"的城市空间，立体化集约空间使居住、生产、休憩、娱乐纳入一种绿色循环体系。

| SHENZHEN/ 深圳 | ULTRA-LIGHT VILLAGE / 超轻村 |

Vision for Yanpaixi village
岩排溪村远景

Regional Construction Studio [College of Architecture Hunan University] / 地方营造建筑室（湖南大学建筑学院）

Node structure
节点构造

Basic Components: Semi-transparent polycarbonate board (30mm thick), Semi-transparent polycarbonate block (40 mm diameter / 30 mm thick), Metal screw suite.

Construction measures: Combination the board and block in series with screws. Fixed both ends with nuts.

基本构成：半透明聚碳酸酯板（厚30mm）、半透明聚碳酸酯块（直径40mm / 厚30mm）、金属螺杆套件。

构造措施：将主体板与垫块依次串接在对应长度的金属螺杆上，两端用螺母固定。

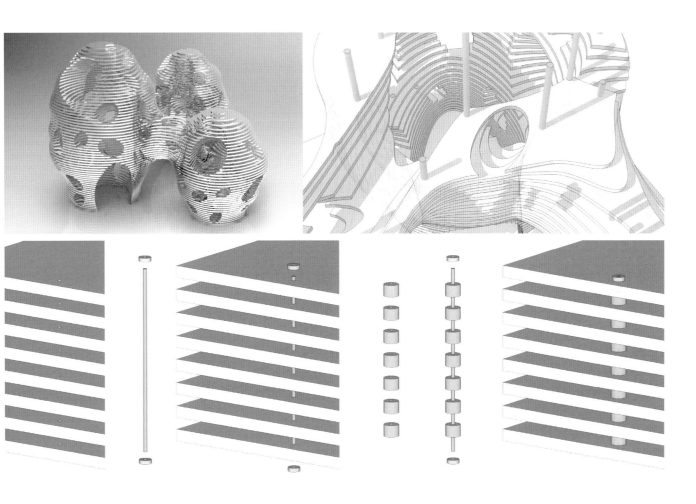

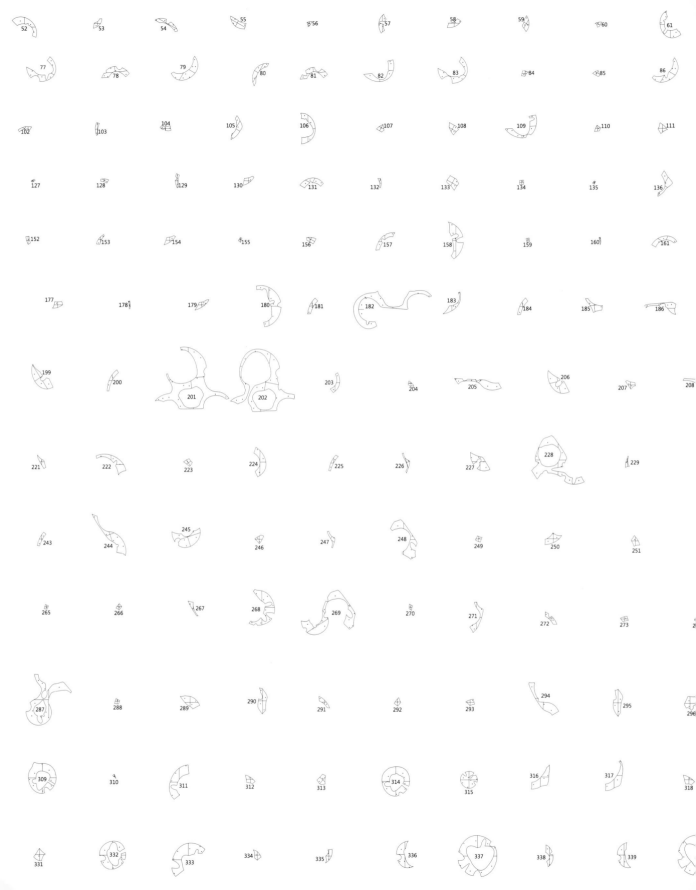

SHENZHEN BUILDS
深圳建造

Curator: Terence Riley　　策展人：泰伦斯·瑞莱

| SHENZHEN/ 深圳 | SHENZHEN BUILDS / 深圳建造 |

In contemporary architecture, the scale of certain projects is so big that the architect becomes not only a designer of buildings but also a city planner and landscape architect, reflecting this Biennale's theme: Architecture creates cities. Cites create architecture. Each of the projects selected, when complete, had a transformative affect on the urban fabric of Shenzhen as well as on the lives and habits of millions of its citizens. Beyond Shenzhen, each of the projects also serve as case study in terms of complexity and scale, reflecting Shenzhen's and China's leading position in the world as a laboratory for urban and architectural experimentation.

In many ways, the buildings being built in Shenzhen today are setting international standards for each of their respective building types. Shenzhen Builds presents five major projects for Shenzhen that are currently in progress—either in design or in construction—and includes the work of both international and Chinese architects: Shenzhen TV Tower by Atelier FCJZ/Yung Ho Chang, the Museum of Contemporary Art and Urbanism by Coop Himmelb(l)au / Wolf Prix, Terminal 3 at Shenzhen Bao'an International Airport by Massimiliano and Doriana Fuksas, the Stock Exchange by OMA/Rem Koolhaas, and Qianhai Metro Plaza by Urbanus. The exhibit focuses on the architect's design process and how it affects environmental, architectural and urban design issues in the city of Shenzhen. The architects have developed their own presentation using models, animations, and drawings, including developmental materials from their offices to demonstrate how these projects were conceived and developed.

Atelier FCJZ - Atelier Feichang Jianzhu
SZ TV Tower

Atelier Feichang Jianzhu (FCJZ) is a leading architecture and design office in China. FCJZ is internationally-honored with most prestigious recognitions in the fields of architecture and art. Projects of FCJZ range from cultural facilities, public buildings, housing, office design, retail and commercial design, lab design to interior design, urban planning, strategic planning and landscape design. Besides, FCJZ has the capacity of product design and exhibition design. The office's major projects include Shanghai Corporate Pavilion at the World Expo 2010, The Bay in Shanghai, Samho Building in Paju Book City, Korea, Ufida R&D Center 1, Split House, Villa Shizilin, Hebei Education Publishing House, Tang Palace Restaurant in Hangzhou, Xishu Bookstore, etc. Recent design projects include JIA teapot, 'FCJZ 1/2' fashion design, TTF Jewelry Design and furniture design for MUJI. In 2008, FCJZ won the international competition for the SZTV tower in Shenzhen.

Atelier FCJZ is endeavored to actively engage in sustainability and interdisciplinary innovation. On-going projects, including Novartis Campus, Jiading Advertisement Park both in Shanghai, and Qianmen Historical Preservation and Regeneration in Beijing, all take on these new directions. Atelier FCJZ received Business Week / Architectural Record China Award in 2006, a WA award in 2004 and China Architectural Arts Award in 2003 and 2004.

The 47-story, 200-meter tall skyscraper is an expansion of the existing facilities of Shenzhen Television and Broadcasting Group in the central area of the Futian financial and administrative district in Shenzhen. Their initial research discovered a similar approach among most high-rise building designs—the universal flat glass curtain wall—from northern hemisphere to the southern one and led them to focus on the following questions: Can a high-rise building have regional qualities? Or, what is the southern skyscraper? These questions helped them to formulate a design strategy that responds to the local climatic and environmental conditions of Shenzhen: We want to design a tower that can simply shade itself so that to reduce the heat gain for the building. As the result of this direction, they designed a skyscraper with a sculpted, multifaceted surface that will cast shadows onto itself during the summer season. The geometrical configuration on the south and north elevations differs from the one on the east and west due to the different sun angles during the day. Building Integrated Photovoltaic (BIPV) technology is also part of the facade system and applied on the upward-facing triangular surfaces on the south. On the roof top are the more conventional solar panels.

For the TV performance and broadcasting hall in the low-rise portion of the project, water is used as the material on the building envelope surface. The intention behind it is to create a pleasant micro climate in a southern urban plaza on the ground level.

Coop HImmelb(l)au
Museum of Contemporary Art & Planning Exhibition, Shenzhen, China 2007

The Museum of Contemporary Art & Planning Exhibition (MOCAPE), part of the master plan for Shenzhen's new urban center, the Futian Cultural District, establishes itself as a new attraction in Shenzhen's fast growing urban fabric.

The project is conceived as the synergetic combination of two institutions, the Museum of Contemporary Art (MOCA) and the Planning Exhibition (PE), whose various programmatic elements, although each articulated according to their functional and performative requirements, are merged in a monolithic body enveloped by a multifunctional facade.

The transparency of this facade and the interior lighting concept allow for a view from outside through the exterior envelope deep into the volume of the space, thereby particularly accentuating the shared entrance and circulation space between the two museum volumes. At the same time, the building skin also allows unhindered view from the inside on the cityscape while giving the visitor the impression of being in a pleasantly shaded outdoor area—an impression enhanced by very wide spans which allow for completely open, column-free and flexible exhibition halls with heights ranging from 6 to 17 meters.

Upon entering the large interstitial space between the two museum volumes, the visitors can, via ramps and escalators, reach the main level, where a kind of public plaza serves as the circulation hub for the whole complex and as orientation and starting point for museum tours. From here also many auxiliary public and private facilities are accessible, including cultural services, a multifunctional hall, several auditoria, a library, a cafe, a book store, and a museum store.

Massimiliano and Doriana Fuksas
Shenzhen Bao'an International airport, 2008

The master plan for Shenzhen Bao'an International Airport intends to offer world-class transportation services. Achieving this means that Shenzhen Airport must operate at the following levels.

T3 at Shenzhen Bao'an International Airport must serve as the global aviation gateway between China and the rest of the world.

As the fourth largest following Beijing, Shanghai, and Guangzhou in China, Shenzhen Bao'an International Airport is a hub for domestic and international flights. Enhancement of international and domestic flights will improve the distribution of wealth among all Chinese cities.

Currently, there are more than 40 cities with population of over one million within 3.5 hours flying time of Shenzhen. This makes Shenzhen an excellent location for a regional hub. Passengers traveling from neighbouring cities could fly to Shenzhen to connect with long distance global flights or flights to other Asian cities.

As the gateway to China and the area of Shenzhen, the Terminal 3 has the unique opportunity to shape the first and last impression of the traveller, national and international. The concept provokes the image of a sculpture with its organic shape. The skin of the airport, consisting of an inner skin and an outer skin with the structural elements in between is enveloping the passenger by surrounding him with the airy structure on his way through the airport, defying classic building concepts. The structure is carefully designed to allow different light experiences while being in constant communication with the outside.

At the core of this concept is the honeycomb, a panel as the unique physical element of the skin. Its design optimizes the performance of materials selected on the basis of local availability, functionality, application of local skills, and low cost procurement. Each panel varies in glass size and in angle of the opening, carefully managing to produce the different light experiences needed for the various functions inside the airport.

The passenger terminal and concourse constitute the major portion of the passengers' perception of an airport. Factors which affect passengers' impressions include processing times, walking distances, ease of orientation, crowding, and availability of desired amenities. Each of these has been carefully considered in developing the concept for the passenger terminal and concourse. The terminal accounts for views to the outside and is planned under a single unifying roof canopy. Natural light is filtering through the double skin. The skin avoids direct sunlight to reduce energy consumption and creates an elegant atmosphere.

OMA
SZ Stock Exchange, 2006

For millennia, the solid building stands on a solid base; it is an image that has survived modernity. Typically, the base anchors a structure and connects it emphatically to the ground. The essence of the stock market is speculation: it is based on capital, not gravity.

In the case of Shenzhen's almost virtual stock market, the role of symbolism exceeds that of the program—it is a building that has to represent the stock market, more than physically accommodate it. It is not a trading arena with offices, but an office with virtual organs that suggest and illustrate the process of the market. All of these factors suggest an architectural invention: our project is a building with a floating base. As if it is lifted by the same speculative euphoria that drives the market, the former base has crept up the tower to become a raised podium. Lifting the base in the air vastly increases its exposure; in its elevated position, it can 'broadcast' the activities of the stock market to the entire city. The space liberated on the ground can be used as a covered urban plaza, and it is large enough to accommodate public events. The Shenzhen Stock Exchange—which will rise to 248m—is planned as a financial centre with civic meaning, located in a new public square at the meeting point of the north-south axis between Mount Lianhua and Binhe Boulevard, and the east-west axis of Shennan Road, Shenzhen's main artery.

The raised base of the SZSE is a three-storey cantilevering podium floating 36m above the ground, with a floor area of 48,000m^2 and an accessible roof garden. The podium and lower tower contain the dedicated stock exchange functions, with the SZSE offices, registration and clearing house, Securities Information Company and ancillary services located higher in the tower. The tower is flanked by two atria—a void connecting the ground directly with the trading floor. Staff enter to the west, the public to the east.

The generic rectangular form of the tower obediently follows the surrounding homogenous towers, but the facade of SZSE is an innovative merger of two conventional building envelope typologies: the window wall and the glass curtain wall. The tower's structure is a robust exoskeletal grid overlayed with a patterned glass skin—the first time such glass has been used for an exterior at this scale. The patterned glass reveals the detail and complexity of construction while creating a mysterious crystalline effect as the tower responds to light: sparkling during bright sunshine, mute on an overcast day, enigmatic at dusk, glimmering during rain and glowing at night.

OMA oversees the design working directly with the client in a shared office on the construction site—an unusual practice for foreign architects working in China. OMA's design and site supervision team is led by Michael Kokora, associate in charge, together with Rem Koolhaas and David Gianotten, partners in charge.

URBANUS
Qianhai Metro Plaza

Shenzhen has expanded and developed its urban environment dramatically in the past few decades, and recently has become an international city with increasing economic power. At the same time, the city has encountered the contradictory situation between shortage of urban infrastructure and limited developable land resources. This problem has been recognized mostly when developing city metro infrastructure, as each metro line would occupy vast amount of land resources in the city center area for the purpose of vehicle and system maintainance. In order to preserve and develop the land resources efficiently, Shenzhen has learnt the precedents from Japan and HONG KONG, incorporated the strategy of "metro bay" concept. If Qianhai metro bay project has progressed and constructed in the future, it might possibly become the first type of mix metro commercial bay system in China, and consequently may become the a new typology of planning future multi-functional metropolis in China.

However, the underground metro maintainance system and metro lines have been confirmed before the process of designing the metro commercial bay on ground level. The solution is to design the column-net within the limited spaces between the metrolines, and naturally raised the mass of above-ground building mass following the flow of underground metro tracks' direction. The result is a dynamic SOHO office loft with the fluent form reflecting the normally-unseen underground urban-scape. the other restriction is structural load bearing limitation, which strictly controlled the above-ground building height to be under 4 levels tall. The design process and complexity are enormous regarding to various aspects of problems, such as the planning and circulation for the 16-meter raised commercial island; the irregular or seemingly random column-net system and the mega scale in designing the building mass. This is no doubt by far the most challenging project far URBANUS.

在当代建筑中，一些项目规模庞大，以至于项目建筑师不仅仅是建筑的设计师，甚至要同时扮演城市规划师和景观建造师的角色，这次双年展的主题亦立足于此：建筑创造城市，城市创造建筑。本次展览中所选择的每一个项目在它完成后，都将会对深圳的都市化构造以及深圳数百万市民的日常生活和习惯产生扭转性的巨大影响。不止是对深圳，这些项目都将成为对规模和复杂性进行研究的重要案例，它们反映出的是深圳和中国作为都市实验场和建筑实验室在世界范围中的重要地位。

今天深圳的建筑物从各个方面为它们各自所代表的建筑类型奠定了国际标准。"深圳建筑"为深圳提供了五个进行中（正在设计中或是正在建造中）的主要项目，为它们工作的既有中国建筑师也有国际建筑师，其中包括：张永和与非常建筑工作室设计的深圳广播电视塔，沃尔夫·普瑞克斯（Wolf Prix）与蓝天组 [Coop Himmelb(l)au] 设计的深圳当代艺术与城市规划展览馆，马西米亚诺 & 多利安娜·福克萨斯（Massimiliano & Doriana Fuksas）设计的深圳宝安国际机场三号航站楼，库哈斯（Rem Koolhaas）和大都会建筑事务所（OMA）设计的深圳证券交易所，都市实践建筑事务所（URBANUS）的深圳地铁前海湾车辆段上盖物业。展览主要聚焦在建筑师的设计过程上，以及它是如何对深圳的环境、建筑以及都市设计问题产生作用的。建筑师们各自发展出了不同的展示方式，包括模型、动画以及绘图，甚至包括来自他们办公室的设计过程中使用的各种材料，由此人们可以了解到这些项目是如何被构思和发展出来的。

非常建筑工作室
深圳广电信息大厦，2009

非常建筑工作室（FCJZ）是中国的一所位于前沿地位的建筑设计事务所，在建筑与艺术领域中都具有相当高的国际认知度。非常建筑工作室的作品的跨度很广，包含文化设施、公共建筑、住宅、办公室设计、零售以及商业建筑设计、实验室设计、室内设计、城市规划、公共建筑物的长远规划和景观设计，同时还从事产品设计和展览设计。工作室的主要项目包括上海2010年世博会的企业联合馆、上海涵璧湾、韩国坡州出版社、北京用友研究中心、北京二分宅、北京柿子林别墅、河北教育出版社、杭州唐宫万象城酒店、深圳唐宫席殊书店等。工作室近期的设计项目还包括为JIA（家）设计的茶壶，'非常建筑1/2'服装设计，TTF 珠宝设计以及为无印良品做家具设计。2008年，非常建筑赢得了深圳广电信息大厦的国际竞标。

非常建筑工作室一直积极致力于可持续发展和跨学科的创造。进行中的项目包括 Novartis 校园、上海嘉定广告公园以及北京前门历史保护和

SHENZHEN / 深圳

再生项目，上述项目都为非常建筑打开了新的方向。非常建筑工作室在2006年获得了商业周刊 / 建筑实录中国奖，在2004年获得了中国建筑奖，并且获得了2003年和2004年的中国建筑艺术奖。

这栋47层200米高的摩天大楼位于深圳福田经济行政区中心，是深圳广播电视集团现有建筑设施的扩展。他们最初的研究发现了大多数高层建筑设计中的共同点：从北侧到南侧都拥有相同的玻璃幕墙。这一发现让他们开始关注这样一个问题：高层建筑是否可以拥有地域性特点？或者说，南方的摩天大楼可以是什么样的？这些问题让他们开始形成一种与深圳本土的气候和环境条件对话的策略。他们希望设计一座可以自我遮阴的电视塔，由此来减缓建筑的增温速度。根据这种设计思维，我们设计了一座雕刻似的、多面的摩天大楼。夏天，这座建筑会向自身投射阴影。南、北立面的几何建构会根据一天中太阳角度的变化而与东、西立面不同。该建筑的外表面系统设计还包括在南部朝上的三角形表面上使用了光电能技术（Photovoltaic）。顶部安装的则是比较常规的太阳能电板。

项目底部的电视演播厅设计中，水被用作建筑曲面的构造材料。这种做法的意图，则是为这样一个南方地区的都市广场的底层空间创造一种宜人的微气候。

蓝天组
当代艺术馆与城市规划展览馆，深圳，中国，2007

当代艺术馆与城市规划展览馆（MOCAPE）作为深圳新都市中心福田文化区的主体规划之一，将成为深圳快速发展的城市肌理中的新焦点。

该项目将两个不同的机构结合在一起——当代艺术馆(MOCA)和规划展览馆(PE)。尽管这二者各自有不同的功能与表现上的需求，但是它们所包含的多样性的主要元素被综合在这一个整体中，被多功能的外表封装起来了。

透明的外表和内部照明的设计理念使得透过外表皮观看内部空间成为可能。由此，两个不同的美术馆共享的入口以及流通空间被特别强调出来。同时，建筑的外皮使内部的观者可以毫无阻碍地观看外部的城市景观，并且获得一种身处于有遮阴的室外空间的感觉。这种感觉正是被这极度宽广的跨距所强调出来的。这种跨距提供了完全开放的、几乎无柱且可变性极强的展厅，这个空间的高度为6~17米。

SHENZHEN BUILDS / 深圳建造

进入两馆之间的广阔的间隙空间后，观者可以通过斜坡或电动扶梯到达主楼层。这里既是一个公共广场又是整个建筑复合体共享的人流流通空间，同时这里还提供导览功能，并且可以成为整个美术馆之旅的起点。这里还为人们提供了许多公共或私密的设施，例如一些文化服务站、多功能厅、礼堂以及一个图书馆、一间咖啡屋、一家书店和一间美术馆商店。

马西米亚诺 & 多利安娜·福克萨斯
深圳宝安国际机场，2008

深圳宝安国际机场的总体规划是要提供世界级的运输服务。要完成这一目标意味着深圳机场必须在以下水准下操作：

深圳宝安国际机场三号航站楼成为中国与世界其他地区之间通行的国际航线的关口。继北京、上海和广州之后，深圳宝安国际机场成为中国第四大国内国际航线的交换中心。由该机场所增强的国际和国内航线将改善中国城市间财富的分布。

目前，每3.5小时就有来自超过40个城市的100多万人口穿行于深圳上空。这使得深圳极好地成为了不同区域进行交换的场所。来自邻近的城市的旅客们可以将深圳作为中转站继而飞往亚洲的其他国家进行更遥远的国际旅行。

作为进出中国和深圳地区的通道，三号航站楼获得了为国内外旅客们留下第一印象和最后印象的机会。这个设计概念展示了一幅拥有有机形态的雕塑景象。机场的表面由外表面与内表面组成，结构成分则在内外表皮之间，旅客们就被包围在这之间。穿过机场时旅客可以始终被这种像空气一样的结构包围着。这与经典的建筑概念背道而驰。这样一种精心设计的结构使得人们在持续行走于机场的时候，与外部空间进行交流的同时也能获得不同的光照体验。

概念的核心是一个蜂巢，一种与表皮使用的独特的物理材料一样的板状结构。对使用的材料的考量基础是本土现成性、功能性、本地施工的便利性以及较低的采购成本，通过设计来优化材料的性能。每一块板的玻璃尺寸不同，并且向着不同的角度打开，根据机场内部的不同功能需求来精心设计以获得不同的采光体验。

旅客候机厅和中央大厅构成了乘客对机场的感受的主要部分。影响旅客印象的要素包含处理时间、步行距离、导视的便利性、拥挤度以及最大程度的舒适性。在候机厅和中央大厅的概念开发的过程中，这些要素都被仔细考量。候机厅的设计使得对外部空间的观赏成为可能，并且被覆盖在一整片统一的顶盖下。自然光通过双层表皮的过滤照射进来，因此避免了阳光直射从而减少了能源消耗，并且创造了优雅的气氛。

OMA 大都会建筑师事务所
深圳证券交易所，2006

几千年来，一直是实体建筑屹立于实体的地基之上。这幅景象一直持续到了现代时期。一直以来，建筑结构扎根于地基，而地基牢牢地将建筑固定在大地上。股市的精髓则是投机，它所依赖的是资本，而不是地心引力。

在这个几乎虚拟的深圳证券交易市场的案例中，象征主义的角色要远远大于项目本身——这个建筑将要代表股市，而不仅仅是物理上容纳这个证券市场。所有这些因素暗示着一种建筑学上的发明：我们的项目将会提供一个拥有悬浮地基的建筑。好似那驱动市场的投机快感将其整个抬起，而传统的地基则沿着塔身攀成一个上升的平台。将"地基"提升到空中将大大增加这个建筑的曝光度。它在这个攀升起来的位置，可以对着整个城市"广播"证券市场的一举一动。被释放出来的地面空间则可以用作一个被覆盖的都市广场，它的面积大到足够容纳许多公共事件。深圳证券交易所——这个将会上升到 248 米的建筑——被规划为市民的金融中心。它坐落于莲花山和滨河大道之间的南北轴线与深圳的主干道深南大道所穿过的东西轴线交汇的新公共广场中。

深圳证券交易所这个悬浮的"地基"使用了三层悬吊式平台结构，"地基"离地约 36 米，拥有楼层面积约 48000 平方米，同时还具备一个可进入的楼顶花园。平台和低处的塔结构主要用于覆盖证券交易的所有功能，包括深交所办公室、登记处、票据交换所、证券信息公司以及位于塔结构高层的辅助服务部门。塔楼底部由两道门廊侧接组成，这样就避免了将地面与交易层直接相连。工作人员从西面进入，而公众从东面进入。

深交所大楼采用与周围的高楼同质的普通的长方形表面。但是深交所大楼的外表则创造性地将两种常见的建筑封装类型结合在一起：窗墙和玻璃幕墙。大楼的框架由强劲的外骨骼格子结构构成，它的表面覆盖着一层刻花玻璃外表——这是第一次将此种玻璃用在这种规模的建筑外表面上。刻花玻璃既解释了建筑结构的细节和复杂性，同时在光线照射时又可制造出一种神秘的水晶般的反射效果：在烈日下闪耀，在阴天里沉默，在黄昏中化作谜，在雨中微微闪烁，在夜里燃烧。

大都会建筑事务所 OMA 直接与客户在施工地的同一个办公室里监督设计——对于在中国工作的外国建筑师而言，这是一种非常罕见的工作方式。OMA 的设计与施工监管小组由 Michael Kokora 领导，由库哈斯协同负责，David Gianotten 作为合伙人负责人。

都市实践
深圳地铁前海湾车辆段上盖物业

当深圳在城市规模上迅速跻身于国际大都市行列，经济力量不断提升的时候，它日益显现出城市基础设施不足以及用地极端缺乏的困境。这个矛盾在地铁建设上尤为突出。例如每条地铁线都有占地巨大的车辆段检修场，它覆盖了大量城市中心地区的土地资源。为了实现土地集约利用，深圳开始采用日本及中国香港等地的地铁上盖物业的做法。前海上盖项目如期建成可能是国内第一个此类工程，成为了一种新的复合性城市多功能体的样板。

在决定加建上盖物业时，本项目所在的车辆段的设计图已经完成。我们只能在现有轨道之间见缝插针摆放柱子，也正是因为有这样的限制，促成我们因势利导地利用与下方轨道吻合的流线型建筑体形，营造出了一个丰富动感的 SOHO 办公区。功能与形式高度统一，外表恰如其分地表现出了看不到的下方的实际情况，这一成果实属难得。然而，建筑体荷载的限制使建筑只能控制在 4 层以下。由于地面升起 16 米导致的流线布局和建筑规范的高难度，使设计面临空前挑战。完全不规则的柱网和超大的尺度更提升了设计难度，而且造价空前高昂。这是 Urbanus 都市实践目前为止所遇到的难度最大的项目。

| SHENZHEN/ 深圳 | SHENZHEN BUILDS / 深圳建造 |

FCJZ / YUNG HO CHANG / 非常建筑 / 张永和
SHENZHEN TV TOWER (SZTV) / 深圳广电信息大厦

In an age when the skyscraper is increasingly questioned for its nearly-exhausted typologies and for being merely a symbol of the vertical race, the super high-rise design for the Shenzhen TV Tower (SZTV) is Atelier FCJZ's attempt to negotiate with the demand for autonomous distinction and desire for integration. It is one of the winning designs of the "4 Towers in 1" competition within an urban complex masterplanning to unify the four new office towers around the new Shenzhen Stock Exchange Headquarters in Shenzhen's Futian commercial business district. With its seemingly generic rectangular form and conventional building envelope typology—the tower will be characterized by its skin and partially cantilevered floor on the eastern front. Its crystalline façade on a glazed system is custom-designed in response to the sun angles and a shading analysis that responses to the southern climate and the subtropical sunlight, while reducing energy needed to cool the building in the frequent summers. The cantilevered floor housing the broadcasting facilities above the ground creates a covered urban plaza for public events, while exposing the insides of the recording studios to the street—becoming the tower's most public interface to offset the insularity of the vertical office tower.

在摩天大厦因竭尽类型而转向纯粹符号性的垂直竞争而受到质疑的时代，非常建筑工作室尝试在超高层的深圳广电信息大厦（SZTV）的设计中将自我特征的需求和整合的愿望达成一致。本设计是"4合1大厦"竞赛的得奖作品之一，这是一个在复杂的城市总体规划下，为统一四个围绕着新证券交易所总部的新办公楼的竞赛。凭借其典型的矩形形态和传统的围护结构类型——大厦将会因其表皮以及向东的悬挑层而极具特色。其在玻璃系统上的晶体幕墙是为回应太阳高度角和遮阳分析而专门定制的，这是为了应对南方气候和亚热带阳光的反应，同时减少在夏季冷却建筑所需的能源。悬挑层承载着演播设施，同时创造了一个有顶的城市广场作为举行公开活动的场地；而且又向外展示录音棚的内部场景，成为了大厦与公共的接合点，打破了垂直的办公大楼与外界的隔阂。

| SHENZHEN/ 深圳 | SHENZHEN BUILDS / 深圳建造 |

FCJZ / YUNG HO CHANG / 非常建筑 / 张永和

SHENZHEN/ 深圳　　　　SHENZHEN BUILDS / 深圳建造

SHENZHEN/ 深圳　　　　　　　　　　　　SHENZHEN BUILDS / 深圳建造

Coop Himmelb(l)au / Wolf Prix / 蓝天组 / 沃尔夫·普瑞克斯

Museum of Contemporary Art & Planning Exhibition (MOCAPE)

The Museum of Contemporary Art & Planning Exhibition (MOCAPE), part of the master plan for Shenzhen's new urban center, the Futian Cultural District, is conceived as the synergetic combination of two institutions: the Museum of Contemporary Art (MOCA) and the Planning Exhibition (PE), whose various programmatic elements, although each articulated according to their functional and performative requirements, are merged in a monolithic body enveloped by a multifunctional façade.

The "urban monolith" of the MOCAPE completes the east wing of the city center's master plan. When seen in combination with the neighboring Youth Activity Hall (YAH), MOCAPE forms the complimentary puzzle piece to the opera and library complex to the west of the central axis. The twisting façade geometry shapes the building in response to the urban surroundings by twisting towards the center of the Futian Cultural District. This rotation establishes a new entry orientation directed toward the axial center of the Futian Central District's main circulatory flows, thus generating a dynamic interaction between the urban, programmatic and pedestrian scales.

With its main level situated at ten meters above the ground—a common feature of the buildings in the Futian Cultural District—the MOCAPE offers a stage-like viewing platform as connecting element to the other buildings.

Upon entering the large interstitial space between the two museum volumes, the visitors can, via ramps and escalators, reach the main level, where a kind of public plaza serves as the circulation hub for the whole complex and as orientation and starting point for museum tours. From here also many auxiliary public and private facilities are accessible, including cultural services, a multifunctional hall, several auditoria, a library, a cafe, a book store, and a museum store.

The transparency of this façade and the interior lighting concept allow for a view from outside through the exterior envelope deep into the volume of the space, thereby particularly accentuating the shared entrance and circulation space between the two museum volumes. At the same time, the building skin also allows unhindered view from the inside on the cityscape while giving the visitor the impression of being in a pleasantly shaded outdoor area—an impression enhanced by very wide spans which allow for completely open, column-free and flexible exhibition halls with heights ranging from 6 to 17 meters.

The skin, which consists of an exterior layer in front of a narrow maintenance space and the actual climatic envelope of insulated glass, is a dynamic surface which is statically independent from the structure of the museum spaces.

The building's technical equipment will reduce the overall need for external energy sources: pollution free systems and facilities utilizing renewable energy sources such as solar and geothermal energy (with a groundwater cooling system) as well as those with higher energy efficiency have been implemented. The lighting concept in the upper level of the Museum of Contemporary Art is enhanced by filtered daylight which contributes to the energy efficient performance of the building by reducing the need for artificial light.

SHENZHEN/深圳 | SHENZHEN BUILDS /深圳建造

© Coop Himmelb(l)au

Coop Himmelb[l]au / Wolf Prix / 蓝天组 / 沃尔夫·普瑞克斯

深圳当代艺术馆与城市规划展览馆

深圳当代艺术馆与城市规划展览馆（MOCAPE，下文简称"两馆"）是深圳福田文化中心区整体城市规划的重要组成部分。建筑实现了两种不同功能要求的整合：两座展馆因其自身的功能、使用要求不同而迥异的种种功能元素，被整合至一个由多功能建筑外表皮围合的独特建筑形态。

当代艺术馆与城市规划展览馆以其城市雕塑的姿态填补了城市中心规划区域东区的最后一块空白。把两馆和深圳少年宫放在一起看时，它与该区域中轴线西侧的剧院和图书馆综合体形成了很好的互动关系。其旋动的外立面几何形体向福田文化区域的方向旋转生成的建筑体块是对周围环境的呼应，还让表皮具有了动感韵律。这个旋转形成了一个新的入口，指向福田中心区域的中轴线，并且暗示了主要交通流线的位置。这是一个在建筑内压以及城市对建筑外压的共同作用下，形成的一个富有动感的雕塑构造的设计理念。

福田文化区公共建筑的共同之处在于其标高的统一，即地上10米处，这亦为"两馆"提供了一个视觉平台，成为了其与周边建筑建立联系的要素。

主入口位于两座博物馆之间，参观者可通过坡道或扶梯进入"两馆"的主层面（距地面10米高），这里俨然是一个公共广场空间，可作为馆内交通流线、导向的枢纽以及博物馆参观的起点。此外，这里还设有各种辅助的公共、私密设施为其参观者服务。其中，公共空间包括文化服务、多功能厅、若干报告厅、图书馆、咖啡馆、书店及博物馆商店等。

建筑通透的外立面及其内部照明理念使得外界可以洞悉博物馆内部的大空间，同时格外突出两座博物馆之间的共享、交通空间。此外，馆内的参观者亦拥有极好的视线，可自博物馆内部观赏室外的城市景观，如同身处有遮蔽的户外空间之中。层高为6至17米不等的大跨度无柱展示空间自由、灵活，为参观者营造了轻松、愉悦的空间氛围。

建筑外表皮的外层为天然石材，内侧为中空隔热玻璃，其间以狭长的围护空间分隔开来。极富张力的外表皮可与静态博物馆空间结构相对独立地存在。

建筑机械设备的选用旨在减少建筑的整体能耗，为了达到这个目标，配置了一系列无排污的太阳能、地源能（包括地源能量制冷）、可再生能源设备以及提高能源使用效率的设备。博物馆采用过滤日光进行照明，减少人工光源的使用，从而大大节约了能耗。

© Coop Himmelb[l]au

| SHENZHEN/ 深圳 | SHENZHEN BUILDS / 深圳建造 |

Coop Himmelb(l)au / Wolf Prix / 蓝天组 / 沃尔夫·普瑞克斯

SHENZHEN/ 深圳   SHENZHEN BUILDS / 深圳建造

Construction site
现场施工照片

PROJECT	SHENZHEN BAO'AN INTERNATIONAL AIRPORT
PROGRAM	Airport Expansion Terminal 3
PERIOD	2008 – 2013 (first phase)
CLIENT	Shenzhen Airport (Group) Co., Ltd.
DEVELOPER	Shenzhen Planning Bureau; Shenzhen Airport (Group) Co., Ltd.
ARCHITECTS	Massimiliano and Doriana Fuksas
GENERAL CONTRACTOR	China State Construction Engineering Corporation, Beijing
AREA	400,000 sq.m.
STRUCTURES, FACADE, PARAMETRIC DESIGN	Knippers Helbig Engineering, Stuttgart, NY
ARCHITECT OF RECORD	BIAD (Beijing Institute of Architectural Design), Beijing
LIGHTING CONSULTING	Speirs & Major Associates, Edinburgh, London
项 目	深圳宝安国际机场
方 案	机场3号航站楼
时 间	2008-2013（第一期）
业 主	深圳市机场集团有限公司
开 发	深圳市规划局，深圳市机场集团有限公司
建 筑 师	马西米亚诺与多丽安娜·福克萨斯
总承包商	中国建筑工程总公司，北京
面 积	400,000 sq.m.
结构、立面、参数化设计	Knippers Helbig 工程，斯图加特，纽约
纪录建筑师	BIAD（北京建筑设计研究院），北京
灯光顾问	Speirs & Major Associates，爱丁堡，伦敦

© Studio FUKSAS

Massimiliano and Doriana Fuksas / 马西米亚诺与多丽安娜·福克萨斯
Terminal 3 at Shenzhen Bao'an International Airport / 深圳宝安国际机场 3 号航站楼

In 2008 Massimiliano and Doriana Fuksas Architects won the international competition for the extension of the airport with the design of Terminal 3. The concept of the project provokes the image of a sculpture with its organic shape. The structure of the building is in steel with a concrete substructure. The skin that envelops the structure, both on the inside and on the outside, shows the honeycomb motive. The 300,000 m² façade is made of metal panels and glass panels of different size that can be partially opened. They follow the honeycomb motive. The building has a roof construction with spans up to 80 m. Through its double layer, the skin allows the natural light to filter and create light plays. Concourse area is one of the key areas at the airport and it is composed of three levels. Each level is dedicated to independent functions: departure, arrival, and services. On the ground floor, the plaza provides access to the departures and arrivals as well as to the cafés and restaurants, offices and facilities for business meetings. Travellers reach the Terminal at level 14.40 m. The spatial concept of the interior is one of fluidity. It combines two different ideas: the idea of movement and the idea of pause. The honeycomb, developed in the suspended structure, is translated into the interior and quoted in different scales throughout the building.

2008 年，建筑师福克萨斯夫妇（Massimiliano & Doriana Fuksas）在深圳机场扩建计划中第三航站楼的国际竞标中胜出。该项目的概念体现了雕塑的有机形态。建筑主体为钢结构，并配以混凝土子结构。把建筑外部和内部都包裹起来的建筑表皮受到蜂巢的启发。30 万平方米的建筑外表由不同尺寸的金属板和玻璃板构成。金属板和玻璃板也受到蜂巢的启发，可以部分打开。建筑的屋顶构造宽达 80 米。通过双层隔板，建筑表皮允许自然光渗入并创造出光线的变化。中央大厅是机场的主要空间之一，共有三层，每一层都对应独立的功能：候机、到站以及服务。在地面第一层，广场提供到达候机厅及到站厅的入口，同时也提供进入咖啡厅和餐馆以及供举行商务会议的办公室和设施的入口。乘客到达的航站楼位于 14.40 米高处。内部设计的空间概念体现了流动性。它结合了两种不同的想法：运动和停止。在悬挂结构中发展起来的蜂巢概念体现在内部设计中，并被引用在该建筑各个不同的层面上。

Construction site
现场施工照片

| SHENZHEN/ 深圳 | SHENZHEN BUILDS / 深圳建造 |

Massimiliano and Doriana Fuksas / 马西米亚诺与多丽安娜·福克萨斯

| SHENZHEN/深圳 | SHENZHEN BUILDS / 深圳建造 |

Image by Zoo Productions / Courtesy OMA
Zoo Productions摄 / OMA 版权所有

SHENZHEN STOCK EXCHANGE

The Shenzhen Stock Exchange is as much an object in the city as it is a product of Shenzhen itself. In OMA's exhibition a linear narrative of the city's history, the stock exchange, and related events since the opening of the exchange in 1991 are revealed and presented. A grid of books with pages of historic images for people to take with them is hung on the wall, thereby collecting and disseminating events both past and present that have taken part in forming of the city of Shenzhen. The collection of images captures moments inside and outside the chronological history. The intention is to represent both the actual physical aspects of the city's relationship to the building and the virtual market history as influenced by local or global events. The projection at the center of the exhibition space displays a continuous loop of videos that documents the Shenzhen Stock Exchange in the context of the city. While the Stock Exchange still has another year of construction its form and relationship to the city can already be interpreted. Rather than representing documents leading to this moment, OMA presents the relationships they have studied between the city and the building and the building and the city. Architecture Creates Cities, Cities Create Architecture.

The Shenzhen Stock Exchange, situated on Shenzhen's main artery Shennan Road, with a floating base, is planned as a financial center with civic importance. Lifting the base in the air increases the exchange's exposure—it can "broadcast" the activities of the stock market to the entire city. The space liberated on the ground can be used as a large covered urban plaza for public events.

The raised 3-storey cantilevered base of the Shenzhen Stock Exchange, floating 36m above the ground, contains three 15,000 m2 floors with an accessible roof garden on top. The platform and the lower tower accommodate the dedicated stock exchange functions, including conference centers, exhibition spaces, a listing hall, and a market watching department. The tower is flanked by two atria—a void connecting the ground directly with the trading floor that opens the large floor plates to natural light. Surrounding the west atrium is a 20,000 sq. m plinth with commercial facilities.

The tower's structure is a robust exoskeletal grid overlayed with a patterned glass skin—the first time this type of glass has been used for an exterior in China. The patterned glass reveals the detail and complexity of construction while creating a mysterious crystalline effect as the tower responds to light.

The SZSE building is designed to be one of the first 3 star green rated buildings in China. It utilizes passive shading through recessed openings that form a "deep" facade reducing the amount of solar heat gain entering the building, improving natural day lighting while reducing the energy consumption. Intelligent lighting systems shut down the interior day lighting when spaces are not in use. Rainwater collection systems are used and the landscape design is permeable to collect water locally and reduce run-off.

| SHENZHEN/ 深圳 | SHENZHEN BUILDS / 深圳建造 |

OMA / 大都会建筑事务所 / 雷姆·库哈斯

深圳证券交易所大楼

深圳证券交易所大楼不仅仅是一座城市建筑,更代表着深圳的精神。大都会建筑事务所的展览对从1991年老证券交易所大楼开张起深圳这个城市的历史、证券交易以及相关事件进行线性叙事,并对此进行呈现。墙上用网格挂着一壁的小册子印有历史图片,可供人们随手取阅,借此收集并传播过去和现在的重要事件,这些事件共同塑造了深圳这个城市。这些图片呈现了编年史中的各种事件,也抓住了编外史的各个重要瞬间。展览旨在表现深圳这座城市与建筑在物质层面的关系,以及由本土或全球事件所影响的虚拟市场的历史。展览的中心空间由各种录像构成一个连续播放的投影,在城市的语境下对深圳证券交易所进行呈现。展览之时,虽然证券交易所大楼还需一年时间尚能建好,但其外在形式及其与深圳这个城市的关系已经被人们进行了解读。与其用各种文档材料带领人们感受建筑建成后的情景,大都会建筑事务所选择呈现他们所研究的城市与建筑以及建筑与城市的关系,而这正契合本届展览的主题——"城市创造"。

以"漂浮平台"为灵感的深圳证券交易所坐落于深圳交通动脉——深南大道,一开始便计划建设成一个有公民参与意义的金融中心。把大厦平台升到空中,增加了证券交易所的曝光度——它可以向整个城市"广播"有关证券市场的活动。因漂浮平台而从地面解放出来的空间则可作为一个大型有顶城市广场来使用,供举行公众活动。

深圳证券交易所3层楼高的悬臂式平台漂浮在离地36米的高空中,包括3层各15000平方米的空间,并包括一个屋顶花园。漂浮平台和较低处的塔座完善了证券交易所的各种功能,包括会议中心、展览空间、上市大厅和市场观察部门。塔座由两个中庭包围两侧——中间的空隙则把地面与交易大厅直接连接起来,交易大厅可以把大型楼板打开让自然光线进入。环绕西边中庭的是20000平方米的基座,里面拥有各种商业设施。

塔座使用粗壮的骨架网格结构,其上覆盖着压花玻璃——这是中国首次在外表结构上使用这种类型的玻璃。这种压花玻璃揭示了这种构造的细节和复杂性,并使其具有通透性,当塔座与光线进行互动时,可创造出一种神秘的晶体效果。

深圳证券交易所大楼是中国首座三星级绿色建筑之一。它通过嵌壁式开口使用被动式遮板,以此形成一个"深度"外观以减少进入建筑的日照热量,在减少能量消耗的同时提高对自然日光的利用。当空间不被使用时,智能照明系统会关闭室内灯光。此外,建筑还使用了雨水收集系统,建筑的景观进行了可渗透设计,用于收集本地雨水并减少流失。

| SHENZHEN/ 深圳 | SHENZHEN BUILDS / 深圳建造 |

| 顺势而为 | 中台城市? | 一半的公司团队 | 几公里长的立面 |
| form follows flow | an inter-mediate city? | half of company as one team | facade length up to kilometers |

| 股道枢纽 | 庞大复杂 | 经济战略物流中心 | 我们需要这么多CBD吗? |
| metro interchange | a grand complex project | new logistic CBD | do we really need multi-CBD? |

| 填,还是不填? | 近两千根地下柱网 | 地下的蛛网动脉 | 要积累,不要归零 |
| to fill or not to fill? | over 2000 existing columns | metro artery weaving underground | to accumulate, not to annihilate |

URBANUS / 都市实践

QIANHAI METRO PLAZA

Shenzhen has expanded and developed its urban environment dramatically in the past few decades, and recently has become an international city with increasing economic power. At the same time, the city has encountered the contradictory situation between shortage of urban infrastructure and limited developable land resources. This problem has been recognized mostly when developing city metro infrastructure, as each metro line would occupy a vast amount of land resources in the city center area for the purpose of vehicle and system maintainance. In order to preserve and develop the land resources efficiently, Shenzhen has learnt the precedents from Japan and HONG KONG, incorporated the strategy of "metro bay" concept. If Qianhai metro bay project has progressed and constructed in the future, it might possibly become the first type of mix metro commercial bay system in China, and consequently may become the a new typology of planning future multi-functional metropolis in China.

However, the underground metro maintainance system and metro lines have been confirmed before the process of designing the metro commercial bay on ground level. The solution is to design the column-net within the limited spaces between the metrolines and naturally raised the mass of above-ground building mass following the flow of underground metro tracks' direction. The result is a dynamic SOHO office loft with the fluent form reflecting the normally-unseen underground urban-scape. the other restriction is structural load bearing limitation, which strictly limits the above-ground building height to be under 4 levels tall. The design process and complexity are enormous regarding various aspects of problems, such as the planning and circulation for the 16-meter raised commercial island; the irregular or seemingly random column-net system and the mega scale in designing the building mass. This is no doubt by far the most challenging project by URBANUS.

SHENZHEN/ 深圳 　　　　　　　　　　　SHENZHEN BUILDS / 深圳建造

Schematic Model Photos
方案模型照片

URBANUS / 都市实践

深圳地铁前海湾车辆段上盖物业

当深圳在城市规模上迅速跻身于国际大都市行列，经济力量不断扩展和提升的时候，它日益显现出城市基础设施不足以及用地极端缺乏的困境。这个矛盾在地铁建设上尤为突出。例如每条地铁线都有占地巨大的车辆段检修场，它覆盖了大量城市中心地区的土地资源。为集约利用土地，深圳开始采用日本及中国香港等地地铁上盖物业的做法。前海上盖项目如期建成可能是国内第一个此类工程，成为了一种新的复合性城市多功能体的样板。

在决定加建上盖物业时，本项目所在的车辆段的设计图已经完成，我们只能在现有轨道之间见缝插针地摆放柱子，也正是有了这样的限制，促成了我们因势利导地利用与下方轨道吻合的流线型建筑体形营造出一个丰富动感的 SOHO 办公区，功能与形式高度统一，外表恰如其分地表达了看不到的下方的实际情况，这一成果实属难得。然而，建筑体荷载的限制使建筑只能控制在 4 层以下。由于地面升起 16 米而导致的流线布局和建筑规范的高难度，使设计面临空前的挑战。完全不规则的柱网和超大的尺度更提升了设计难度，而且代价空前高昂。这是都市实践目前为止所遇到的难度最大的项目。

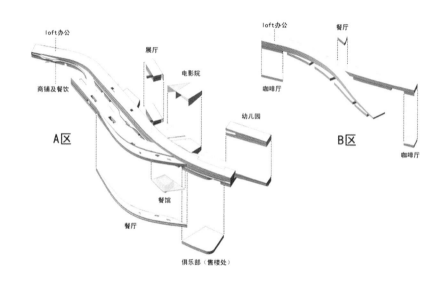

Building Use Program
建筑使用功能

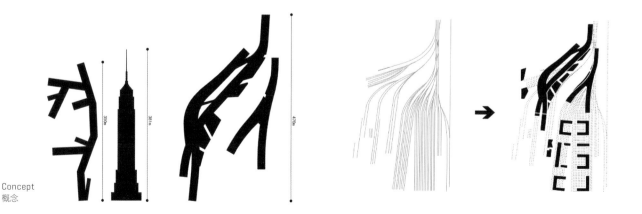

Concept
概念

SHENZHEN/ 深圳

SHENZHEN BUILDS / 深圳建造

BOOM! SHENZHEN
轰隆！深圳

Curator: Mary Ann O'DONNELL
Realization: LEI Sheng
Design Team: CAO Taiming, CHEN Yue, HONG Wudi, WAN Yan, WANG Lechi, ZHANG Chen'ge, ZHANG Yiwei, ZHANG Xueshi

策展人： 马立安
执行： 雷胜
设计小组： 曹泰铭、陈越、洪吴迪、万妍、汪乐弛、张晨戈、张轶伟、张雪石

| SHENZHEN/ 深圳 | BOOM! SHENZHEN / 轰隆! 深圳 |

On July 2, 1979, a section of mountainous, Shekou coastline was detonated. Images of that mushroom cloud circulated throughout China, signaling the beginning of Reform and Opening and the transition from Maoism. Indeed, that first detonation has resonated both literally and figuratively throughout Shenzhen's history. Literally, that explosion initiated construction of China's first export processing zone, the Shekou Industrial Park. Subsequent industrial development both in and beyond Shenzhen also "detonated mountains and reclaimed oceans" to create ports and industrial zones that integrated China's domestic economy with the world system.

Figuratively, the detonation symbolized the aggressive confidence—the willingness to "cut open a road of blood"—of China's leaders to deploy market principals in order to stimulate Chinese society, both economically and culturally. Moreover, for over a decade after the Shekou detonation, Shenzhen symbolized the transformative potential of reform and opening, not just as neoliberal economic strategies, but also and more importantly, as social and individual goals—to reform China by and through opening oneself and the country to the world beyond "the bamboo curtain".

Since the Shekou detonation, Shenzhen has boomed in every sense of the word. In 1979, Shenzhen was a rural area, organized into collective fishing villages, lychee orchards, and oyster farms. From 1979 through 2010, the Municipality's estimated population grew from 300,000 to over 13 million people, its GDP exploded from $US 308 million to over $US 149 billion, and agricultural land vanished, being replaced by international ports, industrial parks, residential areas, shopping malls, and green space. The most notable aspect of this transformation has been the transvaluation of rural life. Under Mao, rural areas were China's revolutionary heart and "villages surrounded the city" was an explicit political, economic, and social strategy for change. In contrast, the Shekou detonation signaled the beginning of a new era in Chinese history—"cities surround the villages". Importantly, the Mandarin expression "surround" might also be translated as "lay siege to" as in "cities lay siege to the countryside". In this context, Shenzhen's boom redefined the scale and intensity of rural urbanization within China and set new standards for developing nations looking to modernize. On the one hand, Deng Xiaoping twice visited Shenzhen, confirming the necessity of deepening economic reforms despite political opposition. Those two visits critically forwarded his agenda to extend reforms to China's coastal cities (in 1984) and the rural hinterland (in 1992). On the other hand, dignitaries and officials from predominantly rural countries as diverse as Cuba and Thailand, not to mention Eastern European and African nations have visited the SEZ to learn from its economic models and sign trade agreements.

Precisely because it announced the country's transition from rural revolution to urban development, fallout from Shenzhen's boom has resonated throughout the lives of everyday Chinese everywhere in China. These changes range from restructuring job opportunities and commodity availability through modifying Chinese property law and education curriculum to dismantling the public welfare system and social entitlements, affecting not only the tastes and lifestyle choices of subsequent generations, but also their moral values and ethical priorities. The generational differences between those who immigrated to Shenzhen in the 1980s and those who came later, as well as differences between those raised before and during Shenzhen's boomtimes have been significant enough to give rise to the SEZ's demographic nomenclature. Old Shenzheners, Old Shekou, locals, new immigrants, post 80s generation, and the rich second generation, for example, are common expressions to describe groups that have been defined by when and how individuals came to inhabit Shenzhen's economic maelstrom.

And yet.

The idea of a Shenzhen boom, is just that—an idea. An increasingly salient feature of world geography, urban booms are statistical artifacts—population growth, market fluctuations, and urbanization levels are created by operationalizing numbers of people, stock prices, and settlement densities and then grinding them through various formulae and arranging the results on graphs. Using the label "boom" to gloss the scale and velocity of urbanization in Shenzhen merely facilitates comparison between the Municipality and its sister cities, Houston and Brisbane, for example. However, the label neither tells us about the excitement a young architect felt upon leaving the known security of a Beijing work unit to "charge the gate" of Shenzhen's nascent Department of Planning, nor does it illuminate the lonely evenings of a teenage migrant worker, who shared a dorm room with seven other, equally but differently displaced country boys.

All this to make a rather simple point. Shenzhen is not what we think it is, rather "Shenzhen" is a linguistic placeholder. The actual city lives and breathes and grows and—yes—booms elsewhere. In other words, to understand and empathize with diverse experiences on the other side of cultural borders and generation gaps, we need flesh out bare statistical bones, creating work that is simultaneously informed by sociological research and poetic license, hard facts and ephemeral truth. Indeed, if we are to become global in any meaningful sense of the word, we must let go of what we think in order to hear what our interlocutor is saying. Thus, the exhibit, BOOM! Shenzhen implodes the idea of a timeline to create fresh perspectives on the lived, environmental, and philosophical meanings of Shenzhen's explosive history. Instead of presenting the SEZ's history as a series of linear events, six young architects from Shenzhen University School of Architecture have designed five pieces that are informed by and comment on different aspects of the city's boom.

The exhibit's centerpiece mushroom cloud Boom! by CAO Taiming and CHEN Yue playfully literalizes the idea of an economic boom, with thousands of tiny figures scaling a mushroom cloud that takes its silhouette from a graph of the Municipality's annual GDP since 1979. In Family Values, a bas-relief rendering of Shenzhen's housing market, Zhang Yiwei humanizes one of the more controversial of Shenzhen's booms—the price of housing, giving viewers insight into how millions of immigrants have inhabited the SEZ. With Futures, a bas-relief of Shenzhen's volatile stock market index Hong Wudi maps the abstract structure of market trading and investment in order to track how capital localization has transformed the cityscape. In poignant contrast, Wan Yan lyrically draws our attention to the uncontrolled and unexpected geological affects of urbanization by having melted a wax candle map of Shenzhen to create Meltdowns, a sculpture based on the SEZ's expanding urban borders. Finally, Zhang Xueshi and animators Wang Lechi and Zhang Chen'ge deconstruct our fetishization of fast track development, by interpreting key dates in the SEZ's history as a greedy snake that "eats" and "shits" history in his flash video Shenzhen Speed.

When viewed together, these five pieces express the visceral contradictions of living in urban boomtimes. Yes, Shenzhen's boom has created a new class of wealthy cosmopolitans, who live in upscale condos and travel to exotic locales. Yes, the source of this wealth has been unequal access to economic resources, deepening contradictions between urban haves and have-nots. Yes, building Shenzhen has entailed the radical reconstruction of the landscape, transforming the coastline beyond recognition and impacting regional climate patterns. And very much yes, Shenzhen's scorching rush to remake and remodel the world has created its own quantum temporality, whereby the city seems to be leaping from one state of being to another every decade. This constantly changing city is the inheritance all young Shenzhen inhabitants—not just the six architects featured in Boom!. Thus, the title of each piece also reminds us of the cultural values and dreams that have informed the Reform era of Chinese world building. Reread the works' titles again, but this time, hear them as poetry: Boom! Family Values. Futures. Meltdowns. Shenzhen Speed. Game Over, reboot. Boom!

Boom! Shenzhen invites viewers to participate in two levels of conversation, one about Shenzhen and one with the young architects. At the level of content, each piece provides viewers with information about China's oldest, largest, and most successful Special Economic Zone. At this level, the exhibit functions like a more traditional timeline. We learn, for example, that in 1997, the Return of Hong Kong deepened cross border integration and in 2004, the last of Shenzhen's villages were incorporated into the Municipal apparatus, making the SEZ the first Chinese city to have no rural residents or areas. However, at the level of aesthetic form, each piece allows viewers to understand Shenzhen beyond economic indicators and demographic statistics. This Shenzhen is creative and messy, playful and abruptly sad. In other words, the young architects' interpretation of Shenzhen's statistical boom allows those of us from other generations and cultures to reach a moment of understanding across our differences and realize something—maybe blatantly obvious, maybe softly elusive, but hopefully more fully human—about boomtime lives.

| SHENZHEN/ 深圳 | BOOM! SHENZHEN / 轰隆! 深圳 |

1979年7月2日，深圳蛇口第一家工业区破土动工，引爆了开山第一炮。被誉为"蛇口开山第一炮"的蘑菇云图片在整个中国传播开来，预示着中国改革开放的开始，也预示着中国从毛泽东时代开始发生转变。的确，这"开山第一炮"不仅在字面上，而且也在象征意义上在深圳的历史中一直回响着。从字面上看，"开山第一炮"预示着中国第一个出口加工区——蛇口工业园的破土动工。随之而来发生在深圳以及其他地区的工业发展也通过"炸山填海"的方式创建了各种港口和经济特区，从而把中国国内经济纳入到国际系统当中。

从象征意义上看，"开山第一炮"象征着中国领导人为了在经济及文化上刺激中国社会发展而部署市场政策的雄心，展示出了中国领导人"杀出一条血路"的意愿。此外，在蛇口"开山第一炮"引爆后的十多年时间里，深圳象征着改革开放的潜力，这不仅仅体现在新自由经济政策上，更重要的是作为社会目标和个人目标来透过"竹幕"改革中国并向世界开放深圳以及整个国家。

蛇口"开山第一炮"之后，深圳也开始了"大爆炸"式的发展。1979年，深圳还只是一个农业小镇，由集体化的捕鱼村落、荔枝果园和牡蛎农场组成。据深圳市政当局估计，从1979年到2010年，其人口从30万增长到1300万，国内生产总值则从3.08亿美元增长到1490亿美元。昔日的农田消失殆尽，被国际港口、工业园区、居民住宅、大型购物中心以及公共绿地所取代。这种转变最引人注目的地方体现在对农村的重新评估上。在毛泽东时期，农村是中国革命的中心，"农村包围城市"是一个明确的政治、经济以及社会策略。与此相反，蛇口爆破则标志着中国历史新纪元的开始——城市开始包围农村。更重要的是，中文的"包围"一词也可以理解为"围攻"，也就是"城市围攻农村"。在此语境下，深圳的"大爆炸"重新定义了中国农村城市化的尺度和强度，并为发展中国家寻求现代化设定了新的标准。一方面，尽管当时的政治并不允许，但邓小平的两次南巡确认了加强经济改革的必要性。这两次南巡使得他在1984年把改革扩展到中国沿海城市，并于1992年进一步扩大到内地的农村地区。另一方面，世界上其他以农业为主的国家，比如古巴、泰国以及东欧和非洲国家都前往深圳特区考察并学习其经济模式，签订了各种贸易协议。

深圳"大爆炸"宣告了中国从农村改革到城市发展的转变，深圳发展的影响也贯穿在中国各地人们每天的生活中。这些改变发生在各个领域，从重组就业结构到消费品的供应方式，从中国物权法和教育大纲的修订到公共福利系统和社会权利体系的解体，影响的不仅仅是后代的生活品位和方式的选择，也影响到了他们的道德价值观以及伦理观念。20世纪80年代移民到深圳的一代人与之后移民到深圳的一代人以及在深圳改革发展之前成长的一代人与改革发展过程中成长的一代人之间的代沟非常大，以至于深圳特区为不同时期的人们赋予了不同的名称，比如"老深圳"、"老蛇口"、"本地人"、"新移民"、"80后"以及"富二代"等。以上不同的名称用于描述在不同时期以不同方式来到深圳经济大漩涡中居住的不同人群。

然而，深圳"大爆炸"的概念，仅仅是一个概念。城市"大爆炸"，从统计学角度上看，是一种人工产品：人口增长，通货膨胀以及城市化水平都是由人口数量、股票价格以及居住密度决定的，用不同的规则对它们进行解读，并在图表上显示各种结果。使用"大爆炸"这个词来解释深圳城市化的规模和速度仅仅是使深圳与其姐妹城市，比如休斯敦和布里斯班之间的对比更加容易。然而，这个标签并没有告诉我们一个年轻建筑师离开北京稳定的工作单位来到深圳新兴的城市规划部门闯荡的兴奋，也没有告诉我们来自不同地区农村的十几岁外来务工者与其他7位室友共享一个宿舍熬过漫漫长夜的孤独。

所有这些都表明了一个简单的观点：深圳不是我们想象中的样子，深圳是一个语言学上的占位符。实际上，其他城市与深圳一样生活着、呼吸着、成长着以及"爆炸式"发展着。换句话说，为了理解并把香港处理文化边界和代沟问题的多种经验移植过来，我们需要除了纯粹的统计数字之外更加鲜活的东西，在社会研究和创新、铁一样的事实以及转瞬即逝的真相的启发下进行创造。的确，如果我们想要变得全球化，那我们必须抛开我们理所当然地认为的，以便于听听我们的对话者在说些什么。于是，这个展览"轰隆！深圳"引爆了用时间表来创造新鲜视角的想法，这些新鲜的视角包括深圳的爆炸式发展历史在生活、环境以及哲学等角度的意义。与用一系列线性事件来展示深圳特区历史不同的是，六位来自深圳大学建筑系的年轻建筑师设计了五件作品，所有作品都受到深圳爆炸式发展的启示，同时也是对这种发展的一种评价。

展览的核心——蘑菇云"轰隆"是由曹泰铭和陈越以一种玩乐的方式对"爆炸"这个词进行的字面呈现，上千个小雕像组成一朵蘑菇云，而蘑菇云的轮廓则由深圳特区政府自1979年开始的年度GDP表格构成。作品《家庭价值观》是一件体现深圳房地产市场的浅浮雕，张轶伟把高价房地产这个在深圳大发展中产生的最具争议性的问题之一进行了拟人化，为观众呈现了几百万的外来人口如何在深圳特区居有定所。作品《未来》则是一件表现深圳股市的浅浮雕，洪吴迪把深圳市场贸易和投资的抽象结构绘制成地图，用以追踪资产本土化是如何改变城市景观的。与之形成明显对立的是万妍的作品《融化》，这是一件基于深圳特区不断扩张的城市边界的雕塑,作者通过融化一幅由蜡烛组成的深圳地图,诗意地呈现出城市化对地理版图所产生的不可控制和不可预料的影响。最后,张雪石与动画师汪乐弛和张晨戈创作的动画影像《深圳速度》把深圳特区历史中的一些关键日子描绘成一条把历史"吞食"掉的贪吃蛇,从而把我们对快速发展所产生的恋物情结进行了解构。

把所有作品放在一起观看,会发现五件作品都表达了生活在城市"大爆炸"时期的深层次的矛盾。是的,深圳大爆炸已经创造了一个新的都市富人阶层,他们生活在高档公寓,喜欢到世界各地旅游。是的,这些人的财富源于他们拥有获取经济资源的特殊路径,这对其他人是不公平的,加深了城市"有产阶级"和"无产阶级"之间的矛盾。是的,深圳的建设已经蕴涵了对自然景观的剧烈重组,使得海岸线的景观面目全非,对区域气候模式也产生了影响。是的,深圳对重建和重组世界的热潮产生了一种量子时序性,似乎每个年代深圳都从一种状态变到另一种状态。这个持续变化的城市是所有深圳年轻居民的遗产,而不仅仅属于展览"轰隆!深圳"的这六个建筑师。所以,每个作品的标题也提醒着我们深圳的文化价值和梦想,其文化价值和梦想预示着深圳会进入到一个在世界范围内设计建筑的改革时代。请重新阅读这些作品的标题,但这次请试着以诗歌的方式来聆听:轰隆!家庭价值观。未来。融化。深圳速度。游戏结束,重启。轰隆!

展览"轰隆!深圳"邀请观众参与到两个层面的对话中,一个有关深圳,一个有关年轻建筑师。在内容的层面,每件作品都为观众提供了有关中国最悠久、最庞大、最成功的经济特区的信息。在这个层面,展览的作用更像是一个传统的时间表,比如我们了解到 1997 年香港回归加深了深圳和香港两个城市的融合,2004 年,深圳最后一个自然村合并为深圳市政的一部分,使得深圳特区成为了中国第一个没有农民或农业用地的城市。然而,在审美形式的层面上,每件作品都允许观众从经济指标和人口统计学之外的角度来理解深圳。这个深圳是创造性的、凌乱的、玩乐式的以及悲情的城市。换句话说,年轻建筑师对深圳在统计学上的"大爆炸"的解读允许我们当中那些来自其他文化或者其他年纪的人跨越彼此的差异来理解深圳,并对深圳"大爆炸"时期的生活有更多认知,这种认知也许显而易见,也许略微难懂,但希望它们是更加人性化的认知。

SHENZHEN/ 深圳　　　　　　　　　　　　　　　BOOM! SHENZHEN / 轰隆! 深圳

1 **Boom! centerpiece mushroom cloud**
CAO Taiming and CHEN Yue playfully literalize the idea of a boom, with thousands of tiny figures scaling a mushroom cloud that takes its silhouette from a graph of the Municipality's annual GDP.

2 **Family Values bas—relief rendering of Shenzhen's housing market**
Zhang Yiwei humanizes one of the more controversial of Shenzhen's booms – the price of housing, giving viewers insight into how millions of immigrants have inhabited the SEZ.

3 **Futures bas-relief of Shenzhen stock market index**
Hong Wudi maps the abstract structure of market trading and investment in order to track how capital localization has transformed the cityscape.

4 **Meltdowns sculpture based on the SEZ's expanding urban borders**
Wan Yan lyrically points to the uncontrolled and unexpected geological affects of urbanization by melting a wax candle map of Shenzhen.

5 **Shenzhen Speed flash video**
Zhang Xueshi deconstructs our fetishization of fast track development, by interpreting key dates in the SEZ's history as a hungry snake that "eats" history.

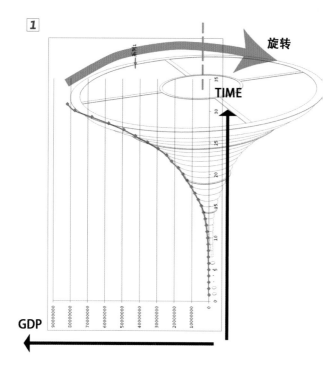

1 轰隆：核心蘑菇云

曹泰铭（Cao Taiming）和陈越（Chen Yue）以一种玩乐的方式对"轰隆"这个词进行字面呈现：上千个小雕像组成一朵蘑菇云，而蘑菇云的轮廓则由深圳特区政府的年度 GDP 表格构成。

2 家庭价值观：体现深圳房地产市场的浅浮雕

张轶伟（Zhang Yiwei）把高价房地产这个在深圳大发展中产生的最具争议性的问题之一进行了拟人化，为观众呈现了几百万的外来人口如何在深圳特区居有定所。

3 未来：深圳股市指数的浅浮雕

洪吴迪（Hong Wudi）把深圳市场贸易和投资的抽象结构绘制成地图，用以追踪资产本土化是如何改变城市景观的。

4 融化：基于深圳特区不断扩张的城市边界的雕塑

万妍（Wan Yan）通过融化一幅由蜡烛组成的深圳地图，诗意地呈现出城市化对地理版图所产生的不可控制和不可预料的影响。

5 深圳速度：动画影像

张雪石（Zhang Xueshi）把深圳特区历史中的一些关键日子描绘成一条把历史"吞食"掉的贪吃蛇，从而把我们对快速发展所产生的恋物情结进行解构。

1979地图

2011地图

SHENZ
AND
HONG
深圳和氵

HEN

KONG
港

COUNTERPART CITIES
对应双城

Curators: Jonathan Solomon, Dorothy Tang
策展人：Jonathan Solomon, 邓信惠

SHENZHEN AND HONG KONG/ 深圳和香港   COUNTERPART CITIES / 对应双城

INTRODUCTION

Hong Kong and Shenzhen are counterpart cities in a single interdependent system: Hong Kong, the former colony turned global finance hub, now a Special Administrative Region of China; Shenzhen, the so-called "instant city" conceived by Deng Xiaoping, now among the nation's most dynamic of cities. Joined by the world's busiest land border crossing, Hong Kong and Shenzhen already form a single metropolis—of sorts. While diverse urban systems in the two cities are already highly integrated many barriers to regional co-operation persist.

Counterpart Cities explores the unique relationship between Hong Kong and Shenzhen through multidisciplinary study of the interdependent infrastructures and natural systems of the two cities. Overcoming historic reluctance to plan across the political and cultural border between them, the project is the first of its kind to bring academics and professionals from both cities to the same table to collaborate on proposals for co-operative action. An exhibition of research and visionary design proposals, Counterpart Cities is also a model for a new way of working across boundaries.

Water—the opportunities and vulnerabilities embodied in it—provides the perfect framework for Counterpart Cities. The effect of both global change and local human intervention combine in complex ways in the ecosystem of the Pearl River Delta. Sea level rise, management of freshwater delivery systems, dredging and effects such as salt-water intrusion, dynamics of the seaport economy, and shore ecologies are examples of systems in which increasingly complexity can only be handled through greater cooperation.

Working on such systems simultaneously allows the project to explore the region's specific challenges and opportunities

© Milkxhake, Courtesy of the University of Hong Kong

in the face of climate change, and to advance the role of urban systems integration beyond contemporary discussions of infrastructuralism or landscape urbanism. Neither a fully natural system nor one in which overall systemic planning is possible, a Hong Kong and Shenzhen megalopolis will ultimately challenge both of these paradigms.

COUNTERPART SITES

The Pearl River is the third largest river system in China and its delta has historically been an important area for agricultural production. Flooding and sedimentation cycles—due to seasonal weather patterns and geological processes—create a shifting landscape between dry and wet, between erosion and expansion. Human occupation, beginning with early agricultural practices and recent accelerated urbanization, has altered this delicate balance through reclamation and excavation within the Delta.

The Pearl River Delta is a major contributor to global greenhouse gasses, and ironically also faces the grave consequences of climate change. Massive gravel excavation from the riverbeds of the Pearl River have effectively lowered the water levels, but invite large levels of salt-water intrusion and dangers of riverbank erosion. This endangers the potable water supply for Shenzhen and Hong Kong as well as the economic activities within the Delta. In addition, the predicted extreme storm conditions exasperate flood related risks in two cities where important urbanization processes have occurred in reclaimed land or narrow flat areas at the foot of steep hills and mountains.

The futures of both Shenzhen and Hong Kong rely on the vast network of infrastructural systems that support their daily operations. Potable water systems, transportation networks, land reclamation processes and real estate dynamics all contribute to the complex interdependencies and competition between the two cities, and will continue to influence their economic and political development. As such, the curatorial team selected three landscape systems that demonstrated the interdependencies between Hong Kong and Shenzhen as the basis for design speculation.

Resource Sharing: Freshwater Networks

A close look at the potable water resources of Hong Kong and Shenzhen reveals the environmental, economic, and social costs of the trans-border infrastructural system. The Dongjiang watershed—a major tributary of the Pearl River—involves a complex set of economies that rely on its water resources. Recent industrial development on the lower reaches of the watershed endangers the safety of this water. An 83km aqueduct diverts water from the Dongjiang to the Shenzhen Reservoir which provides over 70% of potable water for Hong Kong. The city of Shenzhen also draws from the Dongjiang and diverts water to its own reservoirs within the municipal limits and competes with Guangzhou, Dongguan, and other major cities for water resources and rights. Water provision in the Pearl River Delta has long been a contested issue, and is particularly sensitive to the exaggerated drought conditions that climate change will bring. This requires increased collaboration within the region and sensitive solutions to ensure the continued urban growth of the Pearl River Delta.

Collaborative Ecologies: Deep Bay Territory

The Deep Bay, at the border between Shenzhen and Hong Kong, was originally the site of prime aquaculture and ecology in Hong Kong and Shenzhen. Recent urbanization has changed the shores of the Deep Bay drastically. On one side, Shenzhen has reclaimed land and built up the city right to the water's edge; on the other, Hong Kong has retreated from this edge and populated the shore with landfills and power stations; and at the mouth of the Shenzhen River, delineating the border between the two cities, lies Mai Po, an ecologically sensitive wetland with international

| SHENZHEN AND HONG KONG/ 深圳和香港 | COUNTERPART CITIES / 对应双城 |

significance. The polarized conditions along the bay are all vulnerable in the event of climate change. In Shenzhen, the low-lying reclaimed land is subject to flooding, but the reclamation itself also alters the hydrology and is a cause of floods. In Hong Kong, sea level rise would flood much of the land in Mai Po, but also drastically alter the balance between fresh water and saltwater, endangering the sensitive ecological conditions of mangroves, mudflats, and wetlands. Deep Bay is a prime example of the varied models of urbanization between the two cities, and physical manifestation of its political tension.

Economic Symbiosis: Port Infrastructure

The ports of Shenzhen and Hong Kong combined handle almost 10% of TEUs worldwide, and will handle up to 15% in 2030 under optimistic projections. The natural shelter and deep bays of Shenzhen and Hong Kong allowed the two ports to flourish, first under the British rule and later during Deng's opening up policy. Analogous to the relationship between the Port of Los Angeles and the Port of Long Beach, Shenzhen and Hong Kong share geographical proximity but are in direct competition to each other. Shenzhen's larger land area and access to deep waters have prompted Hong Kong to develop mid-stream operations, and their interaction continues to prompt new innovations. However, the two ports are also a major contributor to greenhouse gasses in the region due to lax environmental regulations that have caused significant air and water pollution. Increased storm surge and frequency brought about by climate change prompt design and architectural solutions that protect the waterborne trade and economic viability of the region.

THE DESIGN WORKSHOPS

Counterpart Cities is a collaboration between designers and professionals across the border of Hong Kong and Shenzhen. The project curators worked with 6 teams of to develop strategies to address urban issues in the Pearl River Delta. Three systems: freshwater networks; port infrastructure; and shoreline ecologies; were identified as under threat of climate change, but more importantly, their critical systemic nature require collaboration beyond political and administrative boundaries. Three pairs of counterpart projects emerged and design teams were confronted with the duality and connections between their respective projects and sites, and sustained continuous dialogue with their counterparts across the border to address the impact of climate change through the imagining of architectural futures.

The workshop format of Counterpart Cities is a form of active engagement to address issues in our cities and propose alternatives for their development. The project developed from June 2011 with an intensive research phase at the University of Hong Kong and continued into a six-week design workshop beginning in September 2011. Two reviews in Shenzhen and Hong Kong brought distinguished guests and members of the public into conversation with the six teams.

A PRAGMATIC UTOPIANISM

As a product of this cross-border collaboration, a range of both tactical and strategic approaches characterize multidisciplinary design projects that both communicate contemporary challenges and visualize the possibilities of radical change. A pragmatic Utopianism emerged spontaneously as teams worked to address real challenges of environmental change that Shenzhen and Hong Kong face, but simultaneously attempt to maintain visionary and imaginative futures for the region. As such, all of the teams perceive climate change and environmental degradation as opportunities rather than threats and are opportunistic in their design speculations rather than reactive adaptation or mitigation. In addition, the collaborative nature of the project produced 6 unique projects that differ greatly in scale and program, yet are complementary in their operations and approach.

Stefan Al, Director of the Urban Design Program at the University of Hong Kong, and Doreen Liu of NODE Office in Shenzhen, both acknowledged the problematic nature of shifting sites for freshwater extraction as the sources are eventually compromised and identified wastewater as a potential new source for the region. Al's project, "PRD SEA: Pearl River Delta Special Ecological Area" imagines a reconfiguration of infrastructure across political boundaries to create a networked, distributed, freshwater provision and wastewater treatment system. Treated wastewater from varied land-use types enter in a continuous loop of freshwater and treated wastewater conduits, allowing shortage and surplus in local systems to be absorbed and resources between municipalities to be shared and traded. Liu's project "I-Infrastructure: Water Urbanism" works within this regional framework, and operates at a neighborhood scale, effectively reconfiguring an existing wastewater treatment plant and integrating with a new neighborhood through networked water infrastructure. She structures residential, commercial, recreation, and aquaculture production with the new water treatment infrastructure, recognizing the various stages of water treatment and pairing them with associated land-uses.

Feng Guochuan of Zhubo Architects and Vincci Mak, Assistant Professor of Landscape Architecture at the University of Hong Kong, collaborated to model the effects of natural sedimentation in the Deep Bay and the implications of climate change and flooding in its future. Although both teams base their work on identical assumption and have similar approaches in seeking to manipulate sedimentation patterns as a means to reorganize urban growth along the shores of the Deep Bay, the projects are vastly different. In "Sedimental Urbanism", Feng first utilizes a sensitive analysis of the relationship between indigenous shrimp ponds, mangroves, and mudflats as the basis for a new incremental "reclamation" strategy into the Deep Bay. The new land for agriculture and urbanization is produced while taking advantage of a naturally shifting coastline and ecological processes continue to evolve. On the other hand, in "Symphony of Blades", Mak completes a series of hydrodynamic studies and conceives of a mechanical system that redirects flows and sedimentation, protecting the current shoreline of the Deep Bay but producing new economic program based on sediment collection and infrastructural tourism.

Last but not least, Tom Verebes, Associate Professor of Architecture at the University of Hong Kong and Zhu Xiongyi from CCDI in Shenzhen both reconfigure port infrastructure and its economic activities as a new type of player globally and regionally. In "Future Port", Verebes simulates economic shifts based on port locations and locates a new mega-port at the mouth of Delta and connected by the proposed Hong Kong—Macau—Zhuhai bridge. The new port lands are a series of floating islands that support a new city with new industries. Zones of production are directly correlated to the movement of goods, creating a vision for a new mega-city. Zhu's project takes this a step further in "Mobile Culture, Creative Community" by creating mobile ports that travel around the globe, enacting new modes of creativity and culture where this hybridized port acts simultaneously as a site of exchange, transportation, and production. Existing port zones are adapted to become docks for resources and redeveloped for better utilization of urban land.

All six projects are unabashedly utopian in their idealization of future economic exchanges, political interactions, and environmental processes in Shenzhen and Hong Kong and their envisioning of the development of new frameworks to accommodate them. However, they are also rooted in pragmatic solutions and realities of the region, creating a distinctive and hopeful vision for the future of the Pearl River Delta.

| SHENZHEN AND HONG KONG / 深圳和香港 | COUNTERPART CITIES / 对应双城 |

简介

香港与深圳互相依赖，自成一个系统：中华人民共和国特别行政区香港——过去的殖民地，已然转型成为今天的全球金融中心。被称为"速生城市"的深圳来自于邓小平的构想，现在的深圳已经位居最具活力的城市之列。香港与深圳由世界上最繁忙的陆地边境连接在一起，基本上可算是形成了一个独立的大都会。两个城市中的各种都市系统已经高度整合，地区性合作的多种障碍依然存在。

通过对两座城市互相依赖的基础设施和自然系统进行的多学科研究，"对应双城"探讨了香港与深圳之间的这种独特关系。这两座城市克服了长期的隔离，冲破了彼此间的政治与文化的边界，通过这个项目首次集合了两座城市的学院与专业人士，激发合作，协同行动，并提出方案与意见。"对应双城"是一个兼研究与提供视觉设计方案的展览，也是跨界合作新方式的一个模型。

水，以其蕴涵的机遇和脆弱性，为"对应双城"提供了一个完美的框架。全球变化与当地人为的介入所引发的效应，都被融合进了作用于珠三角生态系统的复杂方式中。海平面上升、淡水资源分配系统的管理、淤积与盐水侵入产生的效应、海港经济的动态情形和海岸生态，都是珠三角生态系统中的重要内容，其逐渐增加的复杂性只有通过更大范围和程度的合作才能得以掌控。

香港和深圳所组成的大都会也将最终挑战这两种模式，以这样的系统工作，项目要求在面对气候变化时深入挖掘这一区域所面对的特殊挑战和机遇，并推动都市综合系统的角色超越当代话语中的基础设施主义或景观都市主义。一个全方位的自然系统抑或是全方位的规划系统，都是不可行的，香港、深圳所组成的大都会也将彻底地挑战这两种模式。

对应地点

珠江，是中国第三大河流系统，珠三角在历史上一直是农业生产的重要区域。洪水与淤积的循环——归因于季节与地理变化——创造了一个在干与湿、侵蚀与扩张之间变化的区域。人类的占领开始于早期农耕和近代不断加速的都市化，因为人类的介入，在三角洲内进行的拓荒与开发改变其原有的脆弱的平衡。

珠江三角洲可谓是全球温室气体排放的主要贡献者，所以也同样地面对气候变迁的严峻结果。大量的河床砾石开挖使得淡水平面下降及海水高位入侵，河岸遭受了侵蚀的危害，这一情形使得深圳和香港的饮用水以及三角洲区域内的经济活动都受到了威胁。除此之外，早已预见的严苛风暴恶化了两座城市的洪水危机，并威胁着在陡峭的山坡和高山脚下进行的土地开垦和窄化平地的城市化进程。

深圳与香港的未来，都依赖于提供日常作业的庞大的基础设施系统网络。饮用水系统、交通网络、土地再利用进程以及房地产动态，都使得两座城市之间的互相依赖与竞争复杂了起来，而且还将继续影响其经济及政治进程。正因如此，策展团队选择了能够体现香港与深圳之间互相依赖的三种景观系统，以此作为设计的基础。

资源公用：淡水网络

仔细研究过香港和深圳的饮用水资源后，就会看到这一跨边界的基础设施系统的环境、经济和社会成本不菲。东江分水岭——作为珠江的主要支流——囊括了依赖于这一水资源的一系列复杂经济。近期，下游的工业发展威胁了水资源的安全，一条长为 83 公里的渡槽将东江水引向深圳水库，并为香港提供了 70% 的饮用水。深圳市也从东江引水以灌入其市内水库储备并与广州、东莞和其他主要城市竞争水资源及其使用权力。珠三角的水资源供给一直是竞争性的问题，而且因气候变化所引致的过分干旱情况尤为让人敏感。这就要求有区域内的紧密合作和灵活的解决方案，以此保证珠三角的城市化进展。

生态合作：后海领域

深圳和香港交界的后海区域，曾是香港和深圳两地的主要水产和生态基地。近来的城市化进程急剧地改变了前海区域的海岸。一方面，深圳开垦了部分土地并且沿水建立起这座城市；另一方面，香港从其边缘撤退，沿海则逐渐布满垃圾填埋场与电站；而在米埔的深圳河入口，则明显现出两个城市之间的边界，这是一片在国际上占有重要地位的生态湿地。后海区域沿岸的两极状况在气候变化的情境下十分脆弱。在深圳，低洼的被开垦土地受到洪水的威胁，然而开垦土地本身也是影响水文和引发洪水的因素。在香港，海平面的上升会淹没米埔的土地，同样会严重地影响淡水和咸水之间的平衡，把红树、泥滩和湿地的敏感生态情境暴露在危险当中。后海也是两座城市之间各种都市化模式的一个重要案例，也是其政治张力的一个物理表现。

经济共生：港口设施

深圳和香港的海港承担了全世界 10% 的 TEU（TEU 是英文 Twenty-feet Equivalent Unit 的缩写，以长度为 20 英尺的集装箱为国际计量单位，也称国际标准箱单位——译注），乐观估计，到 2030 年，该区域可处理的 TEU 将占全世界的 15%。深圳与香港的自然屏障和后海区域，使这两个海港在香港尚未回归和邓小平的改革开放政策之下，一直保持着蓬勃发展的态势。深港两港之间的关系类似于洛杉矶与长滩港，尽管深圳与香港是相邻的地域，但彼此又是直接的竞争者。深圳占有的区域更广并且毗邻深水区，这促使香港发展其中流作业，他们之间的互动继续触发革新。然而，也正是由于松散的环境管制，两个港口成为了该地区温室气体排放的大户，造成了严重的空气和水体污染。为了应变由气候变化引起的不断增多的风暴，团队提出的设计及建筑方案需要保证该地区的水运贸易和经济生存能力。

设计工作坊

"对应双城"是香港与深圳两地的设计师与专业人士的合作项目。该项目的策展人同时与 6 个团队合作，开发解决珠江三角洲都市问题的策略。三个系统：淡水网络、港口基础设施和海岸线生态，都被认为遭受了气候变化的威胁，但更重要的是，这三个系统要求超越政治与行政边界的合作。三组对应项目的出现和设计团队都面临着各自项目同地点的双重性和相关性问题，并且通过与对应的团队 / 对象坚持跨界对话，通过对未来建筑的想象将气候变化的重要性凸显出来。

"对应双城"的工作坊模式要求我们主动介入所在城市，并提出促进发展的议案。这一项目自 2011 年 6 月开始便在香港大学启动了一系列的密集研究，在同年 9 月还开展了为期 6 周的设计工作坊。在深圳与香港的两次评审中，杰出的嘉宾和公共领域的成员都被邀请到与 6 个团队的对话中来了。

实用乌托邦主义

作为跨界合作的成果，一众战术和策略上的方法都具有多学科设计的特征，能够同时传递出当代的挑战，并能将极具可能性的转变进行视觉化。"实用乌托邦主义"是团队针对深港面对的真实的环境变化所迎接的挑战，同时也尝试去保持该地区的视觉和想象的未来。类似地，所有的团队都将天气变化和环境恶化作为机会而非威胁，并如机会主义者般地对待设计构想中的机会而非对应的顺从或缓和。除此之外，项目的合作特性还产生了 6 个在面积和方案计划上都完全不同的独特项目，并在其操作方式上实现了互补。

香港大学都市设计项目主任 Stefan Al 以及深圳南沙原创的刘珩都认可了变动的取水地点的问题所在，作为资源，这些取水点和被确认的污水处理地区都被转化为潜在的新资源。Stefan Al 的项目为 "PRD 海洋：珠江三角洲特别生态区" 想象出了一种穿越政治隔阂的基础设施再配置，创造出了一个网络化的、分类式的淡水提供和污水处理系统。该项目根据不同的土地使用类型对污水进行处理，使其进入淡水和污水处理管的循环，允许当地系统的短缺和盈余进行合并，使城市之间的资源能够得到共享与交换。刘珩的项目"互——基础设施"则在地区性框架中进行，即在邻里范围内开展。该项目有效地重新对已有的污水处理厂进行配置，并且通过网络化的水基础设施将新的邻里联系在了一起。刘珩建构了一种住宅区式的、商业的、再造的、水产业的体系以及新的水处理基础设施，该项目确认了水处理的几个环节，同时，与相关的土地运用进行并置。

筑博建筑的冯果川和香港大学景观建筑副教授 Vincci Mak 合作，模拟后海地区未来的自然沉淀以及气候变化和洪水，制作了模型。尽管两个团队都基于同一个假定展开工作，并且使用类似的方法来分类处理淤积问题，以此方式来重组后海地区的都市发展。他们的方式大为不同。在"淤积都市主义"中，冯果川首先对当地的虾塘、红树和泥潭之间的关系进行了敏感度分析，以此作为后海的新增"回收"策略的基础。农业和都市化所需要的新增土地正是来自于自然变化的海岸线和不断发展的生态进程。在"旋桨交响曲"中，Vincci Mak 完成了一系列水力研究，并构想出一套机械系统，改变了水流及淤积的方向，保护了现有的后海的海岸线，但也能产生出基于沉积物汇集和基础设施的旅游业的新经济项目。

香港大学的建筑系副教授 Tom Verebes 以及深圳中国国际设计的朱雄毅都重新构想了海港基础设施，其相关经济活动也成为了全球和地区性的新类型选手。在"未来港口"中，Tom Verebes 根据海港位置构想了经济转型，并把一个新的超大港口安排在三角洲入海口，用港珠澳大桥将三角洲联系起来。新港口所辖土地是一组漂浮的岛屿，为一个新的城市提供了新型工业。生产区域直接与货物的移动连接在一起，创造了一种新的大城市构想。朱雄毅的项目通过创造全球移动的港口进一步提出了"文化流动，创意社会"，多功能的港口同时作为交换、运输和生产的地域，让新形式的创造性和文化性得以生效。现存的港口区域则可以反身成为资源的码头，再开发成为能更好地利用的城市用地。

上述六个项目都带有不加掩饰的乌托邦性质，充斥在他们对深圳与香港未来的经济交换、政治互动和环境进程的理想化构想中，他们发展新框架来安置这些议题也都弥漫着乌托邦的情节。然而，这些项目也都植根于实用主义的解决方案与该地区的现实情境中，为珠三角的未来创造了一个独特且充满希望的愿景。

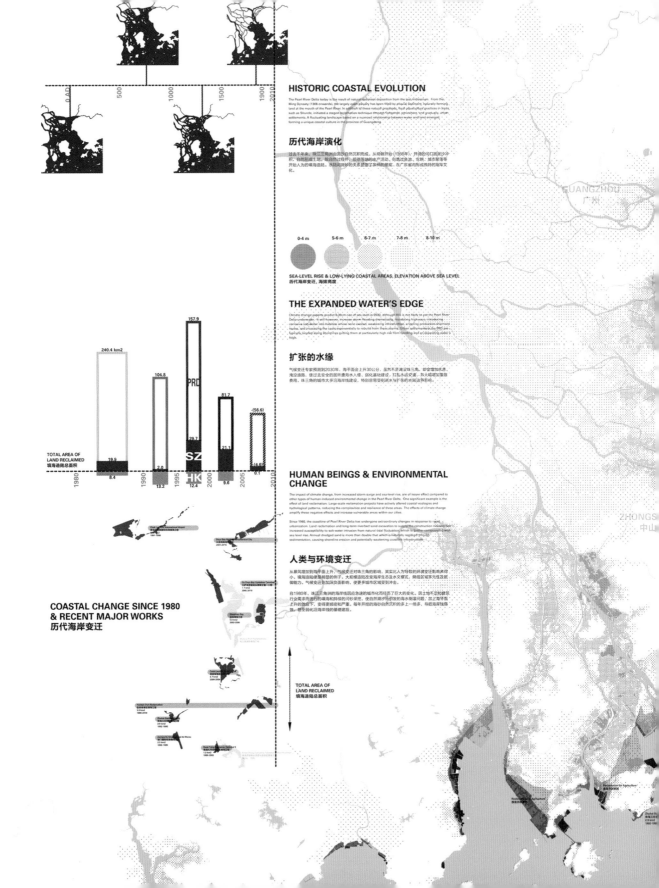

HISTORIC COASTAL EVOLUTION

The Pearl River Delta today is the result of natural sediment deposition from the sea millennium. From the Ming Dynasty (1368 onwards), the largely open estuary has been filled by alluvial deposits, gradually forming land at the mouth of the Pearl River. In addition to these natural processes, fish aquaculture practices in lakes such as Shunde, initiated a staged reclamation technique through fishponds, agriculture, and gradually, urban settlements. A fluctuating landscape based on a nuanced relationship between water and land emerged, forming a unique coastal culture in the province of Guangdong.

历代海岸演化

过去千年来，珠江三角洲由泥沙自然沉积而成。从明朝起（1368年），开敞的河口湾受到冲积，自然形成土地。除自然过程外，顺德等地的水产活动、包括鱼塘、农耕、城市聚落等开始人为的填海造陆。水陆间微妙的关系塑造了多样的景观，在广东省内形成独特的海岸文化。

0-4 m   5-6 m   6-7 m   7-8 m   8-10 m

SEA-LEVEL RISE & LOW-LYING COASTAL AREAS, ELEVATION ABOVE SEA LEVEL
历代海岸变迁，海拔高度

THE EXPANDED WATER'S EDGE

Climate change experts predict a 30cm rise of sea level in 2030, although this is not likely to put the Pearl River Delta underwater. It will however, increase storm flooding dramatically, inundating highways, introducing corrosive salt-water into habitats whose native existed, weakening infrastructure, crippling production shipment routes, and increasing the costs exponentially to rebuild from these disasters. Urban settlements in the PRD are typically located along shorelines putting them at particularly high risk from flooding and an expanding water's edge.

扩张的水缘

气候变迁专家预测到2030年，海平面会上升30公分，虽然不足淹没珠三角，却会增加水患、淹没道路、侵蚀过去安全的沿海水入侵、弱化基础建设。打乱水运交通，将大幅增加重建费用。珠三角的城市大多沿海岸线建设，特别容易受到洪水与扩张的水缘边界影响。

HUMAN BEINGS & ENVIRONMENTAL CHANGE

The impact of climate change, from increased storm surge and sea-level rise, are of lesser effect compared to other types of human-induced environmental change in the Pearl River Delta. One significant example is the effect of land reclamation. Large-scale reclamation projects have actively altered coastal ecologies and hydrological patterns, reducing the complexities and resilience of these areas. The effects of climate change amplify these negative effects and increase vulnerable areas within our cities.

Since 1980, the coastline of Pearl River Delta has undergone extraordinary changes in response to rapid urbanization. Land reclamation and long-term riverbed sand excavation to support the construction industry have increased susceptibility to salt-water intrusion from natural tidal fluctuations, which is further compounded with sea level rise. Annual dredged sand is more than double that which is naturally replaced through sedimentation, causing shoreline erosion and potentially weakening coastal infrastructure.

人类与环境变迁

从暴风增加到海平面上升，气候变迁对珠三角的影响，其实比人为导致的环境变迁影响来得小。填海造陆便是最显著的例子。大规模造陆活改变海岸生态及水文模式，降低区域复杂化及抵御能力。气候变迁也加深负面影响，使更多城市区域受到冲击。

自1980年，珠江三角洲的海岸线因应急速的城市化经历了巨大的变化。因土地不足为建筑行业带来而进行的填海和持续的河砂采挖，使自然潮汐引发的海水倒灌问题，加上海平面上升的效应下，变得更频密和严重。每年开挖的海砂自然沉积的量大一倍多，导致海岸线侵蚀，弱化沿海海岸线的基础建设。

COASTAL CHANGE SINCE 1980 & RECENT MAJOR WORKS
历代海岸变迁

TOTAL AREA OF LAND RECLAIMED
填海造陆总面积

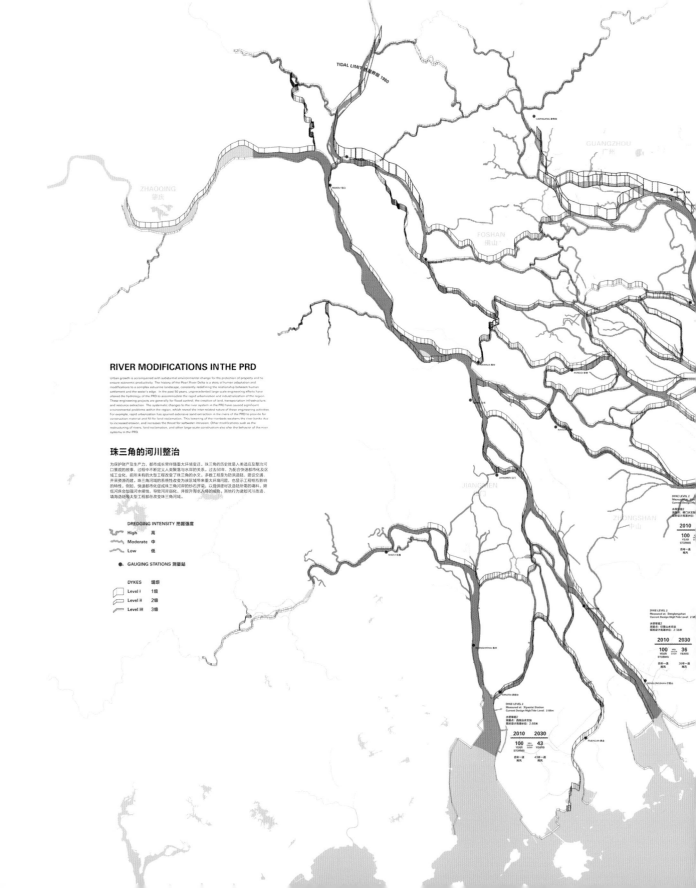

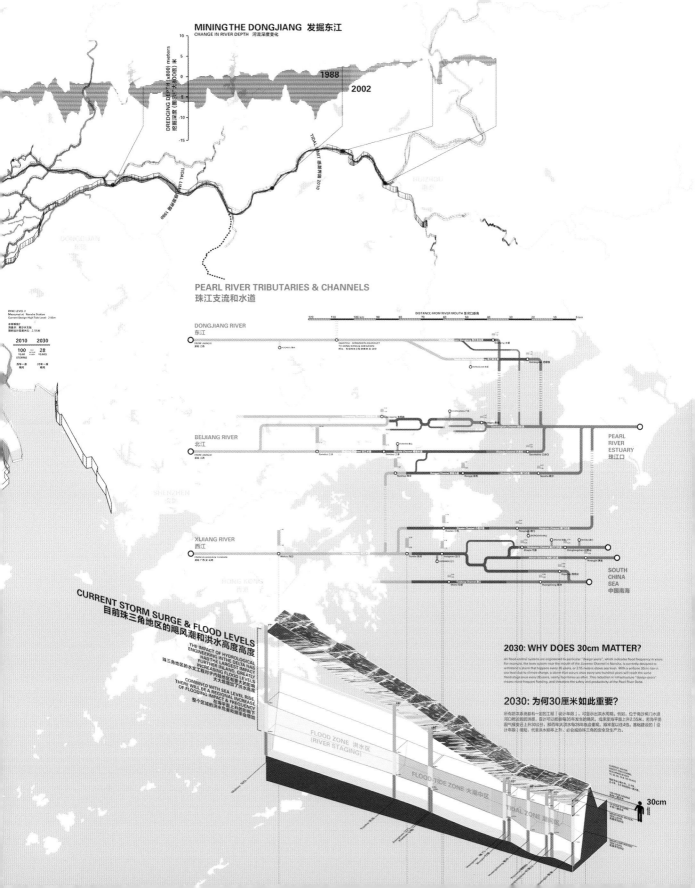

| SHENZHEN AND HONG KONG / 深圳和香港 | COUNTERPART CITIES / 对应双城 |

Resource Sharing: Freshwater Networks

资源共用：淡水网络

The process of urbanization and industrialization has historically relied on the steady provision of freshwater resources. The early growth of Hong Kong required careful watershed planning to increase the yield of freshwater required for the economic development of the city, thus facilitating an expansive network of reservoirs and watershed conservation zones. Today, the urban growth of Hong Kong has exceeded its natural capacity for the collection of freshwater locally, and 80% of its potable water comes from the Dongjiang River through the 83-km Dongshen Aqueduct, a joint effort between Hong Kong and Shenzhen. Shenzhen also regulates urban growth to protect the catchment zones of the reservoirs, but often this is a difficult goal to achieve due to pressures for urban development.

Although the freshwater resources of Hong Kong and Shenzhen are not directly threatened by saltwater intrusion due to climate change, other cities in the greater Pearl River Delta region are potentially effected greatly by the aftermath of sea level rise and extreme drought or flood cycles. The Pearl River and its tributaries form a watershed region that transcends the political boundaries of Guangdong, its five neighboring provinces, and Vietnam. Major coastal cities along the mouth of the PRD have begun to expand their aqueduct system further upstream to avoid saltwater intrusion due to increased hydrological alterations (dredging, sand-mining, sea level rise, etc) and natural tidal cycles. In addition, the Pearl River Water Resources Commission began a regional transfer of fresh water hundreds of kilometers upriver in Guangxi to "flush out" the salty water in 2005. Those transfers occurred twice each month, releasing water from flood reservoirs several days in advance to correspond with high tide. Although the Pearl River has abundant water resources, increased salinity, industrial pollution, and regional water transfer, still shape large-scale water diversion projects that rival initiatives only previously carried out in more arid regions of China.

在过去，都市化及工业化都有赖于淡水资源的稳定供给。为提升城市经济发展效益，香港在发展的早期便注重流域规划，形成了广大的淡水网络，包括水库及水土保育区。现今香港本地所能收集的淡水已不足以支持都市成长。目前 80% 的饮用水来自于东江，经过香港、深圳的共同建立，总长 83 公里的东深供水工程得以启动。深圳试图透过控制都市成长来保护水库的集水区，但发展的压力使目标难以达成。

气候变迁引起的海水入侵，虽然没有直接影响香港和深圳的淡水供应，但大珠江三角洲的其他城市却可能因此受海平面上升、极端干旱或洪水的影响。珠江及其支流流域横跨了政治边界，包括广东与其邻近五省以及越南。珠江口沿岸的大城市已开始将供水工程往上游延伸，以避免水文改变（疏浚、采砂、海平面上升等）导致的海水入侵以及自然潮汐周期的影响。2005 年起，珠江水利委员会开始将淡水引至广西上游数百公里处，试图压咸补淡，每两个月调水一次，会在数天前让水库泄洪以配合满潮。虽然珠江的水资源充足，但持续的盐化、工业污染以及区域调水仍然在型塑大型的引水工程，这类工程以往只在中国更干旱的地方实行。

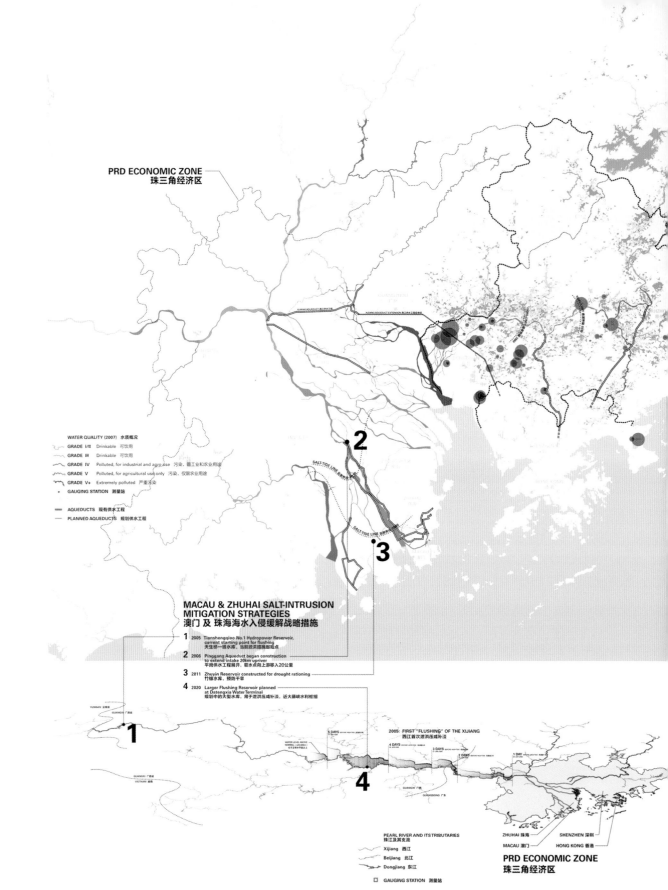

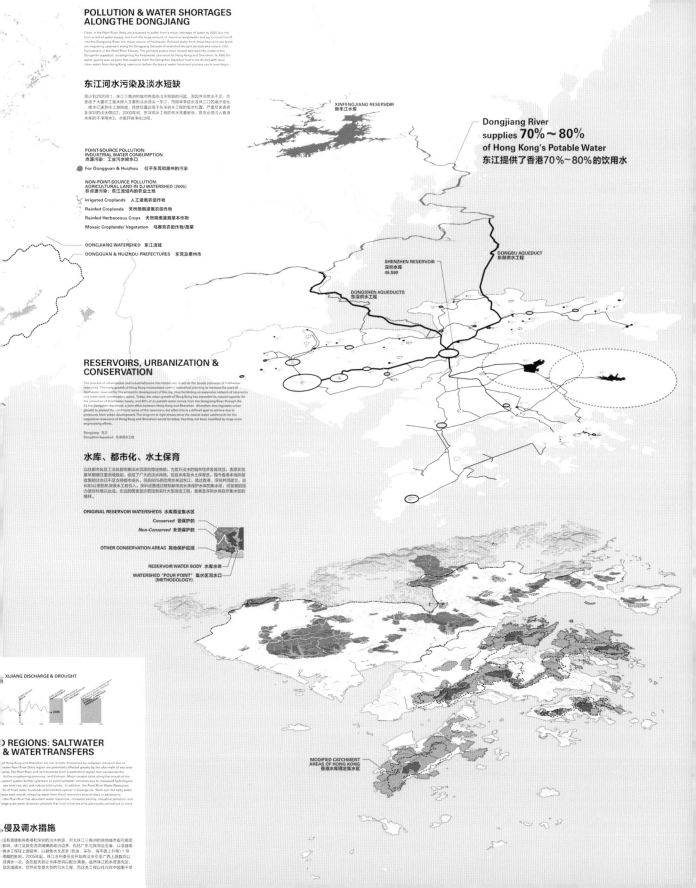

SHENZHEN AND HONG KONG / 深圳和香港	COUNTERPART CITIES / 对应双城
HONG KONG 香港 **Resource Sharing: Freshwater Networks**	**资源共用：淡水网络**

PRD SEA (Pearl River Delta Special Ecological Area)

珠江三角洲特别生态区

What if our administrative boundaries were dependent on ecological boundaries? When freshwater becomes increasingly scarce, could we use watersheds as boundaries to better administer the distribution of water resources?

Climate change, urbanization, and pollution from industrial and agricultural development threaten the steady supply of freshwater to the Pearl River Delta and its cities, while population growth in the region only increases demand. At the same time, Hong Kong and Shenzhen's increasing economic, political and cultural interdependency presents the possibility to tackle these challenges holistically, piggybacking new systems for freshwater delivery onto defunct, existing, and planned infrastructure in the region.

To address the challenges of freshwater provision in the region, we proposed a zone of interconnected water resources: the Pearl River Delta Special Ecological Area (PRD SEA). It adopts four water resource strategies: move from a centralized to a decentralized system; link independent systems to enable resource sharing; piggyback new systems on defunct, existing and planned infrastructure; and finally treat "waste" water as a resource.

The PRD SEA imagines a fundamental realignment of water resources from a centralized, source-to-user system, to one that is distributed and interdependent, relying on a user-as-source network. Traditionally, the collection and storage of fresh water, the distribution of potable water, and treatment of resulting waste water are found in independent systems. Within the PRD SEA, we propose to network these systems in a continuous loop of freshwater and treated waste water conduits, allowing shortage and surplus in local systems to be absorbed and resources between municipalities to be shared and traded.

想象一下，假如我们的行政界限可以取决于生态边界会怎样？在一个淡水资源日益稀缺的时代，我们能否想象由分水岭所构成的新政治界限？

气候变化、城市化、工业和农业发展所造成的污染正在威胁珠江三角洲及其中城市的稳定的淡水供应，而珠江三角洲庞大的人口增长会日益增加对淡水的需求。与此同时，香港和深圳与日俱增的经济、政治和文化上的相互依赖带来了整体上缓解这些威胁的可能性，为本地区已经停用的、现存的和规划的淡水运输设施搭建了新的水利基础设施。

为了应对该地区淡水供应的挑战，我们提议拟建一个相互关联的水资源区域：珠江三角洲特别的生态区（珠三角海洋）。其区域采用4个水资源战略：

(1) 从中央系统变成分散系统；
(2) 连接独立的系统，共享资源；
(3) 在已停用的、现有的和规划的基础设施上搭载新的系统；
(4) 考虑将污水作为一种资源。

"海洋"将水资源从一个集中的由来源到用户的系统，调整到一个分散的、相互依存的和以用户为源的网络。传统处理方法是将淡水收集和储存，将饮用水分配和污水的治理划分在独立的系统里。在"珠三角海洋"，我们提议把这些系统通过连续循环的淡水和处理后的污水的管道连接，使本地水系统中的短缺和盈余尽可能地被吸收，使城市间的资源可以共享和交换。

Team Leader / 设计领队:	Stefan Al (Department of Urban Planning and Design, HKU / 香港大学城市规划与设计学系)
Team Members / 设计团队:	Jason Carlow (Department of Architecture, HKU / 香港大学建筑学系)
	Ivan Valin (AECOM & Department of Urban Planning and Design, HKU / 香港大学城市规划与设计学系)
Consultants / 顾问:	Juan Du (Department of Architecture, HKU / 香港大学建筑学系)
	Iris Hwang (Arup)
	Rowan Roderick-Jones (Arup)
	Kenneth Kwok (Arup)
	Kam-Shing Leung (Arup)
Collaborators / 合作:	Anthony Lam Chuek Wang, Daniel Fung, Peter Lampard
Contributors / 协助:	Audrey Ma, Norman Ung, Dannes Kok, Frederick Li

| SHENZHEN AND HONG KONG / 深圳和香港 | COUNTERPART CITIES / 对应双城 |

SHENZHEN 深圳 — Resource Sharing: Freshwater Networks / 资源共用：淡水网络

I-infrastructure: water urbanism / 互—基础设施

The Pearl River Delta region faces a self-perpetuating dilemma: fresh water shortage caused or exacerbated by serious water pollution as result of rapid industrialization & urbanization. Currently the portable water source for both Shenzhen and Hong Kong is the Dongguan River, and Inter-Infrastructure: Water Urbanism is a proposal for a different approach to conceptualizing the region's potable water source to break from a traditional reliance on engineering and technology. The proposal focuses on the community scale to establish a new water-cycle system through the appropriation of existing urban infrastructures and with the goal of influencing new and less water-intensive ways of life.

A site including a sewage treatment plant and public housing development by the Shenzhen River in Luo Hu is the testing ground for the project. Winding channels and artificial wetlands further purify water after primary treatment and provide public space and amenities to serve the needs of community, and create a self-sufficient water-budget. The design addresses the complexity of city, needs of the people and infrastructure for water through hybridization and inter-relation. The three systems support and intensify each other, working together to maximize the use of limited land and resources. The design draws inspiration from the operations of a circuit-board, a new vertical city is floated above the exiting water treatment plant, reorganizing a fragmented community. This dynamic system of networks connects the existing urban fabric of Shenzhen to the natural wetland ecosystems across the river in Hong Kong, fostering close interactions between the various local communities. In a time of extravagant urban development, this Inter-Infrastructure encourages the re-valuation of our resources.

珠三角正面临一个自我维持的困境，淡水资源短缺、严重的水污染导致了高度工业化和城市化的结果。目前，深圳、香港双城的淡水资源都依赖于东江，我们希望寻求一种新的水源供应概念——有别于对传统工程和技术的依赖。我们在社区尺度内建立一种新的水循环系统，从而生成一种集约用水的新生活方式。

我们选择了一个沿深圳河的现有污水处理厂及相邻的公屋作为项目基地。经过主要的处理过程之后，弯曲的水道和人造湿地能进一步地净化水，提供服务于社区的公共空间和环境质量，并创立一个自给自足的水预算。这个设计通过相互关联来针对城市的复杂程度、人们的需求和水的基础设施，这三个系统互相支持和强化，一起最大化有限土地和资源的使用。设计师从电路板的运行方式中获得灵感，在原污水处理厂上方放置了一个垂直的城市，将原本分散的小区重新组织与加密，并通过动态的系统将深圳的城市肌理与河对岸香港的湿地形态连接起来，形成了各种本地社区之间的紧密互动。在这样一个奢华的城市开发的年代，"互—基础设施"鼓励对我们的资源重新估值。

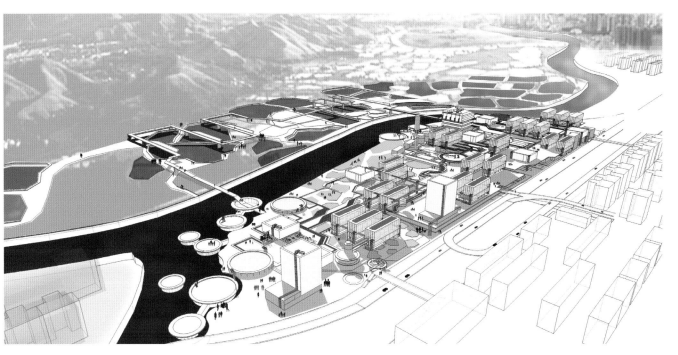

Team Leader / 设计领队： Doreen Heng Liu / 刘珩

Team Members / 设计团队： Yu-Qi Jiang / 姜雨奇
Nan Chen / 陈楠
Miao Liu / 刘苗
Ge Ma / 马戈

Consultants / 顾问： Gu Huang / 黄鸪 (CNMEDRI / 中国市政工程西北设计研究院)
Zhi-Guang Ma / 马之光 (CNMEDRI / 中国市政工程西北设计研究院)
Yuan-Jing Yang / 杨园晶 (CNMEDRI / 中国市政工程西北设计研究院)

Special Thanks / 特别鸣谢： Shenzhen Hydraulic Planning & Design Institutes / 深圳水务规划设计院

SHENZHEN AND HONG KONG / 深圳和香港 | COUNTERPART CITIES / 对应双城

2 Collaborative Ecologies: Deep Bay Territory

生态合作:
后海领域

The Deep Bay, at the border between Shenzhen and Hong Kong, was originally the site of prime aquaculture in both cities. At the mouth of the Shenzhen River lies Mai Po "Marshes" that are not only valued for its natural ecological condition, but an artificial ecologic phenomenon that began in the 1950s when the influx of immigrants constructed "gei-wai's" (tidal shrimp ponds) that replaced the original mud flats, but attract over 55,000 migrant birds in the winter months. The shoreline of the Deep Bay is an apt example of differentiated urban development in Hong Kong and Shenzhen. On one side, Shenzhen has reclaimed land and built up the city right to the water's edge; on the other, Hong Kong has retreated from this edge and populated the shore with landfills, power stations, or other infrastructure. The 5.5km 2007 Hong Kong Shenzhen—Western Corridor connects the two cities over the Deep Bay and suggests a collaborative urban future that is rooted in ecology, economy, and exchange.

后海湾位于香港与深圳的交界，原是两地居首的水产养殖地区。米埔湿地位于深圳河河口，价值不单在于其自然生态环境，更在于其受人为因素影响的生态现象。自50年代起，内地移民开始在后海湾修筑基围（潮汐虾塘）取代原来的泥滩，并在冬季吸引了多达55000只候鸟。后海湾的岸边最能体现出香港与深圳都市演化的不同。深圳不断地填海造地，整个城市沿着岸边发展；香港反而回避后海湾，沿岸皆是垃圾掩埋场或发电厂等基础设施。长达5.5公里的深港西部通道将两城连接起来，象征着以合作的精神尝试将生态、经济与交流作为城市未来发展的策略。

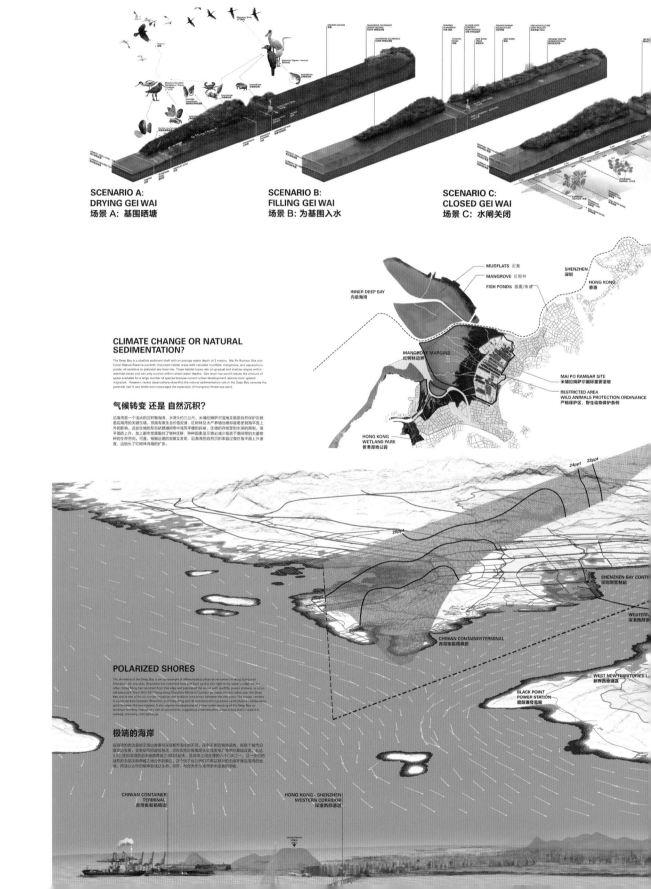

SCENARIO A:
DRYING GEI WAI
场景 A: 基围晒塘

SCENARIO B:
FILLING GEI WAI
场景 B: 为基围入水

SCENARIO C:
CLOSED GEI WAI
场景 C: 水闸关闭

CLIMATE CHANGE OR NATURAL SEDIMENTATION?

The Deep Bay is a shallow sediment shelf with an average water depth of 3 meters. Mai Po Ramsar Site and Futian Nature Reserve are both important habitat areas with valuable mudflats, mangroves, and aquaculture ponds, all sensitive to potential sea level rise. These habitat types rely on gradual and shallow slopes within intertidal zones and can only survive within certain water depths. Sea level rise would reduce the amount of space available for a large number of species because current urban development restricts their upland migration. However, recent observations show that the natural sedimentation rate in the Deep Bay exceeds the potential rise in sea levels and encourages the expansion of mangrove forest sea-ward.

气候转变 还是 自然沉积？

后海湾是一个浅水的沉积物海湾，水深大约三公尺。米埔拉姆萨尔湿地及福田自然保护区都是后海湾的关键生境，但具有生态价值泥滩、红树林以及水产养殖池塘都最受到海平面上升的影响。这些生境的形态依赖潮间带中浅而平缓的斜坡，生境的存续受到水深的限制。海平面的上升，加上都市发展限制了物种迁移，种种因素显示空前心减少泥滩可能得到的大量物种的生存空间。可是，根据近期的观察及发现，后海湾的自然沉积率超过海平面上升速度，这助长了红树林向海的扩张。

POLARIZED SHORES

The shoreline of the Deep Bay is an apt example of differentiated urban development in Hong Kong and Shenzhen. On one side, Shenzhen has reclaimed land and built up the city right to the water's edge; on the other, Hong Kong has retreated from this edge and populated the shore with sensitive power stations, or other infrastructure. The 5.5km 2007 Hong Kong-Shenzhen Western Corridor bridge that crosses the bay offers over the Deep Bay, and is one of the six border crossings that facilitate interaction between the two cities. The bridge corridor is a point where Shenzhen and Hong Kong, and the architectural expressions expressed, collaboratively exist between the two regions. It also shapes the beginning of a new understanding of the Deep Bay an extreme territory instead of a site of oppositions, exaggerating a collaborative urban future that is rooted in ecology, economy, and culture.

极端的海岸

后海湾的海岸线呈现出香港与深圳都市发展的不同，以及截然不同的发展趋势。虽然个城市面朝向的发展，但却有着同的态度和态势；然在深圳岸边城市的建设由海向陆地伸展，而香港岸边的发展呈现向陆地退让之势的态度而岸边则布置诸多较为敏感的基础设施，比如电站等。5.5公里的深港西部通道跨越后海湾之深圳侧和香港侧，是同样之间这六个过境中心之一。这个中心一个不同寻常的存在，见证了深港两地如何以设计合作的态度重新去认识后海湾。而后海湾本身也呈现了着一个不是与对立的极端的地域，也一个预示包含生态，经济，和文化的一个崭新未来的预兆地域。

ARTIFICIAL ECOLOGIES

The Deep Bay, at the border between Shenzhen and Hong Kong, was originally the site of prime aquaculture in both cities. At the mouth of the Shenzhen River lies Mai Po, an ecologically sensitive wetland that is protected under the Ramsar Convention and managed by the World Wildlife Fund. The Mai Po "Marshes" are not only valued for their natural-ecological conditions, but an artificial ecology phenomenon that began in the 1950s when the influx of immigrants constructed "gei-wai's" (tidal shrimp ponds) that replaced the original mud flats. Currently, the 290 hectares of gei-wai's are no longer economically productive, but during the winter months, Mai Po attracts over 55,000 migrant birds to feast on the rich supply of invertebrates exposed by the annual drainage of the shrimp ponds.

人工生态

后海湾位于香港深圳的边缘，原是高地首位的水产养殖地区，位于深圳河河口的米埔。为一个受到拉姆萨尔湿地公约所保护的生态价值湿地。湿地由世界自然基金会管理。米埔湿地的价值不单是为了其自然环境，更在于其受人为影响而成的生态现象。自五十年代起，内地移民开始于后海湾海岸基围吸引原来的泥滩。现在，后海湾的290公顷的基围已失去了原有的经济价值，但在冬季期间，每年吸引超过55,000只候鸟的时候，已经将水的基围成为挥霍的美食。

POLLUTION-CARBON SOURCES
污染源/碳來源

HABITAT MAP
生境地圖

FOOD WEB
食物鏈

EAST ASIAN-AUSTRALIAN FLYWAY
东亚－大洋洲地区飞机航线图

SHENZHEN RIVER 深圳河

BRACKISH WATER 淡水 > 0.5ppt
FRESH WATER 淡水 < 0.5ppt

UNTRAINED 整治前 BEFORE 1995

BRACKISH WATER 淡水 > 0.5ppt
FRESH WATER 淡水 < 0.5ppt

TRAINED 整治後 AFTER 1995

HYPSOGRAPHY SHOWN EXAGGERATED x200
等高线放大200倍

SALINITY CONTOURS
盐度梯度分布

ENERGY / MASS FLOW 能量流

DEEP BAY & SHENZHEN RIVER WATERSHED
后海湾 & 深圳河流域

STREAM FLOW
河流

FLOOD INUNDATION RISK
ELEVATION ABOVE SEA LEVEL
洪水泛滥风险 超海平面高程
- 0 to 2 METER 米
- 2 to 4
- 4 to 6
- 6 to 8

SHENZHEN 深圳
HONG KONG 香港

MAI PO NATURE RESERVE 米埔自然保护区
TIN SHUI WAI 天水围

TO LANTAU ISLAND 至大屿山

SHENZHEN RIVER ESTUARY 深圳口
MAI PO NATURE RESERVE 米埔自然保护区
TSIM BEI TSUI ESTUARY 尖鼻咀河口
TIN SHUI WAI 天水围
HONG KONG - SHENZHEN WESTERN CORRIDOR 深港西部通道
WEST NEW TERRITORIES LANDFILL 新界西堆填区
BLACK POINT POWER STATION 龙鼓滩发电厂

| SHENZHEN AND HONG KONG / 深圳和香港 | COUNTERPART CITIES / 对应双城 |

HONG KONG 香港 — Collaborative Ecologies: Deep Bay Territory / 生态合作：后海领域

Symphony of Blades / 旋桨交响曲

The process of sedimentation is a unique force that singularly shapes the delicate balance of ecologies in the Deep Bay. Although sedimentation occurs naturally, the process has accelerated in recent years due to activities such as rapid urbanization, land reclamation, and other infrastructural projects such as the straightening of the Shenzhen River. Increase in sediment in the Deep Bay reduces its capacity to mediate storm water and increases the risk of flooding in adjacent areas. It also nurtures excessive growth of mangroves around the fringes of the coastline and blocks flood runoff to the sea. If the sedimentation process continues without mitigation, the productive economies and critical urban areas in Hong Kong and Shenzhen would be threatened by increased flooding.

We propose to redirect sedimentation in the bay, by influencing the direction of water current as well as sedimentation patterns. Using principles of hydrodynamics, the project simultaneously limits excessive sediment accumulation in the Bay, while harvesting sediment to create a new local lifestyle economy. It also aims to induce mangrove growths at strategic locations to protect against future storm surges.

Symphony of Blades brings life to infrastructure; proposing industry, education and lifestyle programming amidst a sequence of sediment—manipulating blades at the mouth of the Deep Bay. The coupling of infrastructure and other urban functions fosters a greater awareness about environmental processes. While the blades harvest excess sediment in the Deep Bay, programs such as a sediment-brick factory, energy plant, fertilizer plant and facial mud mask factory, make use of the extra resource of sediment in the Deep Bay to serve various industrial activities which benefit the local community and economy, complimenting the vision of a "sediment-relevant" infrastructural experience.

泥沙沉淀堆积是后海湾自然生态系统平衡不可缺的一部分。虽然后海湾的泥沙沉淀是一个自然现象，但由于愈加频繁的人为活动，例如高速城市化、填海造陆工程和其他基础设施建设项目，如深圳河截弯取直，造成了近年来泥沙沉淀堆积的加快，后海湾的泥沙沉淀堆积不但降低了深圳湾的泄洪能力，同时也大大增加了其周边地区的洪水风险。有机泥沙沉淀也加速了海岸线边缘的红树林植被的生长而影响了降雨泄洪的水流速度。若泥沙继续沉淀堆积，气候变迁所带来的洪水及暴风雨将会威胁港深两地的生产经济和重要都市区的发展。

因此，"旋桨交响曲"项目旨在舒缓后海湾的泥沙沉淀堆积，进而影响水流方向及沉淀堆积的形态。根据流体动力学的原理，此项目一边减缓过度的泥沙沉淀堆积，同时可以收集多余的泥沙以创造一种新的地方经济和生活方式。此项目亦希望能在重点区域引导红树林植被生长来抵御暴风雨及洪水。

"旋桨交响曲"透过一系列的"沉淀物操纵器"和工业生产、教育形式和休闲设施，带给基础设施新的生命力。基础设施和都市功能的结合也可提高市民对环境问题的认识。在"沉淀物操纵器"收集泥沙沉淀的同时也增加了工业生产力，例如制砖工厂、能源工厂、化肥工厂和泥浆面膜工厂，都能将后海湾中额外的泥沙沉淀作为再生资源，这不但对当地的社会和经济提供了帮助，而且有助于市民理解与"泥沙沉淀"有关的基础设施。

Team Leader / 设计领队： Vincci Mak / 麦咏诗 (Division of Landscape Architecture, HKU / 香港大学园境建筑学部)

Team Members / 设计团队： Casey Wang / 王迺慧 (OMA)
Manfred Yuen / 阮文韬 (YS Groundwork: Architecture+Urbanism)

Consultants / 顾问： John Allcock (Asia Ecological Consultants Ltd.)
Terence Fong / 方静威 (ERM)
Dr. Frederick Lee / 李煜绍博士 (Department of Geography, HKU)
Dr. Michael Leven (Asia Ecological Consultants Ltd.)
Dr. C N Ng / 吴祖南博士 (Department of Geography, HKU)
Edward Qiang Shen / 沈强 (Arup)

Research Assistants and Design Assistants / 研究助理及设计助理：
Violette, Yi Peng Chen / 陈奕萍
Dennis Lui Kam Fung / 雷鉴峰
Tracy Xiao Qian Yang / 杨晓倩
Chris Qian Zhang / 张谦

SHENZHEN AND HONG KONG / 深圳和香港	COUNTERPART CITIES / 对应双城
SHENZHEN 深圳 *Collaborative Ecologies: Deep Bay Territory* 生态合作：后海领域	

Sedimental Urbanism

Sedimental Urbanism explores the ecological, social and political future of the Deep Bay and its inhabitants. A vision of the evolving ecology of the Bay is the basis for a proposal to transform the area into a center for sustainable development in the region. Achieving such a vision requires Shenzhen and Hong Kong to work together with natural processes to create an environment where the city and nature become increasingly indistinguishable and mutually beneficial.

Sedimental Urbanism derives its name from the dynamic and sedimentation processes that are rapidly transforming the Shenzhen Bay. It is also a play between the ideas of sediment and sentiment: the physical layers of urbanization that collect to form new ground and the collective unconscious infused into that ground by the people who inhabit it. In its acknowledgement of human satisfaction and happiness as the ultimate measures of value, and its belief in history and culture as essential components of the city, Sedimental Urbanism is openly sentimental. Sedimental Urbanism recognizes that the continual exchange of energy from one form to another—just as in life and death—is as fundamental to the operation of the city as it is to the operation of the universe. Sedimental Urbanism promotes urbanization through erosion, flow, deposition and accretion—manmade and natural.

Sedimental Urbanism acknowledges perpetual change as the defining quality of its context—a context of plants, animals and humans as active players on a stage of natural and manmade assemblages, all of which are temporary.

淤积都市主义

"淤积都市主义"探讨了一系列涉及都市本身及居民的生态与社会政治问题。我们的远景以深圳湾的形态进化作为基础，将其转变成本地区的可持续发展中心，考虑在自然变化影响下的深港双城，提出共创人工与自然系统之间难以区分的互利环境的策略。

"淤积都市主义"的名称来自于流体动力与沉积两种力量所急速改变的深圳湾。淤积这个词有两种含义：它涉及具体的城市化，聚集并组成新土地，在潜移默化的情况下给栖息在那里的人群注入一种新的环境理念，即城市活力是由日常生活经时间沉淀所形成的。"淤积都市主义"有一种毫不隐讳的伤感的意味，坚信历史与文化是人类幸福的重要组成部分，也相信人们的情感与幸福应该是最终的衡量价值的标准。"淤积都市主义"是一种能量的持续转换，就如同生存与死亡，是宇宙基本运转的根基。因此，"淤积都市主义"通过侵蚀、研磨、沉积与增长等方式，促进人工与自然共存的城市化进程。

"淤积都市主义"在定义其环境质量的问题上，也在不断变化。植物、动物与人类，犹如舞台上的演员，活跃于自然与人造集汇的环境里，随波逐流。

Team Leader / 设计领队： Guochuan Feng / 冯果川

Team Member / 设计团队： Aaron Robin / 爱龙
Yujun Yin / 尹毓俊
Qian Feng / 冯茜
Yang Xiao / 肖杨
Siqi Li / 李司祺
Miao Xia / 夏淼
Bo Hu / 胡博
Jie Yang / 杨洁
Cheng Guo / 郭承

Special Thanks / 特别鸣谢： Roger Lee / 李灏 (WWF-HK / Mai Po Marches Wildlife Education Centre and Nature Reserve / 香港米埔沼泽野生生物教育中心与自然保护区)
Leo Lam / 林志权 (HK Drainage Services Department / 香港渠务署)
Jiefan Tan / 谭杰帆 (HK Drainage Services Department / 香港渠务署)
Bin Shen / 沈彬 (CSR Office, China Vanke Co., Ltd. / 企业公民办公室，万科企业股份有限公司)
Shushu Zhou / 周舒舒 (Shenzhen Futian Natural Mangrove Reserve / 深圳福田国家级自然保护区)
Gu Huang / 黄鸪 (Shenzhen Department of China Northwest Municipal Engineering Design Institute Co., Ltd. / 中国建筑西北设计研究院深圳分院)
Yuanjing Yang / 杨园晶 (Shenzhen Department of China Northwest Municipal Engineering Design Institute Co., Ltd. / 中国建筑西北设计研究院深圳分院)

| SHENZHEN AND HONG KONG / 深圳和香港 | COUNTERPART CITIES / 对应双城 |

3 Economic Symbiosis: Port Infrastructure

资源共生：港口设施

Although the Pearl River Delta is crisscrossed by a large system of waterways, natural sedimentation and geological conditions limit efficient waterborne transportation of economic output from numerous industries in the region. Currently, small river-ports scattered throughout the delta region collect goods from their respective locales and feed the deep-water ocean ports of Shenzhen and Hong Kong for transshipment to major trans-Pacific routes. Increased storm-surge and inundation due to climate change are potentially destructive to this economically vital network of port infrastructure, particularly in the western region of the PRD. The Guangdong government has therefore announced plans for large scale dredging projects and port upgrades that would widen and deepen navigational waterways to allow for larger vessels to access river-ports further inland and increase the value of the existing shipment networks.

The ports of Hong Kong and Shenzhen combined handle approximately 10% of the world's TEU throughput. The two ports are simultaneously competitors and collaborators within a complex logistical system of domestic and international shipping networks. This outwardly contradictory relationship is seen in the unique public-private partnerships in the ownership and operations of each of the ports. Additionally, large multinational port operators have stakes in both ports that allows a flexible utilization of available resources and nimble responses to economic and political shifts.

珠江三角洲上水道横贯交错，但区域经济产出所需的水上运输却受限于珠江的地理特性及天然泥沙沉积。目前散落在三角洲各处的小型河港会集合各地货物，运到深圳或香港的深海港口，衔接跨洋运输水路，但气候变迁带来的飓风和泛滥却可能破坏攸关经济命脉的港口设施，尤其是三角洲西边的港口。广东省政府业已宣布了大型疏浚及港口升级工程，以期加深水道，让大型船只能到达内陆河港，提升现有运输网络的价值。

香港及深圳的港口总共处理全球约 10% 的集装箱。两港处于复杂的国内与国际水路运输网络中，既是对手也是伙伴。如此竞合关系，可从港口经营及营运的特殊公私伙伴关联中看出。此外，大型国际港口经营者在两港皆有投资，能弹性利用双边资源，降低政经变动的冲击。

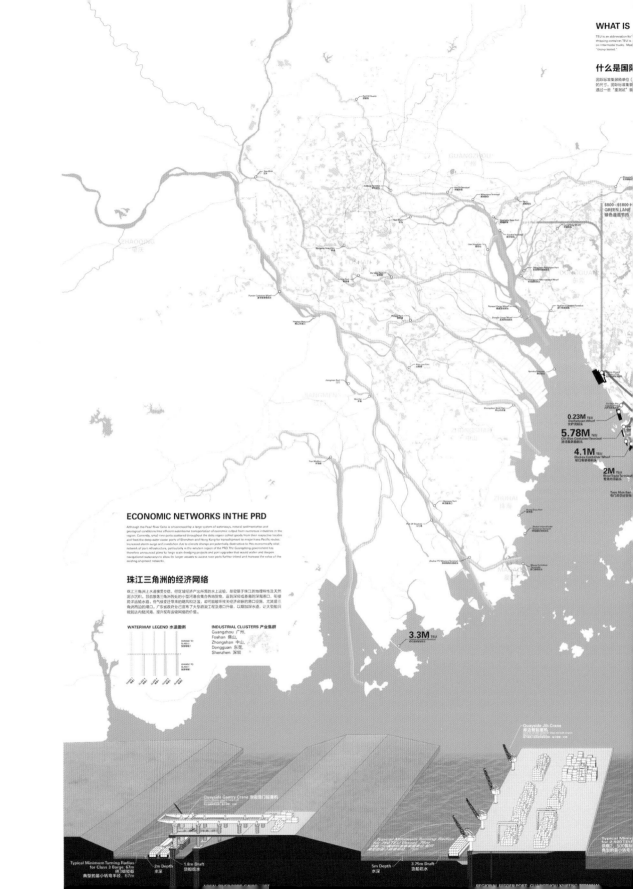

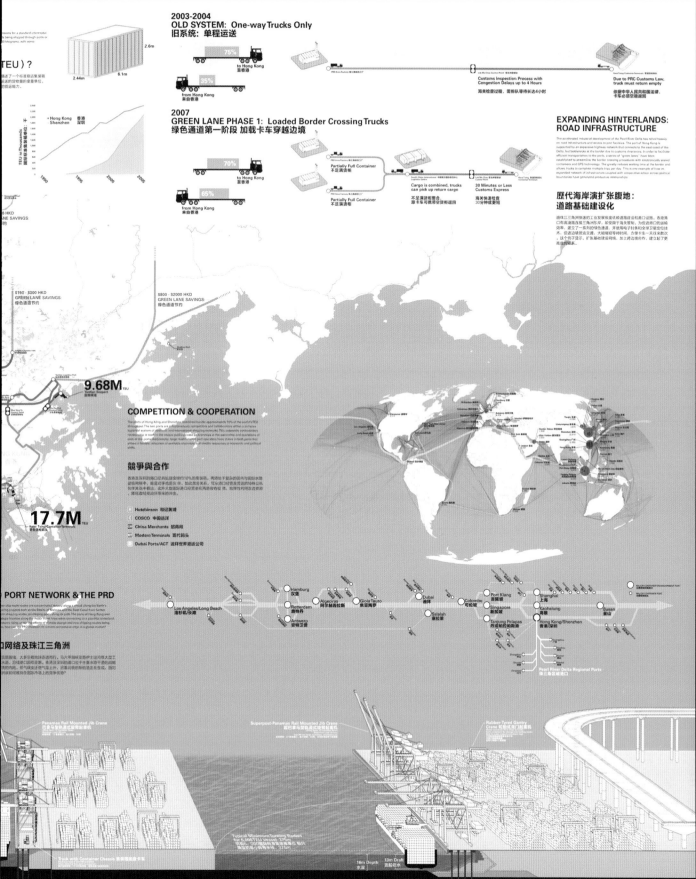

SHENZHEN AND HONG KONG / 深圳和香港	COUNTERPART CITIES / 对应双城
HONG KONG 香港　*Economic Symbiosis: Port Infrastructure*　资源共生：港口设施	
FUTUREPORT	未來港口
The FUTUREPORT project is a design research project which speculates on the future of shipping, land management, manufacturing, environmental sustainability and future urbanisation in the PRD. Through an examination of thirteen ports in the PRD region, there is evidence of an acute limit on the available land for expansion, while older ports have crowded waterways and cannot cater to the largest vessels, due to limitations in the current water depth and logistic movement. As a visionary strategy for the development of a post-industrial economy in the PRD in relation to the future of the SZ-HK ports, a dynamic network of offshore cargo ports is proposed as a new high density urban corridor is proposed in the mouth of the Pearl River, along the future Hong Kong—Macau—Zhuhai Bridge. This new model of urban and industrial growth is conceived as a series of differentiated scenarios documented at instances from 2011 to 2111, linking the PRD region as a continuous city of 50m population by the year 2111. The direct proximity of offshore ports to industrial production is the primary driver of this new model of symbiosis of Shenzhen and Hong Kong. In addition, sources of clean energy harvesting are proposed as integrated wind and solar farms on offshore islands.	"未來港口"是一个对未来进行大胆构想的设计研究方案，探讨珠三角地区的船运、土地管理、制造业。通过考察珠三角地区深圳和香港的13个现有港口，包括在珠江港湾一带的中转码头，我们发现，它们都面临着未来扩建的局限性问题。一些修建比较早的港口，因为水深和物流的限制，很多港口的航道已经相当拥挤，无法停靠大吨位的现代货轮。 这个设计研究项目表现了深港双城未来港口的可能性，在珠江入海口，沿着即将修建的港澳珠大桥，设计了一个动态多变的海上港口城市网络，作为未来从陆地撤出的可能的高密度新城走廊。这个设计提案经过从 2011 年到 2111 年的一系列各不相同的假设情况的考察，展现了一种新的、连接珠三角地区（该地区总人口预计在 100 年后可能达到 5 千万）城市和工业的共同生长。海上港口运输和陆地工业生产的空间关联性也是深圳和香港两个城市共生模型的主要策动力之一。除此之外，大面积的利用风能和太阳能的新能源农场，将保证海上港口网络的自我能源补给。

Team Leader / 设计领队：	Tom Verebes / 汤姆·威尔伯斯 (Department of Architecture, HKU / 香港大学建筑学系)	
Team Members / 设计团队：	Kristof Crolla / 高仕棠 (Department of Architecture, HKU / 香港大学建筑学系)	
	Yan Gao / 高岩 (Department of Architecture, HKU / 香港大学建筑学系)	
	Christian Lange / 克利斯齐恩·朗格 (Department of Architecture, HKU / 香港大学建筑学系)	
Consultants / 顾问：	Ander Chow / 周金松 (Arup Transportation)	
Research Assistant / 研究助理：	Praneet Verma / 普拉尼特·威尔玛 (Department of Architecture, HKU / 香港大学建筑学系)	

SHENZHEN AND HONG KONG / 深圳和香港	COUNTER-PART CITIES / 对应双城
SHENZHEN 深圳 *Economic Symbiosis: Port Infrastructure* 资源共生：港口设施	
Mobile Culture, Creative Community	文化流动，创意社会
With the world growing ever closer through globalization and resources growing ever more in demand, the future of the port lies in efficient and sustainable living. As the land use efficiency of the ports increase, more space becomes available for multi-purpose programs, and port industries are transformed to relate to the human scale with complex functions. Mobile Culture, Creative Community proposes that logistics, business, living and recreation along the shoreline become more closely inter-connected and interdependent, creating a new and robust port economy. To meet the demand of communication and integration between industries in the new port economy, a new "mobile industry" forms a globally productive network of culture and community, which allows for effective communication and creative cooperation. As a prototype, we combine the cultural and creative industries in Dafen with a new port economy for Shenzhen and Hong Kong. Building on the model of art production in Dafen village as an anchor, Mobile Culture, Creative Community creates a large chain of industries which is diverse, integrated, complex, and influential. Mobile platforms float between markets independently. Creative Communities are assembled specifically to the existing conditions and available resources at hand. Being attached to or detached from the land, the Mobile Culture, Creative Community group is continuously transferring and redistributing goods, culture, and knowledge. The Dafen prototype also inspires us, that as an open platform available to the public art becomes part of the community and is no longer restricted to those with artistic training, but becomes a livelihood for any person. Mobile Culture, Creative Community creates a point of exchange at which people from different countries or cultures, with diverse backgrounds, and varied professional skills, can think, talk, create, and work together for the production of culture.	在全球化日趋发达和资源日益短缺的背景下，以货物为对象的港口产业在今后将往高效绿色方向发展。在提高土地利用率的同时，剩余土地可以为城市其他功能的植入预留空间，港口产业由此更趋向多元化和人性化。海岸线的物流区、商贸区、生活区和休闲娱乐区会互相依存、相互促进，最后形成一种全新并充满活力的港口经济模式。此外，结合新的港口经济模式中优质产业之间交流整合的需求，我们构想了"移动产业"的概念。每个港口城市拥有的能够体现城市自身产业特点的移动社区或文化产业，会在世界范围内形成一个互相交换交流、共同创作生产的网络。 我们尝试将大芬的文化产业结合到深圳新的港口经济模式中，以油画产业为核心，逐渐衍生其他产业，形成一个多元、紧密、复杂和庞大的产业链。它们以分离的形式移动到世界各地的港口城市，根据不同地方的岸线条件和资源特点组合成不同的形态，与陆地以及其他"移动产业"发生物质、信息、人才和技术的交换与交流。大芬模式还启示了我们：作为一个向全民开放的艺术平台，这里给没有受过正统艺术教育的人提供了接触艺术并以此为生的机会。在"移动文化产业"这一移动社区的创想中，通过移动性，我们希望不同国家、地区的人，无论是平民大众还是专业人士，能聚集在一起思考创作，互动参与和生产制作。

Team Leader / 设计领队： Xiongyi Zhu / 朱雄毅, Guangsheng Zeng / 曾冠生

Team Members / 设计团队： Qiuru Zhu / 朱秋茹
Jianwei Yang / 杨建伟
Dandan Zhang / 张丹丹
Ruoyu / 陈若愚
Wenyu / 吴问雨
Wei Wang / 王玮
Xue Wang / 王雪
Guirong Yu / 俞贵荣

SHENZHEN AND HONG KONG / 深圳和香港

SHENZHEN: BECOMING A CITY ... FUTURE METROPOLIS OF THE WORLD (Forum Transcript)

深圳：成为一座城……世界未来都市（论坛讲稿整理）

DEC 11th 2011 Shenzhen Biennale Symposium @ Citizens Centre, Futian District, Shenzhen (Multifuctional Hall)

2011 年 12 月 11 日深圳双年展，深圳福田区市民中心多功能厅

Symposium Organizer / Moderator
Laurence Liauw Wie-Wu,
(SPADA Limited / Hong Kong University)

论坛主持人：
廖维武（SPADA / 香港大学）

Session 1:
Shenzhen and Hong Kong (BECOMING A CITY)
Kang Man (ARUP, Shenzhen)
Jiang Jun (Underline Office, Guangzhou / Beijing)
Lee Ou Fan (Chinese University of Hong Kong)
Respondent:
Anderson Lee (Co-Curator, Hong Kong Biennale)

第一部分 / 深圳与香港（成为一座城）
发言人：江民（ARUP，深圳）
姜珺（下划线工作室，广州 / 北京）
李欧梵（香港中文大学）
回应人：李亮聪（香港双年展，联合策展人）

Session 2:
Shenzhen and the World (FUTURE METROPOLIS OF THE WORLD)
Michelle Provost (INTI, Netherlands)
David Giaonotten (OMA, Rotterdam Hong Kong)
Jin Min Hua (Shenzhen Commercial Daily, Shenzhen / Beijing)
Respondent:
Liu Xiaodu (Urbanus, Shenzhen)

第二部分 / 深圳与世界（世界未来都市）
Michelle Provost（INTI，荷兰）
David Giaonotten（OMA，鹿特丹 / 香港）
金敏华（《深圳商报》，深圳 / 北京）
回应人：刘晓都（都市实践，深圳）

Moderator: Laurence Liauw Wie-Wu
论坛主持人：廖维武

SHENZHEN: BECOMING A CITY...FUTURE METROPOLIS OF THE WORLD / 深圳: 成为一座城······世界未来都市

Prologue (Laurence Liauw)

This Biennale challenges us to rethink the question of city making beyond architecture, as a discipline or multi-discipline, and rethink questions of architecture beyond the city. With Shenzhen's rapid rise from fishing village to industrial township to metropolis of todays 14 million within 30 years, what are the real consequences of rapid urbanization? Social, political, economic, cultural, environmental, seen through the lens of architecture reveals what? People come and go, but the city keeps growing. Everyone is an immigrant here. Home is not really here. The conventional notion of place does not make much sense, as the city changes radically from one urban experiment to another. Is architecture merely as side effect, a tool, or a beneficiary in Shenzhen's struggle to define itself, first as an alter ego of Hong Kong, now more as a new urban species for China? Competitiveness of cities in the Pearl River Delta with environmental and social cost begs the question: is more growth the only way forward? Improving quality in a city like Shenzhen is an opportunity for rebirth, with Hong Kong as its Siamese twin neighbor unwilling to change. What are the lessons or influence of Shenzhen on other new towns, and cities (with ambition) in the making? Is Shenzhen's learning over, or could it remain a continuous urban lab for future cities? Can Hong Kong survive without Shenzhen, and what if Shenzhen turns its back on Hong Kong to look northwards? Differences may not be relevant anymore in years to come as we approach 2047. Shenzhen wants to be the most innovative city in China and has UNESCO status, but who does that benefit? A pioneering city full of migrants, high technology, real estate, exports, ecological progress and design, may just be the answer the world is looking for. The re-imagination of Shenzhen could well be Hong Kong's (hope for a) future, possibly China's, and influence to the World.

Symposium Session 1 : Shenzhen and Hong Kong (BECOMING A CITY)

Shenzhen and Hong Kong are famous siamese twin cities born out of accident and political experimentation. Are they merely historical accidents that got lucky and prospered, or is there something in their DNA that distinguishes them from other cities in the making. We investigate in this symposium session BECOMING A CITY the relation between Shenzhen and Hong Kong—in terms or their urbanity, culture, people, economies and politics. Can cities become something overnight, and can they expire slowly as reverse acts of becoming? Shenzhen and Hong Kong thrive on each other and in doing so have both become economic and urban miracles of the 20th century. What were the deciding factors that made this happen, and are these still there? Will one fail eventually or be swallowed up by the other. Will they merge and become a totally different type of city never before seen in history? Will politics get in the way, or will the market prevail? Discussing their differences and similarities in a bid to understand the future of these two projective cities, is key to the future of Chinese cities—as they could become new role models for China or prophets of the end of Chinese urban civilization as we know it.

Introduction

The role of Biennale contributes to how design is communicated with the general public, and how design is taught, and how design is learned and passed to the next generation. Today's first session discusses Shenzhen and Hong Kong, how these two cities developed? This is not really a lecture, more like a conversation, and it encourages people in Shenzhen or in Hong Kong to ask critical questions about these cities and societies. None of our guest speakers are architects, but they know about architecture and cities. Kang Ma, director from ARUP engineering company in Shenzhen. Jiang Jun, critic, publisher, teacher, designer based in Guangzhou and Beijing. Prof. Lee Ou Fan, cultural critic and writer from Hong Kong, who has strong views about culture and education.

Abstracts

Kang Man
I am an engineer with 26 years of professional experience. My

SHENZHEN AND HONG KONG / 深圳和香港

topic today is Shaping of a Metropolis. This is a photo of Shenzhen I took about thirty years ago from the Hong Kong side with much green land in the foreground and the background is the city skyline of Shenzhen. What makes the changes of Shenzhen from a small fishing village with 30 thousand populations to a modern city with a population of 13 million? Looking at population distributions key districts in Shenzhen: Luohu, Futian, we see that connections are very important for a city to develop. Shenzhen's strategy is to achieve development through interaction and linkage, focusing on active boundary links between Hong Kong and Shenzhen. Urban development and city centres are generated from these border connections. Shenzhen Bay area is developing and we certainly need more bridges and expressways. The communications between these two cities is continuing enhanced with the development of technology and workforce in the city. Statistics show the people crossing Shenzhen-Hong Kong per day is the highest (in the world) with people crossing per day including those in vehicles, probably more than one million. Looking at photos of Luohu district taken in 1985, 2008 and 2011, we can see drastic changes between these three periods, especially in infrastructure of rail, ports, bridges and highways, creating a more networked city. The first phase of Shenzhen Metro started in 1998, and was completed in 2004. The second phase with five lines operating was completed in 2011. There a total of 137 stations in service now with a daily flow of people of 3 million. The third phase with 10 lines in total, will be completed by 2016. Line 4 Longhua Line has an elevated railway that will become part of Shenzhen's skyline fusing infrastructure and the city. The current metro lines can let people travel to work, recreation, or meetings within thirty minutes or at most an hour, and also connects Shenzhen with the surrounding cities in Pearl River Delta. There's a concept of a 'one-hour economic circle', using railway to link PRD cities together. Looking at the Hong Kong Metro Map, key is how this network can expand a city and regenerate older parts. This map with red dots are new areas of the redevelopment in the future Shenzhen, regenerating older sub-centers such as Longgang, Yantian, Pingshan, etc. This kind of expansion and regeneration of a city actually requires for advanced technology, skilled workforce and integrated design. Citizens are very concerned about future developments, and more skilled professionals move to the city now. Hong Kong's railway system land development model brought prosperity via station developments. The city itself becomes a hub of the whole region also via connections of rail and metro to airports. This helps Shenzhen and Hong Kong to develop from city to hubs in the PRD, requiring a holistic view and cooperation between engineers, architects, planners and government. A city is not only a place to live or work, but also to exchange creative ideas. Shenzhen and Hong Kong people share the entrepreneurial spirit, having efficient interpersonal networks, which can promote diversity of culture and quality of life. Openness and collaboration of our two cities is very important for the future Hong Kong-Shenzhen Metropolis. Diversity is the most important quality of a metropolis.

Jiang Jun

The characteristic of diversity is very important for a city, which can also be an economic trait. Diversity could mean a lot, including forms and sizes of industries. This is where there is a big difference between Shenzhen and Hong Kong. After 1980s mainland economic reform, Hong Kong moved most industries to the mainland. Shenzhen's growth in 1980s was associated with Hong Kong's capital and skills transfer. Since the 1997 Asia financial crisis impacted Hong Kong greatly, it has not yet identified the new foundation for industries to flourish again. If Hong Kong had not the Chinese Central Government's political assistance, the situation would be even worse. The first differences between the two cities are its industrial base. Hong Kong now has a single-mode economy dependent on Chinese Central Government support. The 2009 global financial crisis also impacted Shenzhen, evacuating lots of low-end industries due to the impact of rising land or labor, and environmental costs, shrinking international markets and RMB appreciation. Shenzhen industries however unlike Hong Kong did not hollow-out, but focused on upgrading its industries quality and diversity. Meanwhile Hong Kong's property prices being the highest in the world affects the survival of small and medium enterprises. Hong Kong recent incoming investment flows ranking in Asia fell from previous year 5th down to 14th now. Shenzhen's low-end industries began to disappear, replaced by high-end IT industry, bio-technology, new materials, renewable energy and creative industries. The new Qianhai district will attract investment and new industries including from Hong Kong. This Qianhai 'special zone', could become an offshore 'Hong Kong Special Administrative Region within Shenzhen Special Economic Zone'. The second difference is the pattern of land development. Shenzhen has expanded to become larger area than Hong Kong SAR. From 2009 onwards high-tech zones have been rapidly developed in Shenzhen, plus extensive upgrading of the old factory areas into new creative industries. In the past, 90% of the factories created 10% of the economy, now the aim is for 10% of high-end industries to create 90% of the future economy of Shenzhen. The third difference is the scale of development, Baoan, Shenzhen's biggest district together with Longgang expanded the SEZ to five times its original size. Shenzhen's center as a whole expanded West from Luohu towards Qianhai with new commercial sub-centres emerging. By comparison last year in Hong Kong, the West Kowloon Cultural District aimed at diversification towards cultural industries. However the result is a masterplan with capital needs that could be developed only by a single owner, underlining Hong Kong's singular property market-led model. By comparison, the economy of Shenzhen is diverse (with the exception of affordable housing development). These differences between diversity in Shenzhen and uniformity in Hong Kong's development, point to different futures for both cities. Hong Kong in terms of science, technology and culture did not learn from other countries. I think Shenzhen could be an alternative for the Pearl River Delta, with its cultural background from the North, while Hong Kong's influence has been declining in recent years. A major question remains : how will Hong Kong adapt to Northern culture (including Shenzhen) and diversify itself from its current urban culture and single-mode economy?

Lee Ou Fan

Let me talk about my impressions of the city's culture. My argument about the tale of two cities starts 40 years ago when the sister cities Shanghai and Hong Kong were important, now more so Shenzhen and Hong Kong. Every time I come to Shenzhen, I feel it is just like 19th century USA with the spirit of 'wild west'. Within the Pearl River Delta, what kind of role does Shenzhen play? It could be centre of creative and design talents, even urban planning talent, but why not Hong Kong? I think Hong Kong is overly bureaucratic, institutional and commercial, and has no space for creative development. Shenzhen offers space to develop at an affordable price. However, if compare Shenzhen's problem with Hong Kong, which is the issue of culture and education. There is only one university in Shenzhen, and we have nine in Hong Kong. We can see the media of the two cities are in competition with each other. The printed media of Shenzhen and Guangzhou is actually quite objective and cover wide topics, while Hong Kong media newspapers have little cultural richness. Hong Kong is very systematic, but its problem is excessive institutionalization and no creative space. I know few Hong Kong books that are influential in the world, on the impact of globalization, or having impact on Chinese culture. There is a crisis of culture (and heritage) in Hong Kong. Lingnan Culture accumulated over hundreds of years are reflected in the daily life of Shenzhen. Many Hong Kong cultural and creative people could not find their own creative space, for example over 100 performance groups cannot find venues to perform. Expectations for West Kowloon Cultural District will offer a wide range of creative spaces, but what about culture of small and medium of creative talent? Out of the 3 masterplans, OMA's design is inspired from local Lingnan Culture, but the Hong Kong Government chose the most conservative masterplan eventually. I was impressed with Shenzhen's cultural contacts. I go to bookstores, and look forward to meeting intellectuals in Shenzhen for conversation. But Hong Kong's intellectuals - they go to Shanghai and Beijing, for what? To earn more money? The diversity of an immigrant city like Shenzhen, its people with all kinds of accents, makes the city more vibrant. When I moved

SHENZHEN AND HONG KONG / 深圳和香港

to Hong Kong in the 1970s, various dialects were spoken. But the generation of immigrants has passed, and Hong Kong is in danger of uniformity now. Northern culture came to Shenzhen, such as Beijing elites in Government, media, technology and education. Southern culture has also influenced the North. This Northern culture is building Shenzhen, seeking freedom in new creative development. I feel positive for the future of both city's cultural exchanges and education, even without big institutions like universities. The largest crisis of Hong Kong education is English standards falling. Bookstores cannot survive high rent in Hong Kong, and publishing costs are too high, so people read less without cultural nourishment. In Hong Kong people are discussing only Hong Kong's problems, without thinking of the relationship between Hong Kong and Shenzhen, between Hong Kong and the World. But Shenzhen's growing too fast also brings problems before urban software development. The first biennial exhibition was sponsored by the Shenzhen Government, but why was the Biennale was not sponsored by the Government of Hong Kong? My question is: if there is now so much innovation and space in Shenzhen, could building better links between hardware and software be solved through innovation?

Respondent : Anderson Lee

First of all, I would like to congratulate you, for bringing these three speakers together as an organization. I can see is hardware aspects of the Kang Man, talking about infrastructure development of both the Hong Kong and Shenzhen City hardware connected together. Then Jiang Jun talked about land use and software development of a city. Then Prof Lee talks about the residents of the two cities meeting and exchanging ideas. How do we then make Shenzhen-Hong Kong become a tale of two cities? I am a Hong Kong Biennale curator, so I believe in order for two city hope to achieve real sister cities, the next Biennale will have only one curator responsible for both cities. Is there still a gap between both cities? I think that Shenzhen has more of a chance, we can see two years ago we had prepared for Biennial venues, we encountered many difficulties in Hong Kong. Professor Lee spoke about what is missing in both cities. I think it is either two-way or three-way city, with tangible and intangible aspects. What should we do about the development of Hong Kong and Shenzhen in the Pearl River Delta, this is what we really should be thinking about.

Symposium Session 2 :
Shenzhen and the World (FUTURE METROPOLIS OF THE WORLD)

Shenzhen defies any rational or conventional model of how a city emerges, booms, survives and transforms itself constantly to avoid expiry. 'Learning from Las Vegas' is simply not enough to describe what Shenzhen has gone through and its impact on the discipline of urbanism. Unique and Frankenstein in its formation, it looks like any other city....or not. Speed, scale and people have to be redefined by Shenzhen and its catalytic effect on neighboring cities in the Pearl River Delta. Has anyone bothered to understand Shenzhen's relation with the world? The fact that it is a UNESCO City of Design and won the UIA Abercrombie award for planning in the last 10 years? The fact that its GDP is highest in China, its population doubled in the past 10 years, and is advancing low carbon city design? The fact that the real estate market of China was born here continues to be redefined by pioneering efforts of its unsung heroes? We will discuss these effects and also the affective potential of Shenzhen on the rest of the World that is still interested in urbanism as a discipline and hope for mankind. Which parts of the World may benefit and learn from Shenzhen's experience and urban experimentation. If nothing else, Shenzhen offers hope for the Future Metropolis of the World in its courage and conviction that going urban is the only way forward.

Abstracts

Michelle Provost

The twin Dutch cities of Amsterdam and Almere are like that between Shenzhen and Hong Kong. Almere was created through land reclamation, for Amsterdam residents to live there in spacious houses with gardens. But Shenzhen is a special economic

SHENZHEN: BECOMING A CITY...FUTURE METROPOLIS OF THE WORLD / 深圳: 成为一座城……世界未来都市

zone with much higher density. Almere, a suburban city with only 190,000 populations has houses built by owners themselves, whilst Shenzhen has a population of 12 million migrants looking for housing. Almere's problem now is how to provide job opportunities for 190,000 residents, who currently work in Amsterdam. So Almere is very interested in Shenzhen's development, as an example to learn from. A variety of urban issues challenge our Dutch political system, such as growing income gap and the destruction of nature, etc. The question for us is: how much would you sacrifice for economic growth? A French animation director drew a comic book about Shenzhen. There are a lot of grey images of factories and warehouses in Shenzhen back in 1997, "...this is going to be lonely, for a modern city next to Hong Kong, Shenzhen has few bilingual Chinese, there's no cafés for me to meet young people interested in the West. After a day or two of meetings, we go back to Hong Kong." As an urban model, it is a pretty strange one. I cannot find the city center; neither can I find typical residential areas, like in the West. It's a city only for business. I felt completely lost in this city before. Nowadays at the Biennale and in OCAT it feels good within a short period of a decade, the entire environment changed. Shenzhen, an industrial city before is changing into a cultural city within such a short period. If we consider Shenzhen the most successful new town in the world, whilst it may not be the most beautiful one, how it has the most energy? It is also known as a place for experiments. This is a city is where you take risks and also validate your hypothesis in real time. Could Shenzhen become a model for the next generation of new towns? In our new book we emphasized three problems which generations of new towns have encountered. The first problem is about ecology and eco-cities. The problem here is doing green for green's sake that does not make full use of green investment. You can also see this in China especially to attract political attention. Secondly are social affects. For whom is this new town built? This is a good question for Shenzhen, where people such as farmers and villagers would not go back to live in the same places. So new towns can change the social structure and the community. Sustainability here can be very practical and social affects. Last, I'll ask what identity do new towns have? They all have unified outlooks with international skyscrapers, twelve-lane highways, high-end residences, highlighting political achievements. How can this kind of standardized new city adapt to local environments in different places? Sometimes China has imported characteristics of other national cities. This loss of local cultural characteristics can also be interesting. What can other emerging new towns learn from the experiences of Shenzhen? Would Shenzhen continue only to be a laboratory of urban development, or become a sustainable place with positive ecology, social structure and identity which can inspire others?

David Giaonotten

This symposium poses an important question. OMA's first engagement in Shenzhen was in 1995, when Rem Koolhaas came here with students of Harvard and wrote a book "Great Leap Forward". It had a profound impact in the West, describing the emergence of new towns in China, the unbelievable real estate boom, unbelievable urban scale and unbelievable skills of people in Shenzhen working on making the city. But the moment we really contributed to Shenzhen was the start of the Shenzhen Stock Exchange project. If you now look at the economic crisis of the world, China's success is one of the successful models remaining. Over 30 years, the people of Shenzhen improved their lives, creating a completely new urban and social invention. How do we build cities and create a social structure for 600 million urban populations expanding? You can't predict what will happen in Shenzhen. Participating in projects is the only way to see what happens next. Mobility is important—free exchange of ideas and knowledge with other cities. The new Shenzhen Stock Exchange reflects the economic freedom Shenzhen and economic connections. It is a new building typology within the CBD. We tried to create a new CBD that has urban social space that creates a new identity for the city. This building form tries to be as transparent as its computerized trading. It is not hiding anything but shows all through its literal and digital transparency. In urban planning for the new Qianhai area competition, OMA said "Let's

SHENZHEN AND HONG KONG / 深圳和香港

not erase urban structure and history, but preserve elements that contribute to the vitality of the city". We analyzed the existing urban fabric and infrastructure, studied what meaningful elements could be added to the plan. Education is important and not internationalized yet, so here is an opportunity. For people's wellbeing we injected Chinese and Western health care systems here. Another sensitive topic is media. We designed an office building for Tencent, a Chinese media company. Initially they wanted a twin tower, but we put everything in a single triangular shaped tower for better communication. After winning the competition, Tencent invited us to design their new headquarters in Beijing instead of Shenzhen. The last issue is a city's identity and how can you create an image of the city? Shenzhen is UNESCO City of Design. Can this be a city's image? Our Shenzhen Eye project (for Crystal Island in CBD), which we collaborated with Urbanus, tried to create this new central image of the city. For the Shenzhen Eye we connected all the different transport stations, stores, offices below ground and we incorporated everything in a big ring structure. If a new town wants to be influential, it has to have imagination. Most new towns lack imagination because they are simply copies of old towns. They can instead be successful by creatively incorporating different elements that define a city's new identity. How can a city like Shenzhen live up to its title of UNESCO City of Design whilst keeping its economic growth? Shenzhen cannot simply copy western cities or other cities in China. Because of its unique economic and political structure, Shenzhen now has an opportunity to become its own unique 'metropolis of the world'.

Jin Minhua

Hong Kong and Shenzhen's relationship with the world has to be viewed over time and space. This is Dafen Village in Shenzhen. After 1980s Reform Open policy, a man from Hong Kong founded a workshop in Dafen, which gradually became the main production site of oil painting for the world. Shenzhen is a city full of immigrants, it has 14 million populations, roughly 60% are from all over the country, 30% come from Guangdong, 10% are indigenous. People from all over the country come to Shenzhen, but few from big cities. People in Dafen village were good at decorative oil paintings. The village community turned into a gallery and then a global industry. Designer Hel Ylyang by virtue of his works gained international awards. The graphic design industry of China was born in Shenzhen. Shenzhen led import and export trading for 20 years, therefore having high demands on product packaging design which also brought in the printing industry. Current printing output is one third of the world market. Shenzhen City of Design was declared by UNESCO in 2008. Established for 30 years Shenzhen has reached a critical point of transition. It used to be a manufacturing city and now is transforming into a creative city. We discuss at local seminars if Shenzhen could be a design based city back in 1999. In 2004 a Shenzhen daily newspaper raised the headline: could Shenzhen become the world's third City of Design? Government at the time supported this ambition, which became reality within the next 4 years. Shenzhen design industry is closely related to the city's development with major innovative industrial design companies born here including BYD, Crystal, Huawei, as well as the interior design and real estate industries. Unesco's evaluation committee recognized that design in Shenzhen as a tool for transformation. In 2004, we joined the creative cities network, divided into 7 areas - design, folk art, cuisine, literature, media arts, music and movie. The first city was Turin in Italy, the second was Seoul. Shenzhen Government plans to setup a design centre / academy, and OCT plans for a design museum. In 2008 December, bidding meeting with UNESCO Shenzhen showed a film and gifts made of local colored paper that reflected Lingnan culture. We wanted to show that Shenzhen was looking to fuse traditional design with modernity. I love this film. It symbolizes Shenzhen from zero to 30. Thirty years in 2008. The zero can be understood as beginning with nothing, starting as a point on the edge of the South China Sea. Shenzhen's most valuable asset is people's ideas and differences. The media respects different points if views as almost a rebellion to conservative ways. We can imagine Shenzhen as a city of dreams. Because it is a city in the sky, so people can fly here. The first generation immigrants in Shenzhen are actually are dissatisfied with the status quo now. Shenzhen used to be like a new kite flying, representing liberal, tolerant, different attitudes. But now the flying of kites is banned here.

Liauw Wie-Wu
Our three speakers have overlapping viewpoints. Michelle Provoost believes Shenzhen is the most successful new town and encourages experimentation. David Gianotten said that Shenzhen can be successful because of its imagination and creativity. Jin Minhua told us about the value of Shenzhen design from an international perspective. At 2010 Expo Pavilion, we discussed how Shenzhen fears losing competitive advantage because its piontering spirit is diminishing and the city is becoming more conservative. OMA's Tencent building is not built in Shenzhen but in Beijing eventually, shows this risk. How then to maintain pioneering frontier spirit in Shenzhen?

Respondent: Liu Xiaodu
I found this forum today very inspiring for me. Michelle Provoost talked about three points to from a historical and an ecological viewpoint, Shenzhen is a completely artificial city. Shenzhen's transformation of social structure is a very important. Developers and infrastructure engineering groups come to Shenzhen, and State capital dominates over private enterprises. A new city from scratch must still go through a process. David Gianotten talked about OMA's experience in Shenzhen, and illustrated some practice problems. The old photo shows Deng Xiao-ping 's Reform slogans. Later we see the Shenzhen Stock Exchange, as symbol or a monument to capitalism. It could also possibly signal the end of an era. Online members heatedly debated about Tencent's architectural design as frightening. Why Tencent is not implemented in Shenzhen, eventually in Beijing? Because of Beijing's cultural and political capital, it can. Self-awareness of this difference is important for the future of Shenzhen. Crystal Island Shenzhen Eye strongly expresses the relationship between Hong Kong and Shenzhen. The high-speed train connects West Kowloon Cultural District to Shenzhen in 14 minutes. Shenzhen design can be the main focus here. Design is complementary to culture, underlining creative relationships between Shenzhen and Hong Kong. Jin Minhua expressed pros and cons of Shenzhen design from Finland to 'Shanzhai copy culture'. I think this label shows a lack of confidence. Applying to be City of Design means we probably lack cultural confidence here. I recently discussed with a Netherlands sociologist that urban development especially in cities like Shenzhen rely on traditional projects, on a capital-led development model. This is manifested by our real estate development. Jiang Jun repeatedly emphasized this single-mode of development in Hong Kong. Shenzhen is close to Hong Kong, so this single-mode real estate development also becomes our capitalist model. But what is the purpose of capital? Western societies are now pursuing knowledge-driven development. Professor Lee says Hong Kong has accumulated centuries of rich culture. This cannot be matched in Shenzhen. Jiang Jun sees status quo in Hong Kong, whilst Shenzhen is getting stronger. From my point of view, this difference is probably an illusion. Hong Kong is looking forward, but does not act, so people were disappointed. Shenzhen has endless possibilities, but in what direction will it go? Shenzhen's problem is losing his pioneering spirit, and our ageing population. Lee Ou Fan repeatedly criticized the suppression of Hong Kong creativity. As the Chinese mainland develops, Shenzhen is slowly pushed into the background. With the decline of Hong Kong's status in the world, Shenzhen eventually will also decline unless it changes.

Closing Comments

Michelle Provoost:
Shenzhen's story inspires a lot. We talked about the hardware and software, maybe a third, called organization. When talking about relationship between Shenzhen and Hong Kong, organization will play an important role. Hong Kong has so many rigid mechanisms, producing a static environment. Shenzhen needs to create new opportunities and avoid the same path. Shenzhen now has so much hardware, but organizations have not matured, as they are in the early stages of development. How then do we keep Shenzhen

SHENZHEN AND HONG KONG / 深圳和香港

a city of innovation?

David Gianotten:

I would like to say that here everything is unique, if you want to compare Shenzhen with Hong Kong. Every city has its own uniqueness. Shenzhen has ceased to be a town, it is a city already existing for some time, having much contact with the world. We decided not to compare and think about the essence of the city itself, traits, as well as future development. Shenzhen is now already an international metropolis, with its own position, and certainly would be better in the future.

Jin Minhua:

The city of Shenzhen is special, and we now see mostly State-owned enterprises from Beijing clustered together, affecting the city. Density is high in Shenzhen with seven or eight dominate industries such as real-estate industry, insurance, banking, logistics and so on. Shenzhen is rebellious by nature. Early on, people came to the so-called liberated zone of Shenzhen, tolerating controversy and opposing views. "I cannot agree with your views, but I can respect your rights". Here the individual actually started to become a reality. Now we are in an era of consumption after accumulating capital, and have begun to weaken our rebellious character. Can other cities in the world still learn from Shenzhen?

Epilogue (Laurence Liauw)

The Shenzhen Biennale was originally setup to showcase new experimental projects in architecture and urbanism, research, speculations and possibilities for China undergoing unprecedented urbanization. This year's Biennale theme set by curator Terence Riley 'Architecture creates cities, Cities create architecture' in an axiomatic endless loop sets up a polemical dialogue for architects and urbanists to think through their own disciplines in order to understand others. This Symposium was able to question the norms and recent history of urbanism as a process and architecture as a discipline, showing that cities are driven by forces beyond their control with unpredictable consequences such as their extreme effects on communities and environments. Design is also not something we can always control in a manner that we expect, when the complexity of cities are extended through time and cultures. Our speakers responded to the Symposium's question with questions of their own in a reflexive manner, sometime defensive, sometimes offensive, always critical. A few brave claims were made about Shenzhen's future, which underscores the belief that our cities will keep providing opportunities (a Western view) rather than problems (a Chinese view). As Hong Kong struggles to come up with a holistic and compelling vision for itself and including its neighbours, Shenzhen is beginning to redefine itself within the Pearl River Delta, in order not to be left behind by the rest of the Mainland (and the World). Hong Kong should heed Shenzhen's wakeup call, not the other way around. Maybe truth lies on the other side of a city's reality. Shenzhen, a global metropolis in the making, poses the important question for China and the World : "what is a city, what makes a city, who makes a city, what could a city become, for whom is the city—and ultimately why do we still need cities".

序言：廖维武

本次双年展促使我们去以一种学科或跨学科的思维去重新思考建筑之外的城市与城市之外的建筑。从一个小渔村到工业乡镇，再到如今拥有1400万人口的大都市，30年来深圳迅速崛起，而快速城市化的真正后果为何？通过建筑的维度又揭示了怎样的社会、政治、经济、文化与环境的真相？人来人往，但是城市依然在发展中。在这个城市中，每个人都是移民而没有真正在此安家。传统的地方观念在此并没多大的影响力，因此，深圳的城市经验彻底转变成了一种新的经验。建筑是否仅仅作为深圳努力界定自我（起初深圳将自我作为另一个香港，而今开始作为新型都市）的过程中的一种副作用、工具或者收益？珠三角在耗费环境与社会成本的竞争中引出了一个问题：更多增长是惟一的出路吗？像深圳这样的城市，自我品质的提升就是一次重生的机会，而香港也不愿放弃与其毗邻的联动相生的机会。深圳对其他新城和正在建设（或筹备建设）的城市有什么影响或者教训？深圳是否已经完成了建设性的学习，抑或它可以成为今后城市建设持续发生的实验基地？没有深圳，香港能否生存？如果深圳背弃香港转而北望又将如何？随着2047年的来临，差异可能再无相关。深圳将成为中国最具创新性的城市，并且已获得联合国教科文组织"设计之都"的地位认可，而届时谁将获得真正的活力？一个汇聚移民、高科技、房地产、出口、生态发展与设计行业的先驱城市，或许就是世界正在寻找的答案。对深圳的重新构想兴许可能会成为香港甚至中国的希望与未来，而这也将影响世界。

研讨会第一部分：深圳与香港（成为城市）

深圳与香港是著名的因机缘巧合与政治实验而产生的双生联动城市，而这只是在历史推动下的偶然状况，还是有蕴藏在二者基因之中的某种因素使其与众不同呢？本次研讨会中"成为城市"这一部分正是在学术或其城市风尚、文化、生活人群、经济与政治层面考察了深圳与香港之间的关系。城市建设是否能一蹴而就？或者城市是否会在反向发展中缓缓灭亡？深圳与香港互惠互利、共生共荣，因此成为了20世纪城市经济发展的契机。什么是至今的决定性因素？其中一个城市是否终将会衰落，或被另一个城市吞并取代？他们是否会最终合并而成为不同以往的新型城市？政治会成为二者发展的障碍还是市场会最终占据主导？借由探讨二者之间的差异与相似，去理解互为映射的城市未来，对中国城市来说是至关重要的——因为正如我们所知，这可能会成为中国城市建设的新型典范或中国城市文明最终的先知。

简介

本次双年展致力于探索设计如何与公众交流、如何传授与传承设计。今天的第一部分将要探讨深圳与香港这两座城市如何发展，这并不是一个真正的讲座，它更多地偏向于一种对话交流，我们鼓励交流有关深圳或香港城市与社会的尖锐问题。在座各位演讲人并不是设计师，但他们都了解城市与建筑。本次演讲人为：江民，奥雅纳工程公司（深圳）董事；姜珺，生活、工作在广州与北京的批评家、媒体人、教师；李欧梵教授，香港文化批评家与作家，而他对文化与教育也十分有洞见。

摘要

江民

我是有着26年专业经验的工程师。今天我要讲的题目是《塑造一个国际大都会》。这是一张大约30年前我在香港拍摄的深圳图景，图片前景有绿地环绕，而背景是深圳城市的天际线。是什么让一个30万人口的小渔村成为今天拥有1300万人口的现代大都市的？在主要人口聚集地区，如罗湖、福田，我们可以看出，正是与香港的紧密联系，成为了这座城市发展的重要因素。深圳的战略是实现与香港的互动联系发展，重点是联络与活跃香港和深圳之间的边界区域经济，而这些边界区域又从发展中脱胎成为城市中心。深圳湾地区的发展需要建设更多的桥梁和高速公路，而这两个城市也随着通信技术的发展和劳动力的增加持续发展着。统计数据显示，深港过境的人员流动量，包括车辆在内，可能超过100万，从而成为了世界之最。从深圳罗湖在1985年、2008年和2011年的照片中，我们可以看到这三个阶段的急剧变化，特别是铁路、港口、桥梁和公路基础设施建设，为联动城市创造了条件。深圳地铁第一期从1998年开始，并于2004年完工，第二期五线同期建设在2011年完工，现共有137个服务站投入使用，日流量也达300万人，第三期共计10条地铁线路将在2016年最终完工。4号线龙华线的高架铁路将成为与深圳城市天际线相融合的基础设施，从而成为城市的一部分。目前，地铁线路可以方便市民在30分钟至一小时内实现工作、休闲或会议出行，同时把深圳与珠三角周边城市连接起来，由此，这个"一小时经济圈"更凸显了城际铁路之重要性。从香港地铁地图来看，关键是该网络如何扩展新城区并重新激发旧城区。地图中的红点所标示出的是深圳将要重建的新区域，如盐田、龙岗、坪山等副中心。实际上，要实现这一区域的扩张和再生，就需要先进的技术、熟练的劳工和一体的规划。市民对城市未来的发展十分关注，因此更熟练的专业人员现在入住城区。香港的铁路系统土地开发模式通过车站的发展为城市带来繁荣，

SHENZHEN AND HONG KONG / 深圳和香港

城市本身成为了整个地区的枢纽并且通过铁路与地铁连通机场。这有助于深圳和香港的发展，并且使二者成为珠三角地区的中心枢纽区域，而这又需要来自工程师、建筑师、规划师与政府间的宏观视角与全面合作。一个城市不仅是人们居住或工作的地方，也是交换创意的场所。深圳和香港市民有着共同的创业精神，同时也有着可以促进多样文化与高品质生活的高效人际网络。对于未来的香港－深圳国际都市区而言，两个城市的开放与合作是非常重要的。多样性是一个国际大都会最为重要的品质。

姜珺

多样性对一个城市而言非常重要，而这也可以成为一种经济性状。多样性的内涵很多，包括行业的形式和规模，而这也是深圳和香港之间最大的差异。20世纪80年代，内地经济改革之后，香港将众多产业迁往内地。深圳在20世纪80年代的成长伴随着香港的资本和技术转型。1997年亚洲金融危机使得香港受到重挫，因此尚未确定新的基础产业是否能在此蓬勃发展，而如果没有中央政府的政治援助，香港的情况会更糟。两个城市之间的第一个差异是二者的工业基础。香港现在有一个依赖于中央政府支持的单一经济模式。2009年全球金融危机也影响了深圳，受到土地、劳动力、环境成本的上升，萎缩的国际市场和人民币升值的影响，深圳疏散了大量的低端产业。与香港不同的是，深圳的工业并不是中空状态，而是注重提升产业的质量和多样性的实体产业。与此同时，香港的楼价是世界上最高的，而这也影响中小型企业的生存。在港投资在亚洲的排名从去年的第5位下跌至第14位。深圳的低端产业开始消失，取而代之的是高端IT、生物技术、新材料、可再生能源和创意产业。前海新区将吸引包括来自香港的新兴产业的投资。前海"特区"可以说是一个身处深圳特区的离岸"香港特别行政区"。两者的第二个差异是国土开发模式。深圳已扩展成为面积大于香港的经济特区，从2009年起，高新区在深圳迅速发展，除此之外，老厂地区也广泛提升为创意产业园区。在过去，90%的工厂创造了10%的经济效益，而现在的目标是10%的高端产业创造深圳未来90%的经济收入。第三个差异是发展的规模，深圳最大的分区宝安与龙岗将经济特区扩展为其原始大小的五倍。深圳的中心作为一个整体继续西进，从罗湖延伸至前海，与此同时，涌现出一批新兴的商业副中心。与此相对应的是去年开始建设的旨在鼓励多元文化、发展文化产业的香港西九龙文化艺术区，但这个方案的最终结果是只能由单一业主注入资金与开发，其反映了香港单一产权主导的市场模式。相比之下，深圳的经济是多样的（除经济适用房的开发）。这些差异在深圳的多样性与香港的统一性城市发展过程中将指向不同的未来。香港在科学、技术和文化等层面并未向其他国家学习。因与珠三角相似的北部文化背景和香港近来影响力的逐步下降，我认为深圳在近几年将成为珠三角地区发展可供替代的选择，而一个主要问题仍然存在：香港将如何从目前的城市文化和单一的经济模式中跳脱出来以适应北方文化（包括深圳）与多样性？

李欧梵

让我谈谈我对这个城市文化的印象。我40年前发表过"上海与香港姐妹双城"的说法，现在深圳和香港的情形更是如此。我每次来深圳，都觉得它有着19世纪美国"狂野西部"的精神。在珠江三角洲，深圳起到什么作用？深圳可能是创意和设计人才，甚至是城市规划人才的中心，但为什么不是香港呢？我认为是因为过于官僚化的机制和商业体制，香港没有创造性的发展空间，而深圳市政府以实惠的价格提倡发展。与香港相比，深圳的问题存在于文化与教育中。在深圳，只有一所大学，而香港却拥有九所。我们可以看到两个城市的传媒行业在相互竞争，深圳和广州的印刷媒体实际上是相当客观的，报道的主题广泛，而香港媒体的报纸却没有文化丰富性。香港是高度体制化的，它的问题是过度制度化导致了创意空间的缺乏。据我所知，香港鲜有享誉世界的、对全球或是中国文化产生影响的书籍。由此可以看出，香港在文化（和遗产）上面临危机，而深圳却在生活中反映了累积数百年的岭南文化。许多香港的文化和创意人现在无法找到自己的创作空间，例如100多个戏剧团体找不到演出场地。西九龙文化区的建设期望为香港提供广泛的创意空间，但中小型文化团体该何去何从？3个总体规划中，OMA的设计灵感源于当地的岭南文化，但香港政府最终选择了最为保守的规划方案。

与深圳的文化交往给我留下了深刻的印象，我去了深圳的书店，并且期待与深圳的知识分子见面交流。但香港的知识分子——他们去上海、北京，为的是什么？为了挣更多的钱？一个如深圳一般具有多样性的移民城市，拥有操着各种口音的人，而他们使整个城市更具活力。20世纪70年代，我移居香港，每个人都说着不同的方言。但移民的一代已经过去，香港现在陷入了均质化的危险之中。北方文化，如北京在政府、媒体、科技教育界的精英影响了深圳，但同时在深圳萌生出的寻找新的创造性而自由发展的南方文化也会影响到北方，即使没有大学这样的大型机构，我对两个城市的文化交流和教育的未来仍保持乐观态度。香港教育的最大危机是英语水平的下降。在高租金的压力下，书店无法生存，过高的出版成本使人们更少阅读，从而丧失了文化的滋养。香港人都只讨论香港的问题，而不思考香港与深圳、香港与世界之间的关系。但深圳的过快增长，也带来了城市软实力开发前的问题。第一届双年展由深圳市人民政府主办，但为什么不由香港政府赞助？我的问题是：如果现在深圳有这么多的创新空间，通过创新是否能创建硬件设施与城市软实力之间更好的联系？

回应者：李亮聪

首先，我想表示感谢，各位成功地将三个演讲人的话题组织在了一起。在我看来，江民所关注的是硬件方面，他谈论的是将深圳与香港连接在一起的城市基础设施的建设与发展。随后，姜珺谈到土地利用和城市软实力的开发。最后，李教授谈到两城居民的会面和意见交换。我们怎样才能使深港成为双生的城市？我是香港双年展策展人，所以我相信，为了这两个城市真正实现姊妹城市的愿望，两座城市的双年展应该由一位策展人负责。两座城市有不可逾越的鸿沟吗？我认为深圳有更多的机会，我们可以看到，两年前在香港的双年展中我们在寻找场地方面遇到了许多困难。李教授谈到的正是这两个城市正在失去的东西。我认为它应该是双向或者三向的，并且有着有形和无形的方面。但在香港和深圳在珠三角的发展中我们应该做些什么是真正值得思考的问题。

研讨会第二部分：深圳与世界（世界的未来都会）

深圳挑战了所有城市如何形成、繁荣、生存及免于消亡而不断自我转型的合理的或常规的模式。"向拉斯维加斯学习"根本不足以用来形容深圳的经历和影响。它的形成是独特而异常的，但它看起来与任何其他城市都类似……抑或不是。速度、规模和人群被深圳及其对周边珠三角城市的催化效应重新定义。是否有人费心去了解深圳与世界的关系？这一答案是否使深圳成为了联合国教科文组织所认可的"设计之都"，同时因过去 10 年间的规划而赢得了国际建筑师协会的阿伯克龙比奖？深圳的常住人口在过去 10 年中翻了一番并在低碳城市规划上遥遥领先，这是否意味着深圳成为了中国国内生产总值最高的城市？深圳是中国的房地产市场诞生之地，并且仍由无名英雄努力开拓、重新定义，这是现实吗？我们将讨论这些影响和其对世界其他的对城市发展规律、人类希望仍有兴趣的地区所产生的潜在影响。世界上那些地区能否从学习深圳的经验和城市实验中获益？然而，除去所有，深圳在它惟一所能选择的城市化进程中体现出的勇气和信念，成为了世界未来大都市的希望。

摘要

Michelle Provost

荷兰阿姆斯特丹和阿尔梅勒是姊妹城市，正如深圳和香港。阿尔梅勒通过土地复垦而建造，而阿姆斯特丹的居民住在宽敞的花园洋房中。深圳是一个有着更高人口密度的特别经济区，阿尔梅勒是一个只有 19 万人口的郊区城市；阿尔梅勒市民自主拥有土地并自行建造房屋，而深圳市有 1200 万正在寻找住房的移民人口。阿尔梅勒现在的问题是如何为 19 万定居在阿尔梅勒但却在阿姆斯特丹工作的居民提供就业机会，因此阿尔梅勒对深圳的发展很感兴趣，并将其作为一个学习的案例。各种城市问题正在挑战荷兰的政治制度，如日益增长的收入差距和对自然的破坏等。我们正在应对的问题是：你会为经济增长牺牲多少？一位法国动画导演画了一本关于深圳的漫画书。退回到 1997 年，深圳还充斥着很多灰色的工厂与仓库，"……这是一个孤独的现代化城市，虽然毗邻香港，但是深圳没有多少人会说英文，那里没有咖啡馆能让我同对西方感兴趣的年轻人们见面。两天会议结束后，我们就回到香港。"作为一个城市的模型，深圳是怪异的，我无法找到市中心，也无法找到像在西方一样的典型住宅区。这是一个只为商业而生的城市。我之前在这个城市曾感觉到彻底的迷失，如今身处双年展 OCT 当代艺术中心，我感觉到了短短的十年内整个环境的变化。深圳从过去的工业城市短时间内就转变成了文化名城。如果我们将深圳认定为世界上最成功的新市镇，虽然它可能不是最漂亮的一个，但它是否会最具有能量？深圳也因其实验性而闻名，它是一个你在承担风险的同时能即时验证自我假设的地方。深圳可以成为下一代新市镇的模型吗？在我们的新书中，我们着力叙述三个新市镇的几代人所遭遇的问题。第一个问题有关生态和生态城市，这一问题就是只为绿色环保而作绿色环保，但并不能充分利用相关投资。这种现象在中国屡见不鲜，尤其是在为了吸引政府关注的情形下。其次是社会影响。这个新市镇为谁而建？对乡村人口不能回到原有土地居住的深圳而言，这是个好问题。由此可见，新市镇可以改变社会结构和生态群落。可持续性发展在这里就变得非常具有社会效用与影响了。最后我想问新市镇具有什么特性？他们都有国际摩天大楼、12 车道的高速公路、高档住宅和突出的政绩等一致的前景。这些标准化的新城市如何在不同的地方适应当地环境？中国已经开始借鉴其他国家城市的特点。这种地方文化特色的丧失也是很有趣的。其他新兴城镇能从深圳的经验中学习到什么？深圳在继续作为城市发展的实验室或有着良好生态环境、社会结构和城市特性的同时，还可以激励他处的可持续发展之地吗？

SHENZHEN AND HONG KONG / 深圳和香港

David Giaonotten

本次研讨会提出了一个重要的问题。OMA 与深圳的第一次接触是在1995 年，库哈斯与哈佛的学生 同来到这里，写了一本书《大跃进》。它对西方产生了深远的影响，书中描述了深圳兴起过程中令人难以置信的房地产热潮、令人难以置信的城市规模和令人难以置信的工作技能。但我们所作的真正的贡献是在深圳证券交易所项目启动之时。如果你现在正目睹全球的经济危机，中国的成功是为数不多的成功案例。三十多年来，深圳市人民为改善他们的生活，做出了一个全新的城市和社会创举。我们如何建设城市并为正不断扩大的 600 万城镇人口创造新的社会结构？在深圳，你无法预知会发生什么事。参与项目是惟一能得知出路的办法。流动性非常重要——这意味着与其他城市的自由的思想与知识交流。新的深圳证券交易所反映出了深圳经济自由和经济发展的关联。它是在"中央商务区"之中新生的构建类型。我们试图建立一个拥有城市社会空间的中央商务区，而它同时能提供新的城市特性。这种建筑形式试图如电脑交易一般公开、透明，它没有任何隐藏，而是完全通过文字和数字展现。在新的前海港口城市竞争规划中，OMA 曾提到："我们不能抹去城市结构和历史，而应保存有助于保持城市活力的元素。"我们分析了深圳现有的城市结构和基础设施，研究了何种有意义的元素可以被添加到该计划中。教育在这里很重要，但是还没有国际化，所以这是一个机会。为了人们的福祉，我们这里引入了中西方的卫生保健系统。另一个敏感的话题就是媒体。我们为中国的传媒公司腾讯设计了一所办公大楼。最初，他们希望建立一个双塔，但为了更好地沟通，我们最终提出了一个独立的三角塔形建筑方案。竞标胜出后，腾讯邀请我们设计北京而不是深圳的新总部。最后一个问题有关城市特性，如何塑造一个城市的形象呢？深圳是联合国教科文组织认定的"设计之都"。这能算一个城市的形象么？我们与都市实践所合作的"深圳眼"项目（位于中央商务区水晶岛的设计）试图创造新的城市中心形象。我们为"深圳眼"将所有不同的运输站、商店、地下办事处等一切设施场所纳入了一个大型环状结构。如果一个新市镇需要有影响力，它必须要有想象力。大多数新市镇缺乏想象力，因为它们仅仅是旧城镇的副本，但他们可以通过创造性地结合不同的元素而成功定义一个城市的新特性。像深圳这样的城市如何才能不负联合国教科文组织"设计之都"的头衔，同时保持其经济增长？深圳不能简单地照搬西方或中国其他城市的经验。因其独特的经济和政治结构，深圳现在有机会成为独特的"世界大都会"。

金敏华

香港和深圳与世界的关系必须从时间和空间上来看。这是深圳大芬村。20 世纪 80 年代改革开放后，一名香港男子在大芬创立了一个车间，尔后其逐渐成为世界油画的主要生产基地。深圳是一个移民城市，它拥有1400 万人口，大约 60% 来自全国各地，广东省的外来人口占 30%，10% 是当地居民。全国各地的人涌入深圳，但是很少从几个大城市而来。大芬村的人们擅长家居装饰油画。村社区变成了一个画廊，随后成为了一个全球性的产业。设计师黑一烊凭借他的作品获得了国际大奖。中国平面设计行业在深圳诞生。深圳市持续 20 年引领国内进出口贸易，因此产品的包装设计也带来了对印刷业需求。当前深圳的打印输出产量占全球市场的 1/3。2008 年，联合国教科文组织宣布深圳为"设计之都"，而深圳改革 30 年来已达到一个过渡的临界点。它曾经是一个制造业城市，现在正步入向创意城市的转型中。我们在本地研讨会中提到，如果时间退回到 1999 年，深圳是否能成为由设计行业支撑的城市？《深圳日报》在 2004 年启用了这样一个标题：深圳是否可以成为世界上第三个"设计之都"？当时政府全力支持这一在之后四年间实现了的雄心壮志。深圳的设计行业与在此诞生的主要创新产业紧密联系，其中包括比亚迪、中信、华为以及室内设计、房地产等行业。联合国教科文组织的评估委员会认为，深圳将设计作为转型工具。我们在 2004 年加入了创意城市联盟，它分为 7 个领域——设计、民间艺术、美食、文学、媒体艺术、音乐、电影。其中第一个城市是意大利的都灵，第二个是首尔。深圳政府计划建立一个设计中心/学院和华侨城设计博物馆。在 2008年 12 月教科文组织的招标会议上，深圳展示出了反映岭南文化的影片与当地的彩纸工艺礼品。我们希望借此表明深圳一直在寻找融合传统设计与现代性的方法。我喜欢这部电影，它象征着深圳从 0 走向 30。30是 2008 年深圳经济特区所经历的 30 年时间，而 0 可以被理解为：深圳从作为一个在中国南海边的点而一无所有的开始。深圳最宝贵的资产是人的思想和差异。媒体尊重从近乎反叛到保守的不同观点。我们将深圳想象为一个城市的梦想，因为它是一个在天空中的城市，人们可以飞到这里。在深圳的第一代移民实际上对现状并不满意。深圳曾经像一只代表自由、宽容与和而不同态度的新的风筝，但现在风筝却被禁止放飞。

廖维武

我们的三个演讲人的观点有所重合。Michelle Provost 认为深圳是最成功的新城并鼓励实验。David Giaonotten 则提出深圳的成功源于它的想象力和创造力。金敏华是以国际视野在谈论深圳的设计价值。在2010 年世博会的深圳展馆中，我们讨论了对深圳因先锋精神的减少与城市逐渐保守的倾向而失去竞争优势的忧虑。腾讯邀请 OMA 的项目不是在深圳而是在北京的最终建立显示了这一危险。那么，深圳如何保持开拓创新精神？

回应人：刘晓都

我发现今天这场论坛对我来说非常有启发。Michelle Provost 从历史和生态的角度谈了三点：深圳是一个完全人工的城市，深圳的社会结构转型非常重要，开发商和基础设施工程组来到深圳并且国有资本主导民营企业。这是一个从零开始的新城市的必经之路。David Giaonotten 有关 OMA 的深圳经验说明了一些实际问题。老照片显示了邓小平的改革口号。随后，我们将深圳证券交易所看作资本主义的象征符号或一座纪念碑，它可能也预示着一个时代的结束。当时在线的参会人员将腾讯建筑的设计作为恐慌而激烈辩论。为什么腾讯不将建筑落户深圳而最终在北京呢？因为北京的文化和政治资本，使它可以容纳这样的建筑。对这种差异的自觉深远地影响了深圳的未来。水晶岛的"深圳之眼"强烈表明了香港和深圳之间的关系。高速列车可在 14 分钟内贯通深圳与西九龙文化艺术区。深圳的设计能成为焦点。设计与文化相辅相成。金敏华的话题从大芬村到山寨文化，阐述了深圳设计的利弊。我这样的对比标签显示出了信心的缺乏。申请成为"设计之都"，意味着我们可能缺乏对自我文化的信心。我最近在与荷兰社会学家探讨城市发展，尤其探讨了如深圳等依靠传统的以资本为主导的发展模式的城市发展。我们的房地产开发恰恰体现了这一点。姜珺反复强调香港的发展模式单一，而深圳毗邻香港，所以这种单一模式的房地产开发也成为了我们的资本运作模式。但是资本的目的是什么？现在西方社会追求知识驱动型的发展。李教授说，香港已积累了几个世纪的丰富文化，而深圳却无法与之匹配。姜珺承认香港的地位，而深圳也在日渐强大。以我所见，这种差异可能是一种假象。香港只是期待着但不采取行动，所以人们会感到失望。深圳有无限的可能性，但它会去往何方？深圳自己的问题是开拓进取精神的丧失与人口的老龄化。李欧梵多次批评香港对创意的抑制。随着中国内地的发展，深圳正慢慢被推入背景中。香港的世界地位正在下降，而除非深圳有所改变，否则它最终也将随之下降。

收盘评论

Michelle Provost

深圳的故事给我们很多启发。我们谈论硬件和软件，也许还存在第三种条件，那就是组织机构。在谈到深圳和香港之间的关系时，组织机构发挥着重要的作用。香港有这么多的刚性机制，因此产生了一个静态的环境，而深圳需要创造新的机会并避免重蹈覆辙。深圳现在有很多硬件，但组织机构尚未成熟，因为它还处在发展的初期阶段。那么，我们如何使深圳保持为创新型城市呢？

David Giaonotten

我想说，如果你想比较深圳和香港，这里的一切都是独一无二的。每一个城市都有其自己的独特性。深圳已不再是一个小镇，它成为一个城市已经有一段时间，并且已与世界有很深的联系。我们决定不去比较与思考城市的本质、特点以及未来的发展。深圳现在已经是一个有着自己的立场的国际大都会，未来肯定会更好。

金敏华

深圳市是特殊的，我们现在看到的大多是来自北京的国有企业在此集结而影响着这座城市。房地产、保险、银行与物流等主导产业在深圳高密度聚集。深圳天性反叛，早先人们来到深圳所谓的"解放区"，容忍争议和反对的意见："我不能同意你的看法，但我能尊重你的权利。"在这里，个体事实上开始成为一种现实。现在我们处于后消费时代，资本积累已开始削弱我们性格中的叛逆。世界其他城市能否依然向深圳学习？

尾声：

廖维武

深圳双年展的最初设置是用以展示中国在建筑与城市规划、研究、猜想与可能方面的前所未有的城市化实验项目。这一届双年展的主题"建筑创造城市，城市创造建筑"由策展人泰伦斯·莱瑞先生拟定，这个主题处于不言自明的无限循环中，是为建筑师与城市规划专家通过自己的学科去理解他人而设立的。本次研讨会可以质疑规范、城市化过程新近历史以及建筑学学科，它显示了在无法控制的力量的推动下，难以预料的城市发展后果，正如其对社区和环境的极端影响。设计也是因城市的复杂性贯穿于时间与文化之中而不能按照我们所希望的方式行使。我们的演讲人以自反的方式回应了他们自己的问题与专题讨论会的质询，这些回应有时是防守型的，有时是进攻型的，但是总能切中要害。关于深圳的未来，城市将继续提供机会（西方的观点）而不是问题（中国的观点）。香港正努力为自己提出一个全面的和令人信服的愿景，而包括其邻居深圳为了不落后于内地（与世界）的其他地区，也已在珠三角范围内开始重新定义自己。香港应留心深圳的觉醒，而不是相反。也许真理隐藏于城市真实的另一面。作为正在酝酿中的国际大都会，深圳对中国和世界都提出了一个重要问题："什么是一个城市？是什么构成一个城市？一个城市由谁构成？城市会变成什么？为什么并且为了谁我们最终仍需要城市？"

SHENZ
AND
CHINA
深圳和中

HEN

8 URBAN PLANS FOR CHINA
八个城市项目

Curators: Jeffrey Johnson, Xiangning Li
策划人：Jeffrey Johnson, 李翔宁

SHENZHEN AND CHINA / 深圳和中国　　8 URBAN PLANS FOR CHINA / 八个城市项目

Urban China

China is without question one of the most vital places in the world for tracking trends in urbanization. During the past three decades since Deng Xiaoping initiated the Open and Reform policies, China's unprecedented rate of urbanization has forever changed the way we think about and engage with the city. With the influx of foreign investment fueling economic growth, mass populations moved from the countryside to urban centers seeking economic opportunities and a better way of life. Never before has the world experienced such a mass exodus. It took China just a single generation to urbanize at a rate that took America a century to accomplish and Europe before that almost two. Just over thirty years ago, only 19% of China's population was urban equaling less than 200 million people. Today, China is nearly 50% urban with an urban population of over 665 million.

Even more striking is that this trend is expected to continue. According to McKinsey Global Institute, by 2025, 350 million additional people will move to urban centers creating more than 220 cities with 1 million or more population (the United States currently has only 9 cities of 1 million or more and Europe 35). 170 new mass-transit systems could be built to support 40 billion sqm of new floor space, 50,000 of these buildings could possibly be skyscrapers—equivalent to ten Manhattans. Without dispute, the world has never experienced an urban project of this scale and magnitude.

Due to this accelerated urbanizing process, the Chinese city exists in perpetual change experiencing extreme cycles of destruction, construction and reconstruction. Existing city centers are transformed overnight and vast rural landscapes are usurped and rapidly urbanized. New cities form where only villages existed just years before. What was new only a decade ago is now often deemed obsolete and replaced with newer, bigger and many times bolder constructions. This continuous physical and spatial transformation yearns for radical new urban proposals. Architects and planners of the past could have only dreamed of having the opportunity to conceive and realize projects of this size, complexity and significance. The urban project has experienced a resurgence in China as its relevance extends beyond the proverbial drawing board towards the real domain of the city.

8 Urban Projects

For the exhibition "8 Urban Projects" eight contemporary projects were selected that together we felt represented a broad snapshot of urbanism in China today. To exhibit a diverse sampling of approaches we chose projects from a varied range of Chinese and international studios. Each project was selected based on a set of critical issues or themes we felt define the challenges of the urban project in China today. Additionally, each of the projects was selected based on their location within a specific spatial condition that characterizes the unique contemporary urban terrain that has emerged over the past thirty years in China. Due in part to these perpetually shifting conditions, we felt that each of the projects challenged the conventions of what seems to be outdated and obsolete ways of thinking about the city, yielding exceptional and, at times, radical new urban forms and strategies. Viewed together we feel the eight urban projects construct a conceptual Chinese city that begins to define a new urban paradigm. Some of the critical themes that we found consistent across many of the projects and that we feel are highly representative of the urban condition today in China are expanded on below. Though far from comprehensive, we feel the list below illustrates how the unique conditions and challenges have generated new urban modalities reflected in the work of the selected projects.

Human-centered/Quality of Life

The forces associated with accommodating the millions of new urban inhabitants hastily has resulted in the construction and multiplication of standardized urban models that seldom consider the quality of the place or the community it is creating and/or replacing. Even with a recent trend of promoting "quality of life" and improved "lifestyles" many are still forced to live in environments that lack essential public amenities and social spaces, and convenient access to public transportation and daily necessities. The physical and mental well-being of the inhabitants is of paramount importance and the recent focus on the "health" of cities has inspired many more human-centric proposals that cultivate diversity and social interaction through well-considered public spaces and cultural amenities, programmatic diversity, "healthy" buildings and environments, and a more symbiotic relationship with nature and landscape. In each of the 8 urban projects selected the intent to propose living environments that privilege quality of life over mere accommodation is clear.

Steven Holl's Shan-Shui Hangzhou project proposes to create an exemplary quality of life by creating a living environment that balances scenic landscape, architecture and culture. Also in Hangzhou, Woods Bagot's have planned leisure, culture, entertainment and recreation programs around the existing Xixi wetlands to offer a place of respite from hectic urban life. Yung Ho Chang/Atelier FCJZ's Jiading Ad Base promotes a "humanistic" urban scale with small mini-blocks with interior courtyards and colonnaded commercial streets that will encourage social interaction. OMA/Urbanus' Shenzhen Eye proposes to vitalize a vast and lifeless urban area with a new urban landmark that links together through public parks and landscapes new cultural facilities and creative industries. And, James Corner Field Operation's Qianhai specifically proposes "4 Public Realm Typologies: Parks, Pedestrian System, Streets, Landmarks" to create a maximum quality of life in a new CBD.

Ecology/Environment

Perhaps the most critical issue confronting urbanism in China today is the environment. The rapid urbanization over the past thirty years has caused a drastic decrease of arable land and critical strains on natural resources like water. Additionally, when you consider that the built environment consumes the majority of the world's energy (construction + operation), how cities are designed and planned is of utmost importance. Increased density, orientation, infrastructural efficiency, mass-transportation, building technologies and materials, and strategic adaptive reuse of existing structures can all provide for more "sustainable" urban strategies. China has a great opportunity to provide the prototypes for more intelligent and sustainable cities in the future.

Concerns about sustainable issues are prevailing in China, especially at the government level. Reports on ecological impact of all public projects are required to be submitted to the planning bureau for approval. Although the idea of sustainability is widely accepted, its real application remains more or less on a paradoxical level and the actual performance of the "Green" projects is still questionable.

Even more promising is that we are seeing some recent projects reach beyond the performance-driven aspects of "sustainable" design towards a re-centering or re-balancing of nature with the built environment. James Corner Field Operation's Qianhai project is presented as an urban ecology, with landscape and nature providing the essential linkages between the built environment, and the built environment with the bay. Landscape maintains its more traditional scenic and recreational role but also becomes a social catalyst through the organizational system of spatial networks. In Steven Holl's Shan-Shui Hangzhou a "fusion" of landscape and architecture is proposed and the project finds inspiration from the existing natural landscape and the historical Hangzhou water culture. And, the Xixi Wetlands projects by Woods Bagot and WSP demonstrate how development can play a critical role in the preservation and promotion of a natural wetland area.

Preservation/History/Nostalgia

Over the past thirty years cities in China have lost vast numbers of historical buildings and districts due to demolition, deterioration and indifference. Seldom considered in the propagation of the "new" is the positive role preservation can play in defining a city's and its inhabitants' identity and sense of place. With the market forces seemingly adverse to preservation and rehabilitation, we have seen a shift in this thinking with recent success that have been able to demonstrate that preservation can be both culturally and commercially lucrative.

At the same time, nostalgia in contemporary China, as a cultural phenomenon also blends with the nostalgic consumption of historic images, and in many cases leads to "fake reality" of tradition and heritage in architecture and urbanism. Overt nostalgia also results in an obsession with recreating historical contexts, and oppression of creativity for an imaginative future. The projects selected for exhibition generate a real sense of place and history rather than creating theme parks with historic motifs.

David Chipperfield's Rockbund project located along Shanghai's historic Bund demonstrates the cultural value of architecture and preservation. The once dilapidated northern section of the historic Bund has been converted into a cultural and commercial district ingeniously adapting the historic colonial architecture to new contemporary uses. By contrasting the colonial architecture of the past with new infill structures the project creates a narrative of multiple histories, reinforcing Shanghai's identity as a cosmopolitan city. Steven Holl's Shan-Shui Hangzhou master plan also reinforces the value of preservation through the adaptive reuse of existing industrial buildings that will compliment the new, and with the restoration and revitalization of the water canal infrastructure.

Scale Diversity—Program Diversity

The fabric of most Chinese cities can be said to exist at predominately two scales: micro and mega. From the informal villages and urban settlements to the courtyard houses and the lilong, the historical trace of life in many cities is found at the small scale. The majority of the new development is of mega-scale creating a huge gap between the mega and the mini. As a response to the predominance of large-scale development many recent proposals advocate a broader spectrum of scales that allow for diverse programs and flexibility in use.

Yung Ho Chang/Atelier FCJZ's Jiading Ad Base proposes a grid of mini perimeter blocks with public courtyards creating a high-density/low-rise alternative to the superblock, which enables more diversity and interaction—a finer-grain urbanism. As opposed to multiplication of similar blocks it aspires to encourage uniqueness and difference within the constraint of the small block. We see this resistance to superblocks also in James Corner Field Operation's Qianhai proposal. Their proposal of mixing diverse programs is enabled through the use of smaller urban blocks with mixed-use podiums and commercial programs at the street. The Ordos 20+10 project proposes to create individually unique architectural projects that at an urban scale collectively construct an urban experience with diverse scales and programs.

Connectivity/Networks/Porosity

In part due to the current land policies many of the newly developed and planned projects are conceived as large autonomous and isolated 'chunks' of land disconnected and isolated from the adjacent urban context. Many in fact are gated enclaves with limited and controlled access. Urban districts are comprised of isolated "urban islands" that are completely disconnected from the city. Although the walled enclave has historic precedent in Chinese urbanism, many recent proposals advocate alternative strategies that promote intensified social interaction through more porosity and connectivity.

OMA/Urbanus' Shenzhen Eye is conceived as an infrastructural "ring" linking disparate urban elements and discrete zones together and with the city and its surroundings. James Corner Field Operation's Qianhai project promotes connectivity and interaction through "social landscapes" not only within the project area but importantly linking the revitalized waterfront with the urban core. Steven Holl's Shan-Shui Hangzhou is conceived as a spatial link to adjacent projects by David Chipperfield Architects and Herzog & de Meuron and to Hangzhou through water infrastructures encouraging the public to permeate and flow through the project.

Place Identity

With the radical transformations that have occurred in the urban and natural Chinese landscape the ability for urbanism and architecture to shape and enhance place identity is critical. It reinforces architecture's role and responsibility as a cultural artifact that helps shape the image of place and the identity of its inhabitants. With a complete saturation of overly branded projects and the intense competition between iconic urban structures, some recent projects have attempted to transcend the superficial to create more lasting and meaningful proposals, all the while maintaining their commitment to extend beyond the norm and aspire towards uniqueness.

In the Civic Center district of Shenzhen, OMA/Urbanus' Shenzhen Eye proposes a new civic landmark across from city hall emphasizing the role of the public in the city. David Chipperfield's Rockbund reinforces Shanghai's cosmopolitan identity by combining its colonial architectural past with new contemporary cultural facilities. Ordos 20+10 projects utilizes iconic architecture designed by a group of international and Chinese architects to generate a unique urban experience which showcases the city's identity as an economic center and social and cultural enclave in the vast northwest region of China.

Spatial Politics

Historians often lament that democracy and power politics may both have their advantages and disadvantages. However, the trend of today's politics in China has an obvious determinant affect on architecture. The production of space becomes a process for the solidification and exemplification of ideology. Architecture and urbanism, from the perspective of Plutonomy, are the physical expression of the political and economic power underneath. However, with appropriate guidance, politics can also become a positive influence on architecture and urbanism.

In the special case of Qingpu—Jiading, an enlightened mayor has commissioned numerous architects of international notoriety, including Yung Ho Chang/FCJZ's (Jiading Ad Base), to showcase their experimental architecture, making the town a famous stage for cutting-edge contemporary architecture and urbanism. In the case of the master plan project of Cui Heng New District of Zhong Shan, the hometown of Dr. Sun Yat-sen, the master plan was prepared at the 100th anniversary of Dr. Sun's revolution, which encouraged cultural and economic linkages between mainland and Taiwan. And, David Chipperfield's Rockbund project responds positively to the challenges of Shanghai's colonial heritage.

SHENZHEN AND CHINA / 深圳和中国　　8 URBAN PLANS FOR CHINA / 八个城市项目

城市中国

中国无疑因其迅速的城市化进程而成为了全世界最具活力的地方。邓小平提出改革开放政策的30年来，中国史无前例的城市化进程永久地改变了我们对于城市的理解和介入方式。随着外资的涌入拉动经济增长，大批人口从乡村移民到城区，寻找致富机遇和更好的生活。世界历史上还从未出现过如此庞大的迁移现象。中国只用了一代人就实现了美国用了一个世纪、此前的欧洲几乎用了两个世纪才完成的城市化。就在30年前，中国只有19%的城市人口，相当于不到两亿人口。今天，中国将近50%的地区实现了城市化，城市人口超过了6.65亿。

更为显著的是这种趋势有望持续。根据麦肯锡公司的数据，到2025年，又将有3.5亿人口迁入城市，创造220个人口规模在百万以上的城市（美国目前只有9个人口过百万的城市，欧洲有35个）。170个新的大众运输系统将建立，以支持400亿平方米的新陆地空间，其中5万个会是摩天大楼，相当于4个曼哈顿。毫无疑问，世界从未经历过如此规模和强度的城市化进程。

在这种加速的城市化进程中，中国不断经历着摧毁、建造、重建的极度循环。现有城市中心一夜之间被改造，广阔的乡村风景被侵占并迅速城市化。新城市所在地在几年前还是乡村。十年前还是新的建筑，现在通常被认为过时了，要被更新、更大、更大胆醒目的建筑取代。这种持续的物理和空间的转换要求彻底的、新的城市化方案。过去的建筑师和规划者做梦才能想到设计和实现如此规模、如此复杂和重大的项目。城市项目在中国经历了一次复苏，它超越了宣传板上的口号，延伸进城市的现实领域。

8个城市项目

我们为"8个城市项目"展览挑选了8个当代项目，我们觉得它们是一是中国今日都市主义的辽阔景象的一个缩影。为了展示不同的样式，我们在国内外的工作室中进行广泛挑选。选择每个项目的基础都是我们认为能够明确今天中国城市项目挑战的重要问题或主题。此外，每个项目的选择也都基于一个特定空间下的当地情况，这些空间体现了中国过去30年里涌现出的独特当代城市地域的特性。部分地由于这些持续变化的状况，我们感到每个项目都挑战着我们如何去判断什么是过时的、挑战着我们对于城市的陈旧思考方式，制造特例和彻底更新的都市形式和策略。放在一起来看，我们觉得八个城市项目构建了一个开始定义新的都市范式的概念中国。我们发现的一些关键主题贯穿着许多项目，我们认为它们能高度代表今日中国的城市化状况，这些主题将在后文展开。尽管不够全面，我们还是觉得这个清单表明了特殊的条件和挑战如何催生了这些项目所反映出的新的城市形态。

以人为本 / 生活质量

为了适应几百万的新城市人口，标准化城市模型的建设和繁殖迅速产生，却鲜少考虑地方的质量或这个地方正在形成或更新的社群。尽管在今天经常倡导"生活质量"和进步的"生活方式"，许多人仍旧在缺少基本公共设施、社会空间以及便利的公共交通和日常必需品的环境里求生存。居民的身心幸福是首要的，最近对城市"健康"的关注启发了更多以人为本的建议，这些建议认真考虑了公共空间和文化设施的内容多样性、"健康"建筑和环境问题以及自然和景观的共生关系，培育多样性和社会互动。被选出的8个城市项目的目的都十分明确，倡导注重居住质量的生活环境。

斯蒂文·霍尔事务所的"山水杭州"计划提出创造一个在风景、建筑和文化中取得平衡的居住环境，建立一个生活质量的典范。也是在杭州，伍兹贝格在西溪湿地规划了休闲、文化、娱乐和创意项目，为狂热的都市生活提供一个休憩之所。张永和/非常建筑事务所的嘉定中广国际广告创意产业基地，提出"人文主义"的城市标准，微型楼群中的庭院和廊柱商业街将促进社会的互动。大都会建筑事务所OMA和深圳市市实践建筑事务所的"深圳眼"，提出激活一个面积广大但是缺乏生气的城市区，用新的城市地标将公园和风景、新的文化设施和创意产业连接在一起。James Corner Field Operation的前海规划特别提出"四个公共地带的拓扑学：公园，步行区，街道和地标"，在一个崭新的CBD里创造最高质量的生活。

生态 / 环境

环境问题大概是今天中国城市化所面临的最严峻的问题。过去30年来极速的城市化造成了可耕地的急剧减少和水等自然资源的临界应变。此外，只要想一下环境建设消耗着大量的世界能源（建设+运行），就会想到城市如何被设计和规划是极为重要的。不断增长的密度、方向、基建效能、大众运输、建筑技术和材料以及现有结构的战略性再利用，都能提供更加"可持续的"都市策略。中国有着一个重大机遇，在原有基础上创造更加智能和持续性的未来城市。

考虑到可持续是中国的普遍问题，特别是在政府层面，所有公共项目生态影响的报告必须获得规划部门的批准。尽管可持续的观念已经被广泛接受，其真正的应用仍旧有些矛盾，最近的"绿色"工程的实际表现仍旧存在问题。

更为肯定的是，我们看到最近一些项目已经超越了"可持续"设计的效果驱使层面，走向自然与建筑环境的重新定位和重新平衡。James Corner Field Operation的前海计划是一种配合了景观和自然的都市生态学，以建筑环境和海湾之间的基本联系为前提。景观保持了它更为传统的风景和娱乐功能，同时通过空间网络的组织系统成为了社会催化剂。在Steven Holl的山水杭州中，风景和建筑的"融合"被提了出来，项目的灵感来自于现有的自然景观和历史上的杭州水文化。同时，伍兹贝格和维思平的西溪湿地计划展示的是土地开发如何在一个自然湿地的保存和推广中扮演关键角色。

保存 / 历史 / 乡愁

过去的 30 年里,中国的城市由于拆迁、恶化和漠不关心失去了许许多多的历史建筑和区域。在"新"的繁殖中很少考虑保存在定义城市及其居民的身份和地点认知方面的积极作用。尽管市场形势看上去不利于保存和复原,但我们在最近的一些案例中看到了这种想法的变化,这些成功案例证明了保存在文化和商业上是双赢的。

同时,当代中国的乡愁作为一种文化现象,同样与对历史图像的乡愁消费联系在一起,在很多情况下导致了建筑和都市主义上对传统的继承的"伪现实"。公然的乡愁也引起了对重新创造历史语境的痴迷,并抑制了想象未来的创造力。本次展览的项目生成了对地点和历史的真正感知,而不是用历史图案制造主题公园。

戴卫·奇普菲尔德的"洛克·外滩源"计划选址在富有历史韵味的上海外滩,证明了建筑及保存的文化价值。一度荒废的北外滩地区被改造成文化和商业区,巧妙地将历史上的殖民建筑与新的当代用途结合了起来。通过旧时殖民建筑和新填充结构之间的对比,这个项目创造出了一个多重历史的叙述,强化了上海作为国际界大都市的身份。斯蒂文·霍尔的"山水杭州"主体规划同样强调保存的价值,将新部分与已有工业建筑进行有效再利用,修复和激活水道基础设施。

规模多样性——项目多样性

大部分中国城市的肌理可以说基本在两个极端上:极小的和极大的。从非官方的乡村到城市定居点、庭院洋房和里弄,许多城市的历史遗迹都已成为凤毛麟角了。大多数发展项目都是超大规模的,这在极大和极小之间造成了巨大的缺口。作为对占主导地位的大规模发展的回应,最近许多方案都提倡更宽泛的规模,可以容许项目的多样性和应用的灵活性。

张永和 / 非常建筑事务所的嘉定中广国际广告创意产业基地提出了一个微型建筑圈,配有公共庭院,为大街区提供高密度 / 低高度的替代物,容许更多的差异与互动——微都市主义。与繁殖相似街区相反的是,它渴望在小街区的限制下鼓励独特性和不同。我们在 James Corner Field Operation 的前海项目中也同样看到了对大街区的抵制。他们的方案是融合不同的项目,使用更小的城市板块,配合多用途的裙房和街边的商业项目。鄂尔多斯 20+10 项目提出了个性化的独特建筑项目,在城市范围内共同打造不同规模和内容的都市经验。

连通性 / 网络 / 多孔性

部分地由于现今的土地政策,许多新近开发和规划的项目都被设计成庞大的自治而孤立的地"块",与周边的都市语境脱节和隔离,其中许多实际上是在限制和控制下围隔的封闭之地。城区由一个个孤立的"城市岛屿"组成,与城市完全脱节。尽管墙内之地在中国都市主义历史上已有前例,许多近期的方案还是倡导替代策略,通过更多的空隙和连接推广密切的社会互动。

大都会建筑事务所 OMA 和深圳市都市实践建筑事务所的"深圳眼"项目,构想的是一个结构"环",将分散的都市元素和散落的区域与城市及其周边连在一起。James Corner Field Operation 的前海项目倡导用"社会景观"推进联系和互动,不仅在项目范围内,复兴的海滨地带与都市核心的连接也同样重要。斯蒂文·霍尔的"山水杭州",与戴卫·奇普菲尔德事务所、赫尔佐格与德梅隆的项目以及杭州市在空间上相连接,通过水利基础设施鼓励公众参与项目并流动于其中。

地方身份

随着中国的城市和自然发生急剧的变化,都市主义、建筑塑造和加强地方特征的能力成为关键。这就强化了建筑作为文化产品的作用和责任,它有助于塑造一个地方的形象和其居民的身份。随着过度标签化项目的完全饱和与图标性城市结构的激烈竞争,最近的一些项目不再只做表面文章,而是去创造更持久更具意义的方案,同时恪守承诺,超越成规,追求独特。

在深圳的市民中心,大都会建筑事务所 OMA 和深圳市都市实践设计建筑事务所的"深圳眼"提出了一个坐落在市民中心对面的新城市地标,强调公众在城市中的作用。戴卫·奇普菲尔德的"洛克·外滩源"计划通过结合过去的殖民建筑与新的当代文化设施,强化上海国际大都市的身份。鄂尔多斯 20+10 计划利用一群国际和中国建筑师设计的图标性建筑,产生一种独特的城市经验,将城市特征呈现为经济中心和中国西北的社会、文化重镇。

空间政治

历史学家经常感慨民主政治和极权政治都有各自的优缺点。尽管如此,今天中国政策的趋势在建筑上有着显著的决定性影响。空间生产成为意识形态凝固和塑造的过程。从政治经济学的角度看,建筑和都市主义是其下层政治和经济权力的物理体现。不过,通过适当的引导,政治也能成为建筑和都市主义的积极影响。

在青浦-嘉定特别项目中,开明的区长委任了一大批国际知名建筑师,包括张永和 / 非常建筑事务所(嘉定中广国际广告创意产业基地),让他们展示各自的实验建筑,让这个地区成为前沿当代建筑和都市主义的展示舞台。翠亨为了纪念辛亥革命一百周年而准备的孙中山故乡中山新区主体规划,鼓励大陆和台湾之间的文化和经济往来。此外,戴卫·奇普菲尔德的洛克·外滩源计划积极回应了上海殖民遗留物提出的挑战。

| SHENZHEN AND CHINA / 深圳和中国 | 8 URBAN PLANS FOR CHINA / 八个城市项目 |

James Corner Field Operations

Qianhai / 深圳前海规划

WATER FINGERS + WATER FRONT

Qianhai Water City recognizes and re-establishes "water" as the defining element in Qianhai's territorial identity. Water is deployed as an infrastructure and an amenity in this two-fold re-structuring of the Qianhai harbor, to improve its ecological performativity, social life and physical characteristics. First, we propose 5 mega-scale filtration fingers that function not only as remediation infrastructure for entire site, processing both the water flowing through the channels and rivers, as well as the stormwater run-off of the entire 1804 hectares; but also act as urban parks to activate social life and add value to new urban development. The second move is a continuous waterfront that along with the 5 fingers are the new open space framework and social infrastructure of Water City. 4 programmatically distinct waterfront parks that respond to the edge conditions of their adjacent districts; forest park, performance & art park, play park and port park construct a varied yet cohesive waterfront experience.

FABRIC + DISTRICTS

The 5 water fingers orient the enormous site into 5 districts, each with its own "primary" programmatic disposition—commerce and business; trade and logistics; research and innovation; civic and culture; and lifestyle and mix. These thematic qualities also help to drive the architectural massing and typologies that make up the fabric in each district to create distinct, legible destinations at the territorial and neighborhood scale. As such, we have defined maximum building heights and programmatic distribution for each subdistrict that is calibrated to it specific geographic context and particular programmatic make-up. Development frontage along the waterfront and public spaces is maximized, and these are typically areas of highest activity and density within each district. The pilot district (Guimiao commerce & business district) is sub-divided into 7 specific sub-districts with different programmatic emphases: commercial, business, event, life, art, culture creation and daily life. Though each sub-distincts has its own mode of street, architecture and public space, they're still connected by main street and traffic hub. Moreover, the infrastructure is fused into the urban design.

水廊道 + 水岸公园

前海水城方案以"水"为重要元素体现前海片区的地理特色，并将水作为基础设施和环境质量来重构前海湾，从根本上改善该前海区域的生态、社会生活和景观质量。首先是净化处理从现状河渠流入海港的水体来改善湾内水质。前海水城方案通过引入5条超大尺度的净水廊道，来处理河渠水体和1804公顷的新建城区的雨水排放。同时，5条大尺度水廊道也结合了城市湿地公园的功能，来改善周边的城市生活并增加相邻片区的开发价值。其次是构建一个环绕前海湾的连续的水岸公园体系。4个彼此相连的特色水岸公园提供给前海湾一个连贯的水体验：水岸森林公园，演艺公园，游乐公园和海港公园。

肌理 + 区块

肌理和5条分区水廊道的引入将整个前海区域分为5个特色分区。每个区域的城市活动彼此不同，并以此来形成区域功能特色：金融商业区，贸易物流区，科技创新区，文化艺术区和生活区。这些分区的主要功能划分使前海片区结构清晰，易识别。我们通过结合地理位置，区别性地进行高度控制和功能细分，并在沿水和靠近公共空间的区域提高建筑物密度和强化公共生活的强度。在前海水城的启动区（桂庙金融区）内，我们将整个片区通过城市功能的侧重进一步细分成7个亚级市区：金融，商业，会展，文化，艺术，创新和生活片区。虽然每个亚区有各自不同的街区结构、建筑类型和公共空间分布，但各个亚区被精心地通过城市主要街道以及交通枢纽联系在一起。现状的城市基础设施也被很巧妙地结合进了城市设计中。

| SHENZHEN AND CHINA / 深圳和中国 | 8 URBAN PLANS FOR CHINA / 八个城市项目 |

Yung Ho Chang / Atelier FCJZ / 张永和 / 非常建筑事务所

JIADING AD BASE / 嘉定国际广告创意产业基地

This project takes on a 'mini-block' approach to planning a low density office zone in Jiading, a satellite town outside Shanghai's city center. Resisting the current Jiading fabric of autonomous buildings surrounded by landscape and parking lots, it takes on low density planning looks at creating intimately scaled urban streetscapes through the introduction of a network of scaled down mini-blocks. Measuring just 40x40 meters, each of the 18 hollowed out perimeter block is separated from the next by a 10 meter street grid. A 3-meter wide ground level setback of the blocks on all sides, then creates a continuous network of pedestrian arcade through the whole site.

Through a catalog juxtapositioning different perimeter conditions (floor plate area, room thickness and edge porousity), different prototypes are selected that give a wide spectrum of mix of office spaces (open office mix with office-in-rooms) to meet different office requirements. A wide variety of courtyards interspaced throughout the 18 blocks result from this across a spectrum of varying privacy, ranging from those for exclusive tenant usage to the ones serving as an extension of the streetscape with full public access.

Each architecture applies 5 materials in construction, brick, stone, concrete, grid and each explored a new mode of usage, even the same material can be used in different ways. So far, those facades are unified in a way, but each keeps its own nature..

整个广告创意产业基地的规划思路为"水乡步行城市"，故本项目将其延续并演化。建筑物底层与纵横的城市步行道路相互咬合，形成互动的有机关系。运用小尺度街道构思，运用10米、20米的街道尺度，形成不同尺度的人性活动空间。在街道尺度上，首层退后3米，设计适合南方多雨地区的骑楼，同时创造丰富的沿街商业立面。

平面设计的概念是"生活化庭院"和"不同尺度办公空间"的结合。根据每个建筑在规划中的地理环境和周边景观以及潜在的服务人群和企业（如严谨的行政办公类型和自由的创意办公类型），为每个建筑安排了位于不同层，具有不同尺度的庭院，形成了庭院体系，如城市广场（城市庭院）、大型中庭、为会议室配备的景观庭院、为个人准备的休息阳台和室内的共享大厅。办公空间与不同尺度的庭院相结合，引入室内楼梯和室外楼梯，形成相对开放的办公空间与相对封闭的办公空间，从而适应不同的办公要求。A1-A18具有以不同比例混合的大小办公空间，平面柱网与机房等功能性用房的布置充分考虑了办公设备的尺寸和使用者的舒适性。

在18栋建筑中，运用了5种材料：砖、石材、混凝土砌块、格栅和金属网材质营造围合空间。每栋建筑探寻一种材料的做法，同一种材料在不同的建筑中做法也不相同。至此，创意园区中，建筑立面既获得了统一的效果，又保证了每栋建筑的个性。

SHENZHEN AND CHINA / 深圳和中国
8 URBAN PLANS FOR CHINA / 八个城市项目

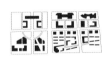

Typical Jiading Superblock Infill:
Buildings are placed as autonomous objects within superblocks surrounded by landscape and parking lots.

建筑独立存在，体量过大，缺少步行联系，无法形成合理的交通系统。

Proposed Superblock Infill:
Fills superblock with a network of mini-blocks to create human scale urban streetscapes.

确定设计机理。
建筑体量适中，建筑之间为10m或者20m街道，街道及是分割又是联系，适合步行。

Site Plan/总平面
- 10M VEHICLE AND PEDESTRIAN STREETS /10m 人行+车行
- 20M PEDESTRIAN STREETS /10m 人行
- 15M STREETS WITH CANALS /15m 沿河
- 50M STREETS WITH CANALS /50m 沿河

Typical Jiading Superblock:
- Buildings are placed as independent islands
- Buildings are placed as independent objects within superblock and do not form a coherent streetscape.

建筑独立存在，体量过大，缺少步行联系，无法形成合理的交通系统。

Proposed Jiading Block:
- Buildings are integrated blocks
- Perimeter of the buildings are designed to create continuous pedestrain walkways on all sides of the building.

建筑四面同时退让3m，形成挑檐，创造出适宜步行的空间，使每一条街道由建筑立面围绕，每一个建筑立面都有街道相临。

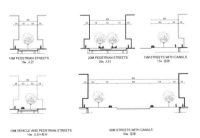

10M PEDESTRIAN STREETS / 10m 人行
20M PEDESTRIAN STREETS / 20m 人行
15M STREETS WITH CANALS / 15m 沿河
10M VEHICLE AND PEDESTRIAN STREETS / 10m 人行+车行
50M STREETS WITH CANALS / 50m 沿河

 → →

filled block 　　perimeter block 　　ground level arcade
满铺　　　　　　环楼　　　　　　　骑楼

由于铺满生成经济，创造丰富的建筑内部使用空间，在底层的基础上生成骑楼不仅使立面造型得到丰富，同时创造了丰富的步行空间。

Masterplanning
规划

Yung Ho Chang / Atelier FCJZ / 张永和 / 非常建筑事务所

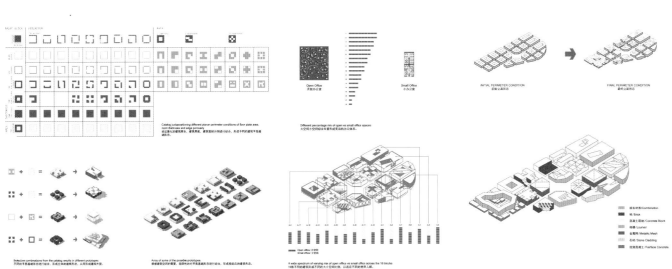

Massing prototypes
体块样本

Streetscape
街景

| SHENZHEN AND CHINA / 深圳和中国 | 8 URBAN PLANS FOR CHINA / 八个城市项目 |

David Chipperfield Architects / 戴卫·奇普菲尔德建筑事务所

ROCKBUND PROJECT

Following the establishment of international trading relations in the nineteenth century, Shanghai became a commercial and cultural centre of East Asia and home to a large number of European commercial offices and consulates. Examples of Shanghai's Art Deco style—European building styles combined with Asian elements, characteristic of the city's early twentieth century architecture, are strung along the Bund, Shanghai's boulevard on the west bank of the Huangpu river.

An ensemble of historic buildings, now called the Rockbund Project, reflects the diversity of the colonial architecture and forms the Northern part of the Bund. By restoring the existing buildings and planning new ones, a team of international architects is helping to revitalise the Rockbund Project, which will accommodate office complexes, hotels, retail and apartments. David Chipperfield Architects has been commissioned with the restoration and conversion of eleven buildings.

The aim of the restoration concept is to present buildings that have aged with dignity and style: The façades have been carefully cleaned and repaired without destroying the original fabric. Conversions from a later date have been removed and the buildings returned to their original state to the greatest extent possible. Existing structures within the roof area of some buildings have been expanded in reaction to contemporary changes in usage. The expansion of the Andrews & George Building to create a 60-meter-high office tower, Rockbund 6, will see its historic fabric being combined with contemporary architecture: The listed three-storey façade of the existing building that marks the southern edge of the planning area will be renovated and eleven storeys added in the form of a stacked construction, creating a landmark building that assumes the scale and materiality of the surrounding architectural environment.

The National Industrial Bank (N.I.B) and the Royal Asiatic Society Building (R.A.S) have gained new extensions oriented to the inner courtyard, located in the south-west of the block, Museum Square. The new facades were rendered using 'Shanghai Plaster', the same material as used on the adjoining buildings. The former R.A.S. Building, once China's first public museum, now houses the Rockbund Art Museum—a museum of contemporary art. Inside this Art Deco style building, the newly created flexible areas enable a range of different presentation concepts. A new atrium links the upper floors, creating a triple height space.

The whole Rockbund Project will be accomplished with the completion of Rockbund 6 in 2014. The restoration works on the façades and interiors of the historic buildings were completed in spring 2011. The Rockbund Art Museum was opened on 4 May 2010 with an exhibition of works by Chinese artist Cai Guo-Qiang.

PROJECT	Rockbund Project
ARCHITECT	David Chipperfield Architects
MODEL PHOTO	Roman März
MODEL KINDLY SPONSORED BY	Shanghai Rockbund Master Development Co. Ltd.
项　　目	洛克·外滩源项目
建筑设计	戴卫·奇普菲尔德建筑事务所
模型摄影	Roman März
模型赞助	上海洛克外滩源综合开发有限公司

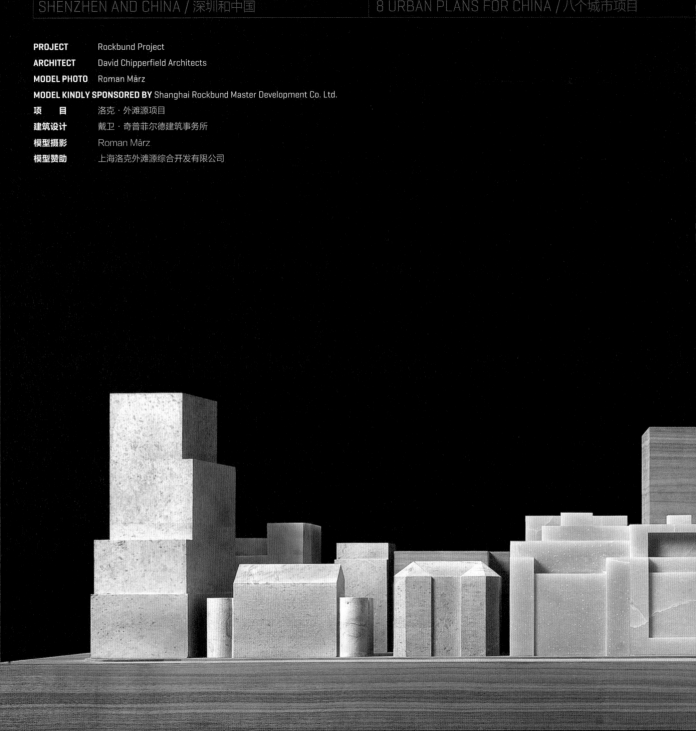

David Chipperfield Architects / 戴卫·奇普菲尔德建筑事务所

洛克·外滩源

自 19 世纪开放通商之后，随着大量跨国公司和领事馆的进驻，上海一跃成为远东地区最受瞩目的经济、文化中心。20 世纪初，融汇中西的"上海式"装饰艺术风格（Art Deco）风靡一时，在浦江西岸的外滩建筑群中占据着独特的地位。

外滩源项目恰位于外滩的最北端，街区内的一系列保留历史建筑充分体现出了上海殖民地建筑在类型上的多样性。在一支国际建筑师团队的共同努力下，外滩源项目将通过对老建筑的保护，对新建筑的规划和对高档办公、酒店、零售与居住功能的引入，实现街区的整体复兴。而戴卫·奇普菲尔德建筑事务所则在其中承担着 11 栋历史建筑的修缮和改造设计工作。

对外滩源历史建筑的修缮，以展现建筑在历经岁月沧桑之后的绰约风采为宗旨：在保护其原始材质不受破坏的前提下，对建筑立面进行精心的清洗与修复并尽可能地拆除所有后期改建，力求在最大程度上恢复各历史建筑的原始设计。同时，为了满足新的功能需求，在设计中还对部分历史建筑的屋顶进行了小规模的扩建。作为外滩源项目的亮点之一，原有的美丰洋行将被扩建成一栋 60 米高的办公楼；保留的三层历史建筑立面将与其上部增建的逐级退进的十一层建筑体量紧密结合，形成位于街区南端的标志建筑——洛克六号楼。

中实大楼 (N.I.B) 和亚洲文会大楼 (R.A.S) 的扩建部分均面向位于街区西南的内广场——博物院广场。为了保持其与周边历史建筑的协调，二者选择了与之质地相同的水刷石作为外立面材料。原亚洲文会大楼，曾是近代中国最早向社会开放的博物馆，现已改建成一座当代艺术博物馆——外滩美术馆。在这栋装饰艺术风格的历史建筑内部，改造后的展厅空间为各种展览提供了灵活的布展空间，而改造后的中庭空间则将建筑上部的三层空间连成一个整体，为人们提供了全新的空间体验。

整个外滩源项目，包括洛克六号楼在内，将于 2014 年全部完成，而历史建筑的外立面和室内修缮则已于 2011 年春季先期完成。2010 年 5 月 4 日，外滩美术馆以中国艺术家蔡国强先生的展览为开幕展，率先对外开放。

| SHENZHEN AND CHINA / 深圳和中国 | 8 URBAN PLANS FOR CHINA / 八个城市项目 |

Steven Holl Architects / 斯蒂文·霍尔建筑事务所

Hangzhou Shan-Shui

Steven Holl Architects (SHA) won the master plan for this international competition in 2009. In the current development, the SHA site brings together an urban constellation with David Chipperfield Architects's design for the reuse of the boiler factory buildings on the southern portion of the site, and Herzog & de Meuron's reuse of the former oxygen factory buildings to the north.

The heart of the "bowtie" plan is a "Water Tower" and a "Mountain Tower," alluding to the spirit of Hangzhou. The Water Tower branches into tributary forms connecting to the north, while the Mountain Tower connects via landscape forms to the south.

There are five large scale elements in the SHA design, which hover between landform and architecture, connecting to the factory buildings at each end of the site.

1. WATER TOWER

The round water tower rises from a sheet of water, connecting with a curved bridge crossing Dong Xing Road to the north. This diffused glass tower houses serviced apartments and offices with retail space at the base, and a restaurant and event space at the top.

2. CANAL SPREADERS

An existing canal feeds five new canals, lined with new hybrid buildings. These "Canal Spreaders" characterize a new zone of living by the water. They offer a variety of housing types as well as offices, cafes, restaurants and shops along the public paths at the water's edge.

3. LANTERN TOWERS

The lantern towers take inspiration from the old stone lanterns in West Lake, setting "fire over water." Photovoltaic glass curtain walls gather the sun's energy during the day. At night, one elevation of each tower glows via special Fresnel glass, reflecting the day's energy in the water. One loft apartment per floor is connected by an elevator to collective lobbies below the pond. Health club, spa, retail shopping and parking levels connect the lower levels.

4. MOUNTAIN TOWER

At the center of the site, the Mountain Tower is joined via an escalator bridge to an event space at the top to the Water Tower. This tower of translucent ceramic skin and green roofs branches to the landscape of the 3D Park.

5. 3D PARK

A fusion of landscape and architecture in the 3D Park yields public green-space roofs with openings; as "gardens within gardens," bringing nature and light to the lower levels. It is flanked by a 200-room hotel, served by a spa and restaurants opening to roof gardens, and bisected by a pedestrian link, lined with shops.

Hybrid buildings, the mix of functions, the merging of architecture and landscape, and the invigorating programs inserted into the re-used factory buildings, characterize this unique urban constellation.

Aspiring to a twenty-first century urban vitality in balance with landscape, Shan-Shui will be a magnificent and inspiring new section of the great city of Hangzhou.

| SHENZHEN AND CHINA / 深圳和中国 | 8 URBAN PLANS FOR CHINA / 八个城市项目 |

Steven Holl Architects / 斯蒂文·霍尔建筑事务所

杭州山水

斯蒂文·霍尔建筑师事务所（SHA）的规划设计方案是在2009年的国际竞赛当中获胜的。SHA地块目前的方案中汇集了戴卫·奇普菲尔事务所于基地南部的杭锅厂房的改建方案及赫尔佐格与德梅隆事务所于基地北部的杭氧厂房改建方案设计，形成了"城市之星"。

"山之塔"和"水之塔"伫立在蝴蝶结形基地的中心，暗喻了杭州山水的神韵。"水之塔"分岔成了不同的支流向北区延伸，而"山之塔"则通过山峦般的形态连接至南区。

SHA的设计包括了五个大规模的建筑元素，这些建筑在地景与建筑间延展，并连接基地两端的厂房。

1. 水之塔

圆柱形"水之塔"从犹如镜面的水中升起，她曲线形的天桥跨过东新路，向北伸展。这个漫射玻璃塔楼内有服务式公寓、办公室，低层的裙楼设为商业空间，而顶层则安排了空中活动空间。

2. 延展水巷

一条现有的运河连接五条新建的水巷，水巷边上并排着综合性的新建筑。这些"延展水巷"建筑形成了一个全新的河岸生活。延展水巷不仅提供了不同类型的居住空间如大型别墅，还设置了咖啡厅、商店、办公区等，让河边的人行道富有蓬勃的生机。

3. 湖心灯塔

湖心灯塔自西湖三潭印月的石灯塔中汲取灵感，构成了"水中火"的景观。光伏幕墙的蓄光作用在白天发挥采集太阳能的功效，到了晚上，湖心灯塔通过安装在每栋塔楼其中一面的特制菲涅耳玻璃发出的微光，将白昼的能量倒映在湖里。湖心灯塔每层复式公寓由电梯直接通达湖面底下连通大厅组成的结合厅。地下层由健身房、水疗院、商业及停车库连接起来。

4. 山之塔

"山之塔"在基地的中心点，通过电动扶梯连接到水之塔顶部的活动空间。塔楼的外墙面设计成半透明瓷质墙板，塔楼向南伸展，与3D公园屋面上的空中花园相连。

5. 3D公园

"3D公园"将绿化与建筑融为一体，屋面形成了公共花园；屋面上的天窗开洞为低层空间带来了光线及生气。3D公园被设计为拥有200间套房的目的地酒店，它包括了空间宽敞的水疗院及若干通往屋顶花园的餐馆。主要人行道与开放式的斜坡走道两侧排列着高端商店，给3D公园带来了活力。

综合性建筑物与各功能的混合、建筑与绿化的融汇及通过新功能激活旧厂房构成了这个独特的城市汇聚点的别具一格的色彩。

怀着对21世纪都市活力与自然生态均衡的憧憬，"山水杭州"将是杭州市的一个壮观又创新的新区域。

SHENZHEN AND CHINA / 深圳和中国   8 URBAN PLANS FOR CHINA / 八个城市项目

SHENZHEN EYE

Shenzhen is a fast developing city depending on an intelligent infrastructural network. Our proposal organizes Shenzhen centre vertically into three new connection layers which can be experienced in different speeds of displacements, by train, by car, by bicycle or by foot. The new layers generate different notions of time, and co-exist over each other, encouraging exchange, interaction and merging. The new connections transform the void in the center of Shenzhen into a dynamic and active public space.

OMA and Urbanus competition entry for Crystal Island in 2009 addressed the problems of isolation and inactivity in the Central plaza of Shenzhen. Despite Shenzhen city's award winning urban planning, the heart of its center has remained disconnected and void of people. We wanted to construct a space with a cultural atmosphere and different activities, changing and improving the quality of public space.

The Shenzhen Eye project introduces a scheme with 'three levels, one path'. A composition of an elevated Ring, a ground level Plaza and underground Connectors was designed to address the lack of people, identity and connections at Civic centre. The three components enhance the viability of each part while unifying the overall design of the development. The policy of 'privately owned public space' (POPS) allows the contribution of funds and expertise from the private sector, merging the spaces for leisure and commerce.

In order to meet the challenge of bringing more people and activity into the current park, the hovering Ring will join the existing elevated network of Shenzhen and lead the pedestrians to the Plaza and beyond. The Ring undulates, touching the landscape forms when access is needed, or lowers itself down to create an access point to the Underground Connectors. Loosely defined public programs unify the entire plaza, cultivating exciting new activities while maintaining the initial concept as a physical connector.

The Underground Connectors link all public transportation hubs in the area, as well as the Book City, the new Ganxia development, Shenzhen Sheraton Trade Centre, and Shenzhen Yijing Commercial Centre. The Connectors are activated with retail, restaurants, gyms and movie theatres, creating a new typology of commercial and public space as layer below ground. The underground space has direct access through ramps, patios and elevators to a large Green Park and Plaza above, making the shopping and dining experience unique compared to all surrounding retail areas.

The landmark, "Shenzhen Eye" is an active functioning centre of culture and creativity. The current proposal is to enhance the void with program related to an information centre and exhibition space, while maintaining its physical and symbolic status as the focal point of the region.

深圳眼

凭借着智能的基础设施网络,深圳正处于迅速发展的历程之中。我们在市中心新建起三层垂直连接体,它将给人们带来不同速度的出行体验,如火车、汽车、自行车或步行。新连接体扩大了本区域内人们活动的时间跨度。不同层面间互为补充,为信息交流、人员互动和空间融合提供了便捷的平台。新连接体将会把深圳中心区的空地打造成充满活力的公共空间。

在2009年水晶岛竞赛中,OMA与都市实践的联合方案获胜。获胜方案反映了深圳中心广场缺乏互动性与活跃性的问题:尽管深圳在城市规划方面曾荣获诸多奖项,城市中心却向来空旷,缺少与人们的连接。因此,设计师希望在此打造一片具有文化氛围的活动空间,改变并改善城市公共空间质量。

"深圳眼"方案采用了"三层连接、一条通路"的设计理念。整个连接体系由架高环线、地面广场及地下连接构成,以解决当前市民中心人流稀少、活动有限、连通不足的问题。三层连接体相辅相成,共同呼应了整体开发构思。"私有公共空间"(POPS)的政策吸引了私营企业在此投入财力、人力,为空间增添了休闲和商业氛围。

为了能将更多的人流和活动引入现有的市民公园,架高环线将与深圳现有的空中道路网络连通,将行人引导至广场和其上方空间。高低起伏的大环或与景观连接,构成通路,或向下与地下连接体相接。整个广场中的公共活动统一规划,既涵盖各种新功能,又维持其原有的"连接体"设计理念。

地下连接体将区域内所有公共交通枢纽以及书城、新岗厦、深圳贸易中心和深圳怡景商业中心连通。零售、餐饮、健身和电影院等为连接体带来活力,形成了一种全新的地下商业公共空间。地下部分与地面绿地公园和广场之间的坡道、天井和电梯的连接,为人们带来了与众不同的购物和用餐体验。

名为"深圳眼"的地标具有文化和创意两大功能。现有方案旨在利用信息中心及展示空间增强"深圳眼"的功能性,并维持其作为区域核心和焦点的地理地位和象征意义。

| SHENZHEN AND CHINA / 深圳和中国 | 8 URBAN PLANS FOR CHINA / 八个城市项目 |

Fang Zhenning / 方振宁

ORDOS 20+10/Shenzhen

Curated by independent curator Fang Zhenning, "ORDOS 20+10/Shenzhen" is a mini version of a bold group architecture project, "ORDOS 20+10", in Dongsheng District, Ordos, Inner Mongolia Autonomous Region.

"20+10" states the objectives of the project planned and designed by 20 Chinese architects and 10 international firms involved in this project. The project plan worked out by the planning team specifies the land area, the scale and scope of work, and every architect has been assigned two plots for the development of 20,000 sqm. office floors.

This project is unprecedented in the sense that the structures of the architectures and city are planned by the architects independently. Aiming to be the home to various small-and-mediums sized high-end enterprises, "ORDOS 20+10" intends to build a mini city of "sustainable business complex."

ORDOS 20+10 / 深圳

"ORDOS 20+10／深圳"是独立策展人方振宁根据中国内蒙古自治区鄂尔多斯市东胜区的大胆的集群建筑项目"ORDOS 20+10"策划的浓缩版。

"20+10"的意思是由20位中国建筑师和10位外国建筑师共同进行规划设计而设立的有关建筑师的行动目标的标志，从参与本项目的建筑师中诞生的策划组，在制定的"项目任务书"中对用地、规模及设计范围作了明确的规定，每位建筑师被平均分配到两块设计用地，并要求两块用地上的所有建筑物的总建筑面积控制在2万平方米左右。

像这样以建筑师自主规划的方式来决定建筑和城市的结构体系，在造城史上也属前所未有的实验。该项目定位为中小型高端企业聚集的办公园区，有意把它打造成一个"企业生态综合体"的迷你城。

| SHENZHEN AND CHINA / 深圳和中国 | 8 URBAN PLANS FOR CHINA / 八个城市项目 |

WSP and Woods Bagot Asia / 维思平和伍兹贝格亚洲

XIXI PROJECT

Hangzhou's Xixi National Wetland Park is a precious and well-preserved wetland in the west of the city centre, 5km away from the West Lake. Its unique location close to Hangzhou's center provides the city with a beautiful greenspace/ecosystem. Both Woods Bagot and WSP 's great designs in different scales for XIXI Wetland are displayed in this Biennale.

Woods Bagot

The Strategic Plan for the Xixi Region aims to establish a number of preliminary recommendations for infrastructure, planning and design that will make the region attractive place for workers, residents and tourists. The conceptual masterplan is intended to inform large scale urban economic and real estate decisions to be made. The Xixi park region strategic plan offers an unique opportunity to create a place that is vibrant city within a city—a city of culture and life.

WSP

WSP's approach is to return the landscape to nature. Since ancient times, Xixi Wetland has been a refuge for its ancient culture, unique landscape and charming natural scenery—qualities the design aims to keep intact. The planning, site strategies and architecture all adjust to and follow characteristics of traditional Chinese architectural spaces - maintaining a temperament of harmony, peace and serenity. This also includes a traditional approach neighborhood relationships, attitudes to life, and so forth.

First Natural Protection: To help the local ecology flourish, it is important to protect existing materials: natural levees, vegetation and animal habitats. To enhance these aspects, the use of willow stumps—for example—helps protect the water's edge and provides ideal natural conditions for the loach (an eel-like fish). The stumps encourage growth of plants in shallow water and maintains the current landforms. WSP's plan also calls for replanting 'reed'. All of this promotes growth of additional local plants and provides breeding environments for many types of insects and birds. By creating these conditions, the West River example shows a view of recovery and water purification.

Second Natural Protection: Over thousands of years, artificial ponds have been created and repaired, paddy fields, gardens and bamboo planted—all making a habitat for ducks, geese, fish and other animals. It is important to maintain this natural form of agriculture and allow visitors to enjoy the idyllic setting.

Third Natural Protection: The design respects the atmosphere and romantic quality of the current village setting. To maintain this scattered, natural texture, the new village structures are planned as small, self-contained, discrete blocks and courts. This multicenter arrangement takes advantage of the environment without overpowering it. The paths, bridges and trails between the village blocks help move visitors slowly and casually through this protected landscape. The reconstitution of the Xixi Wetlands reveals the beauty and strength of the land, rivers and lakes.

| SHENZHEN AND CHINA / 深圳和中国 | 8 URBAN PLANS FOR CHINA / 八个城市项目 |

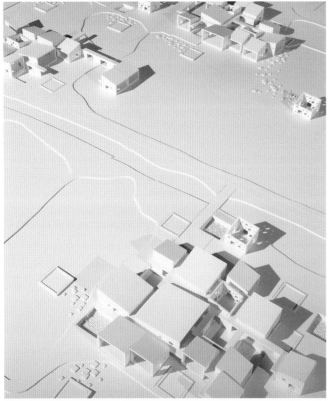

WSP / 维思平

Woods Bagot Asia

WSP and Woods Bagot Asia / 维思平和伍兹贝格亚洲

西溪国家湿地公园

西溪国家湿地公园位于杭州市区西部，距西湖不到5公里，是国内罕见的保存较好的城市外围湿地型生态资源。西溪湿地在杭州市绿地生态系统中具有其独特的地位及作用。本次展览展出了伍兹贝格和维思平两所设计公司为西溪湿所作的不同规模和等级的精彩设计。

伍兹贝格公司

伍兹贝格为西溪湿地所作的战略规划旨在为城市基础设施、规划和设计建立基本准则，从而使这一地区能够吸引工人、居民和游客。概念性总体规划计划为大型城市经济和房地产提供决策。西溪湿地公园的地区性规划提供了一个独特的机会，这个机会可以创造一个充满活力的城市——一个具有文化和生活的地方。

维思平建筑设计

维思平的整体规划及景观设计目标：回归自然，回归原始——西溪自古就是隐逸之地，有古朴的人文风范、江南独特的湿地景观、旖旎的自然风景、浓郁的田园水乡风情，希望规划、景观及建筑设计能调整与自然的关系，回溯中国传统建筑空间形态，传递出中国文化所有的和谐、平和、清寂、幽宁、冲淡的气质，也包括了传统的邻里关系、生活态度、处世理想和人生境界等。

保护第一自然——原生植被、动物及自然堤岸、湿地形态等。采用柳树桩的生态护坎，非常有利于泥鳅、黄鳝类生物的生存；营造更多水面、浅滩和水草地，增加水、陆关系的变化形式，使更多的植物、昆虫和鸟类能找到合适的生存、繁殖地；补种芦苇，它不仅是西溪河渚芦花景致的一种恢复，更有利于水体的净化。

保护第二自然——经过千百年人工修葺的鱼塘、水田、荷塘、人工栽植的早园竹等与饲养的鸭、鹅等家禽动物构成的农业自然形态。体现渔耕文化、休闲文化的田园风情。

尊重第三自然——散落具有浪漫气息的村落肌理及原有道路。新规划建筑以原有村落宅基地为基础，采取功能群组分散原则，以形成各地块自成体系又足够"离散"的多中心聚落；充分利用原有小桥、巷道、田间小径，作为主要的人行步道，重构西溪固有的隐逸风骨和水乡形态。

| SHENZHEN AND CHINA / 深圳和中国 | 8 URBAN PLANS FOR CHINA / 八个城市项目 |

CUIHENG NEW DISTRICT PROJECT / 中山翠亨新区

Zhongshan Municipal Committee of the Communist Party of China and People's Government of Zhongshan Municipality, have assembled an international research and design team, which is lead by professor Wu Zhiqiang, the chief planner of the World Expo 2010 Shanghai and vice-president of Tongji University and features the coordination of DHV, Kuiper Compagnons and Mckinsey Company for the plotting of the future blueprint of Cuiheng New District.

Beijing in 2003, Prof. Wu Zhiqiang has been the planning study and development of Zhongshan City, and the achievements of this planning combined the 8 years' thinking and research and proposed the overall development though of "Cultural Zhongshan", as well as such main ideas and innovative concepts as constructing the spiritual homeland to overseas Chinese around the world, the cluster urban areas and the third-generation new towns, which can be summarized as the following:

2—TWO sides cooperation experimental district
Based on the realistic basis of Taiwan-Hong Kong link, Zhongshan shall focus on common construction by the mainland and Taiwan, and interaction of the mainland, Taiwan and Hong Kong, and undertake cultural industry, green high-quality domestic industry, cutting-edge intellectual industry and talent exchange and cultivation, etc.

0—GLOBAL Chinese spiritual homeland
Mr. Sun Yat-sen proposed the great slogan "rejuvenate the Chinese nation", and Cuiheng New District (C.H.N.D) will hold the flag of "it will gather the Chinese and Western, surpassing the great powers" high. Gather the force of the Chinese in the whole world, build the landmark for national rejuvenation, and create new knowledge and new civilization that stem from China and lead the future.

1—ONE new generation metropolitan
C.H.N.D With innovative industry, advanced culture and high-quality environment, Cuiheng New Area will comprehensively improve the overall quality of the bay area, and create a new center of Zhongshan city that develops from scale advantage to energy advantage, from leading quantity to leading quality and from energy consumption to energy production, so as to be in line with the world high-end service platform to rejuvenate the whole metropolis, and drive the city community with networked and efficient communication network.

1—ONE manmade waterfront with wisdom sea-use
C.H.N.D will take more advantage of natural wisdom (ecological reclamation, environment- friendly development and conservation etc.) to achieve harmony between human and ocean, and become the model of "smart" sea-use.

The total land area of the actual planning is 232km², construction land areas is 96km² and the starting phase region is about of 50km². The contents for the participation of this exhibition are partial achievements made in the planning of Cuiheng New District, dividing it into two sections of "Spatial Strategic planning" and "Conceptual Urban Design".

中山市委市政府组织了以世博会总规划师、同济大学副校长吴志强教授为核心，荷兰德和威环境工程有限公司、荷兰高柏伙伴规划园林建筑顾问公司和麦肯锡公司共同协作的国际研究和设计团队，共谋翠亨新区未来蓝图。

吴志强教授从2003年开始研究中山的规划发展，本次规划成果融合了8年的思考与研究，提出了文化中山总体发展思路，在翠亨新区建立全球华人精神家园、建构群体都市、第三代新城等主体构思和创新理念，可以概括为：

2——两岸共建更紧密合作示范区： 基于台港纽带的现实基础，中山应以两岸共建、三地互动为抓手，适于承接文化产业、绿色优质生活产业、前沿型智慧产业和人才交流培育等相关产业。

0——全球华人精神家园： "振兴中华"是中山先生提出的伟大口号，翠亨新区将以伟人"融汇中西，超越列强"的宏愿为旗帜，集全球华人的力量，共建民族复兴的地标，创造源于中国、引导未来的新知识、新文明。

1——新一代群体都市示范区： 翠亨新区将以创新的产业、发达的文化、优质的环境，全面提升湾区的整体质量，创造一个从规模制胜到能量制胜、从总量优先到质量优先、从消耗能源到生产能源的中山滨海新中心，以接轨世界的高端服务平台提振都市整体，以网络化、高效率的交通网络带动城市群落。

1——新一代智慧用海实践区： 在高效用海的基础上，翠亨新区更将利用自然的智慧，生态填海、用利消害、以用促育，实现人与海洋的和谐共生，实践新一代智慧用海的发展理念。

翠亨新区规划范围约为232km²，建设用地面积为96km²，近期启动区面积约为50km²。本次参展内容是翠亨新区规划的部分成果，分为空间发展战略规划和总体城市设计两部分。

SHENZHEN AND CHINA / 深圳和中国

8 URBAN PLANS FOR CHINA / 八个城市项目

住宅
Residence

住宅首先是一座微型的城市。开敞的庭院或天井体认了建筑和自然的交接，房屋的位次致意着神明和祖先，进深和行止明确着长幼与尊卑、同或、住宅后的园林还隐喻了中国思想中法度与人情间的阴阳谐和，在城市的尺度上居住的组织并不总那么秩序井然，而现代城市的规划理念并不能完全适用于这样"演化出"的城市。

一份未定的地图
AN INDEFINITE MAP

什么是城市？对于不同的历史和地区，"城市"的涵义是不同的。一方面，由建造城市的建筑类型和符号不同，给成城市的建筑风貌有一种近乎天然的冲动；另一方面，作为城市组织的单元空间了个体经验与一个更大的世界，这个世界不仅包容了复杂的、种明和社会体系的总体特征，它还框定了人置身于这个广阔之中的意像。换种意义更言，从文象意味上来说，要识别一个城市，城市不过是建筑类型的阴影而已。

地標
Landmark

所谓地标其实是个西方人的说法，在中国传统中，特别廉耻的工程总是易于引起争议，更不要说"引人注目"这个给有歧义的词。大多数中国建筑较乏可以识别的特明特征，只有特别的端装物才能长久地在城市的天际线上一枝独秀下去；除了建筑物本身惊人的尺度，依靠有利的地形，一些地标建筑也可以收到江山壮丽的效果。

园林
Garden

CHINESE CITIES IN TWO VIEWS
双城记

CURATOR: TANG KEYANG
TEAM: XI'AN UNIVERSITY OF ARCHITECTURE AND TECHNOLOGY
WANG ZHAOZONG   HUANG WENTAO
DESIGN: YANG JIYU
项目策展：唐克扬
项目团队：西安建筑科技大学建筑学院 王兆宗 黄文滔
平面设计：杨济瑜

SHENZHEN AND CHINA / 深圳和中国

CHINESE CITIES IN TWO VIEWS / 双城记

A Chinese City in Two Views aims to present the Chinese urban history in two correlated approaches. In one view, a city might be the sum of all its historical fragments and is often represented by its culminating stage. Observers of such cities usually turn to general typological principles that generate 'identikit' of them. In another view, Chinese cities are constantly changing entities with specific causes for their transformation. For such cities the purpose of our show is not only to provide established perspectives of a city but also to examine how it was transformed through time.

To juxtapose the two views is not only to highlight the methodological gaps in examining historical cities, but to reveal that the tension between such views constitutes part of the urban histories. For this purpose, two group of cases are tactically selected from a large reservoir of research materials, showing both "canonical" Chinese cities and slices cut into their genealogy. For example, the Chang'an of the Sui and Tang dynasties can be called a 'city of cities', which embodies both a 'typical' construct of classical Chinese urban spaces and a particular moment of transition. This exhibition visualized the aforementioned cities in both time and space. The design of the show pays special attention to two aspects of its presentation: how specific socio-historical context makes a certain representation of cities inevitable; and how the convention of representing urban spaces accordingly prescribes the circumstances of our modern practice.

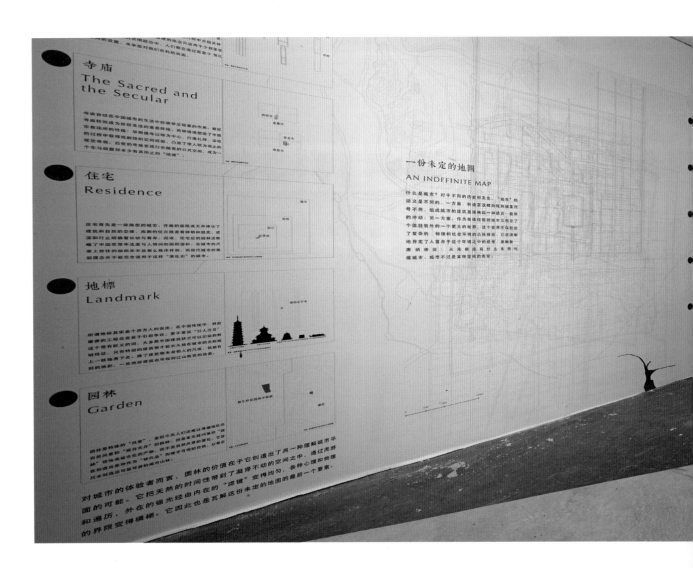

"双城记"意在以两种不同的视角来理解中国城市的发展。在其中一种视角里,城市是它的所有历史碎片的总和,通常呈现为它的极盛状态,依据一般的类型原则,如此打量这座城市的人生造出了一种"特征标准像"。在另一种视角里,中国城市则是变化的个体,盛衰都自有其情境,对于这些城市而言,我们的展览不仅仅希望秀出它最为人熟知的一面,也想揭示它历时的变化。

将这两种视角置于一处可勾勒出不同城市图景间的冲突,同时,我们试图说明这种冲突本身构成了城市历史的一部分。为此,展览从浩如烟海的研究资料中选择了两组不同的例子,既展示那些"经典"的城市的常态,也呈现它们富于意义的断片。例如隋唐长安可以称为"城市的城市",它包含了中国古代都城的一般构造(空间原则),同时,它又是承前启后的关节点(历史变迁)。无论展现"空间"还是"时间"本身已隐含

着对中国城市的认知。展览设计特别注意了两方面的情况:首先是特定的社会情境凸显了哪些城市的表现要素;其次,如此再现城市的传统又是如何影响设计师对于当代城市的理解的。

陸 宮殿和官署
Palaces

发掘和研究传统中国城市时，人们关注得最多的可能就是宫殿等"国家建筑"了。上古的"宫殿建筑"比如半坡的"大房子"没有礼仪和起居功能的明显分离。宗庙有时也不妨合一。后世宫殿制度的确立，为构造多少类似的建筑物带来不同的功能，这些功能以各种不同的方式组合在一起，形象化地解释了皇权和国家的定义和

柒 寺庙
The Sacred and the Secular

寺庙曾经在中国人的生活中扮演举足轻重的作用。在宗教狂热席卷朝野，政教难以彻底分离的中古，曾经出现过举国"伽蓝"舍宅为寺的盛况。后世宗教退从于政体，佛唱间杂看"俗讲"，寺庙转而成为世俗生活的重要陈地，东京大相国寺门前的集市成了今天庙会的源头之一。

捌 住宅
Residence

住宅首先是一座微型的城市，它在重复着我们描述中国古代城市时常常用到的那些方面，只不过尺度都小了一大号：它常常是自我对称的闭合院落，院落的布置同时体现着"形象"和"场域"，开敞的庭院或天井体现了建筑和自然的交融，房屋的功能位次敬重着神明和祖先，进深和行止明确着长幼与尊卑，间或，住宅后的园林

玖 地標
Landmark

所谓地标其实是个西方人的说法。在中国传统中，特别靡费的工程总是易于引起争议，更不要说"引人注目"这个带贬义的词。大多数中国建筑和它的用途并没有必然的关联，缺乏可以识别的鲜明特征，所以只有特别的建筑物才能长久地在中国城市的天际线上一枝独秀下去。

拾 园林
Garden

园林是特殊的"风景"，直到今天人们还难以准确地区分自然风景和"宛自天开"的园林。但是毫无疑问某些"园林"恰恰是城市化的产物，而不是自然风景的演化，它旨在创造出某种作为"替代品"而唾手可得的自然，以拳石尺水创造出可居可游的闹市山林。

一份未定的地图
AN INDEFINITE MAP

什么是城市？对于不同的历史和文化，"城市"的涵义是不同的。一方面，和语言这样的纯粹抽象符号不同，组成城市的建筑直接唤起一种感官一肢体的冲动（比如面前一段连续的台阶，不仅在你的脑海中传递一个"梯子"的讯息，更多的，还使得行人得到向上行走的意念）；另一方面，作为集体经验的城市又包含了个体经验外的一个更大的世界，这个世界不仅包含了复杂的、物理和社会环境的总体信息，它还清晰地界定了人置身于这个环境之中的感受。詹姆斯·唐纳德说：从来都没有什么东西叫做城市。城市不

壹 中心和内外 / Center and Depth

即使只从平面上观察也有好几种理解中国传统城市的可能性。首先相当一部分历代都城是左右(东西)对称的，这带来了今天人们津津乐道的"中轴线"。但是事实上并不存在这样一条"中轴线"，首先"中轴线"在物理上并不贯通，明清北京宫城中央的门道，即使皇帝本人也难得走几次；其次，"中轴线"在视觉上也并不

贰 墙和门 / Gates and Doors

中国传统城市的城墙已经成为这些城市本身的象征，但是关于墙有几个误解。首先，墙不只有守御功能，因此它并不一定很高。隋唐长安城的城墙就比明清北京矮得多，关于它的高度人们有着不同的推测。但是它的面积之大却是人所首知的，这样大的城市实际没发生过什么真正的城头攻防战。这些历史事实帮助人们重新思考城

三 地形和水系 / Built with Nature

有人说中国城市中习见的正交网格来自于"井田制"，不管怎么说，将自然地形和水系引入城市，算是打破了这种陈规。大兴长安城的设计者宇文恺在"九二六高坡"上营曲江池，北京西城的网格里出现水边的"烟袋斜街"，都构成了某种意外。

肆 市井 / Urban Market

"市"在城市的叫法里占据了一半的位置，足见它的意义了。但是市场的物理形式本身好像没什么定谓。它首先有随风聚散的"事件建筑"的简易特征，宋朝城市的商业场所"瓦子"的含义，据说讲的就是"来时瓦合，去时瓦解"；其次，市场的建筑讲究的是成规模有规律的排列组合，而不是某种特出的个性，"行业"的"行

伍 道路和城市 / City Street

中国历史都城的道路给人们留下深刻的印象，首先是因为"大道如青天"。朱雀大街宽达150米，成为许多东亚城市效仿的典范。然而这样超尺度的街道除了展示威仪还有什么意义呢？如此宽阔的街道有时已经接近于广场了。常常有政治和娱乐性的活动，《太平广记》中的歌舞竞赛和"甘露之变"后的流血示众都在此举行——

0 2500 5000 10000m

Fake White-Collar

越境偷渡 **Escaping Borderline**[32]

特区 **SEZ**[35 >>> 55]

(Special Economic Zone)

一村一品 Br
named after the produc

Shenzhen Becoming A Global Enterprise", a diagram showing how the hinterland "in Shenzhen, including the 1st frontier and 2nd borders, work as a mechanism behind Shenzhen's urbanization." (Diagram Kyle H. Olsen 2007)

The socially designed nature of Shenzhen is an expression of the combination of control and openness. It is a city located between two "gaps" — the first from between SEZ and Hong Kong and the "2nd front" between SEZ and the interior. The gates of the first front were set up by Mao between 1950s and by Deng Xiaoping in the 1980s. In the early days, The first lines are set up to ensure China's separation from the outside world, the second they form are set up to allow for a relative opening of the first lines as the outside world. Shenzhen becomes a threshold of space. The inner economic gap between the interior and Hong Kong, as well as a preferential policy on business duties that SEZ enjoyed, that once, existed into a big and corrupt town city extending along the boundary between Shenzhen and Hong Kong as commercial premises between the two lines. (Reference Duan: A history of Growth & Out of Central, BANG Lei, 2014)

Labor Insurance Mask Underwear

计生 **One Child Policy**[33]

超生 **Chaosheng**[34] (Illegal Child beyond the Plan)

农民工 **Migrant Workers**[36 >>> 16]

中国制

The story of "Taishan suits" become and the textile industry etc. comparative advantage of Chinese textile industry with the sewing machine communication in 1980s is the rest than the "manufacturing era" and Chinese industrial period cycle in Chinese elaboration of the Factory

DIY Sanitary Pad

URBAN CHINA / INFORMAL CHINA
城市中国 / 自发中国

Curators: Jiang Jun, Su Yunsheng
策展人：姜珺、苏运升

State and Society from the Perspective of Space - On "Urban China / Informal China"

Jiang Jun

China, as well as its governance, has become one of the most important issues of our age. Examining the evolution of the relationship between state and society in China is the key to understanding these issues. For some of its historical clues and spatial representations, I used to briefly outline in my essay Informal China: A History of Control and Out-of-control (2006), which included a series of cyclically reappearing events in China's history and examined the significance of the basic relationships that are still significant in understanding today's China, such as the relationships of central plain—frontier, central—local, centralization—market, planning—market, etc. Some of the chapters have been edited into special topics of Urban China magazine when I was the chief-editor. Urban China / Informal China is basically the extension of the thinking process. As a piece of exhibited work, it focuses more on spatial representation—the spatial forms such as city, village, architecture, objects, graphics and text—the cause-effect product of the state-society interaction, in which space could be not only the consequence of the interaction, but also the impetus for their further interaction. For exactly this reason, space possesses its contradictory attributes being both incompetent and omnipotent. The interpretation of space hence seems both obscure and popular. Despite the fact that this work cannot wholly or deeply express the variants and the invariants of China's spatial patterns in a macro—history, it can at least, within a limited length, demonstrate for the audience a sort of spatial perspective to untangle the complexities within China's history.

There was a long wall in the exhibition. The exhibited works were therefore designed as a sheet of wallpaper, which is close to a spread-out historical reel. The chronological division roughly covered China's historical periods including feudalistic time, semi-colonial time, wartime, and post war period, in which the post-war period was the emphasis. China's urbanization, as well as most of the massive changes with urbanization as the carrier, had taken place during this period. The "urban revolution" during the market economy reform after 1980s was, to a certain extent, the continuation in the form of space of the social and ideological revolutions of the previous 30 years. The format of the work, on the other hand, is close to those of "tabloids", "clippings", or the "big-character posters" presented in the form of "dictionary". Most of the contents originated from the relevant topics of Urban China magazines, such as immigration, Chinatown, Shenzhen, made in China, etc., with others originated from my previous discourses on these topics. Most of the sections were preceded by keywords, such as "People's Commune", "Special Zone", or "Urban Village", etc. The method of word-coinage was sometimes more explanatory than the actual meanings of the words themselves here. There are also a few sections headed by two related keywords, such as "Great Leap Forward" — "Small Furnace", or "One Child Policy" — "Chaosheng (illegal child beyond the plan)". The contrast and interweaving of the two words reflects the interaction and tension between state and society. In term of the colors: red and blue represent the two driving models of space production—the top-down model dominated by the state, and the bottom-up model invigorated by the society. In this way we can roughly tell from the wallpaper the extent of state-domination during the first 30 years of the post-war era, and the rise of social forces during the latter 30 years.

As for the spatial configuration of the contents, the wallpaper has also roughly displayed from top to bottom a series of dimensions from large to small. The Tense City series focus on the coexistence and density of diversified tenses within the process of rapid urbanization, such as "new city" (progressive tense), "demolished cities" (negative progressive tense), "abandoned cities" (future perfect tense), etc. The Dirtitecture series focus on the flexible uses of architecture in the irresistible transitions of context or mutations of content, such as the individualized communism mansion or the stranded boathouses. The Social Products series focus on the nameless designs that are produced spontaneously in the social contradictions, such as the underwear made of labor-insurance gloves or working-class-cloth during the shortage economy era, or the cheap plastic "leather shoes" wore by migrant workers when they have to work in the city. The Urban Images & Texts series focused on the silent confrontations among diversified discourse subjects with the tools of slogans, graffiti, or other public writings in specific urban spaces, such as the confrontation between the official "civilized demolition" and the individual "return my homeland" in the urban demolition sites. This spatial series have basically covered the categories of majors under the tradition of Bauhaus education: urbanism, architecture, products and graphics, but the emphasis has been shifted from the internal ontologies to the external driving forces. In my personal educational practices I refer the latter as the "spatial sociology". However, sociology is just one of the aspects. A complete external spectrum should at least include geography, economy, society, politics, culture, etc. In the interweaving of interdisciplinary theories, today's education is still far from enough.

Basically, "Informal China" reflects the instinctive repercussion of a developing society against the absence of state power or governmental functions, such as the characteristic "fake certificate" services in the absence of a legal environment, the private lending system in the absence of public services orient to medium & small-sized enterprises, the illegal extension in the absence of urban planning and social management, the informal social housing role of urban villages to the vulnerable groups in the absence of welfare policies. It's important to note that it's not only the savagery of social economics, but also the vitality of local government, that have been demonstrated by the instinctive repercussion. Both have integrated the dual forces of rent-seeking and autonomous creation from bottom-up. On the one hand, the state-dominated marketization within the framework of plan economy is a process of stimulating the vitality of local society; on the other hand, it's also a process of guiding the positive energy from the chaos. In terms of the central-local relationship, "the centralized China" will gradually change into "a special-zoned China" with regional functional clusters as sub-centers, so as to potentially maximize the vitality of the local governments within the framework of a centralized country. In terms of the state-society relationship, the state will promote the formation of civil society through the regulation and improvement of legal mechanisms, social welfare, criteria of public services, so as to maximize the vitality of the society. From this perspective, the social creations emerge from the historical reel of "Informal China" will also test the wisdom of the governance of "Urban China".

SHENZHEN AND CHINA / 深圳和中国　　　　URBAN CHINA / INFORMAL CHINA / 城市中国 / 自发中

空间视野中的国家与社会
——有关"城市中国 / 自发中国"

姜珺 / 著

中国及其治理方式已经成为我们这个时代的重要课题，而读解国家与社会在中国的关系演变，则是深入这一课题的关键。有关其中一些历史线索及其空间呈现，我曾在《自发中国：一部控制与失控的历史》（2006）一文中作过点到即止的勾勒，包括一系列在中国历史上反复重现、对理解今日中国之走向依然有镜鉴意义的基本关系，如中原与边疆、中央与地方、集权和分权、计划与市场等，其中一些局部章节后来还成了我担任《城市中国》主编时的专题。"城市中国 / 自发中国"基本上是这一思路的延续。作为展品，它更侧重于空间呈现，即包括城乡、建筑、物品和图文等尺度在内的空间形态，在国家与社会互动作用下的因果产物。空间一方面是二者互动的结果，另一方面又可能成为二者进一步互动的原因。正因为此，空间才具有了无能与万能的矛盾属性，对空间的读解才显得既深不可测，又通俗易懂。尽管这个展品不可能完整或深入地呈现中国空间形态在大历史中的变与不变，却至少可能在有限的篇幅下，为观众呈现一种在中国历史中梳理杂陈的空间视野。

展览提供了一面长墙，展品因此被设计成一面墙纸，其内容接近一幅展开的历史卷轴，断代上大致涵盖了中国帝制、半殖民、战时、战后等几个时期，重点在于战后，中国的城市化以及以城市化为载体的各种剧变都发生在这一时期，尤其是 80 年代后的市场经济改革时期，这一时期的"城市革命"某种程度上是前 30 年社会和思想革命在空间中的延续。形式接近"文摘"、"剪报"，或以"词典"形式呈现的"大字报"，大部分源自《城市中国》杂志的一些相关专题，如移民、唐人街、深圳、中国制造等，另一部分则源自我之前对这些专题的一些论述。大部分版块都被冠以一个关键词，如"人民公社"、"特区"或"城中村"等，这些关键词的造词法有时比它们的词义本身更有说明性；有些版块则被冠以两个相对的关键词，如"大跃进"和"小高炉"，或"计生"与"超生"等，二者的反差和交织反映了国家与社会的互动和张力。红色和蓝色分别代表空间生产的两种驱动模式——自上而下的国家主导和自下而上的社会能动，因而在墙纸版面上我们大致能够直观地看到战后前 30 年国家主导的程度以及后 30 年社会力量的兴起。

在内容的空间形态上，墙纸从上至下也大致呈现了一个从大到小的尺度系列。"时态城市"系列着重于呈现剧烈城市化进程中多种时态的共存及其密度，如"新城"（进行时）、"拆城"（负进行时）、烂尾城（未来完成时）等；"脏筑"系列着重呈现建筑在其难以抵御的语境变迁及内容变化中产生的弹性使用，如个体化的共产主义大厦[1]或搁浅登陆的疍民船屋等；"社会产品"系列着重呈现在一定社会矛盾中自发产生的无名设计，如短缺经济时期由劳保手套或工装布改造而成的内衣，或农民工进城所穿的廉价塑料"皮鞋"等；"城市图文"系列着重呈现不同话语主体以标语或涂鸦等公共书写方式在特定城市空间中的无声对抗，如拆迁现场官方"文明拆迁"与民间"还我家园"的对抗。这一空间序列基本上涵盖了包豪斯传统下对设计教学的专业分类：城市、建筑、产品和图形，但重点更多地从内在的本体转向了外在的动力。我在自己的教学实践中将后者称为"空间社会学"，但事实上社会学只是一方面，一个完整的外部谱系应该至少包括地理、经济、社会、政治和文化等方面。在跨学科理论的融合方面，我们今天的教学做得还远远不够。

基本上，"自发中国"反映了一个发展中社会在国家力量或政府职能缺失下的本能反弹，如在法律规范缺失下出现的特色办证服务，在面向中小企业的公共服务缺失下出现的民间借贷，在规划管理缺失下出现的违章乱建，在福利政策缺失下城中村对弱势群体起到的廉租房供给作用。需要说明的是，这种本能反弹不仅表现为社会经济的野性，也表现为地方政府的活力，双方都自下而上地结合了自发寻租和自主创新的双重力量。国家在计划框架下进行的市场化，一方面是对地方社会活力的激发过程，另一方面也是对这种活力中的正面能量加以疏导的过程。从中央 - 地方关系上看，"集权中国"将逐步转变为一个以区域功能组团为次中心的"特区中国"，从而可能在大一统国家的框架下实现地方政府的活力最大化；从国家 - 社会关系上看，国家将通过法律制度、社会福利、公共服务的规范和完善促成公民社会的形成，从而实现社会活力的最大化。从这一角度看来，在"自发中国"历史卷轴中呈现的民间创造也将考量着"城市中国"的治理智慧。

1　1960 年起为集体设计的城市人民公社。编注。

文明 Civilization[1]
蛮夷 Barbarism[2]

作为史上最大的农耕文明之一，中国曾有着自我中心的天下观，视其中心边缘为蛮夷[2]。而正是这些蛮夷[2]的相互征服产生了另一种类型：绿洲文明[1]——一种以水源为节点嵌入戈壁沙漠的城市网。沙漠商人们因其在这一终极迷宫中的生存能力，而成为丝路[3]贸易的特许经营者。[《新丝绸之路：旧大陆的宿命》，姜珺，2010]

As one of the biggest agriculture civilizations[1] in the history, China used to have a self-centric conception of the world – a civilized center surrounded by barbarism[2]. However, it was the conquest of barbarism[2] that had generated another typology – the oasis civilization[1], which is a city network imbedded inside the gobi with water sources as nodes. The exclusive franchise of the trading network was hereby granted by Allah to the desert merchants due to their unique ability in surviving in the ultimate labyrinth. [New Silk Road: The Fate of the Old Continent, JIANG Jun, 2010]

丝绸之路[3]是古代国际贸易体系的纽带，贯穿起欧亚大陆的一系列贸易城市
The Silk Road[3] is a route across the Eurasian continent along which the ancient international trade system and a series of trading cities were generated

丝路 Silk Road[3]

丝绸之路[3]因其对欧亚大陆各大天堑的渗透而知名。它如此漫长，以至于可以轻易被地方势力割据，控制丝路[3]者也就控制了贸易的利益。丝绸之路[3]的各个片段为历史上的不同帝国交替控制，这些帝国的此消彼长决定着欧亚大陆上的地缘政治变迁，以及丝路[3]城市的竞争与浮沉。[《新丝绸之路：旧大陆的宿命》，姜珺，2010]

The Silk Road[3] was famous for its penetration over the natural blockade of Eurasian continent. It was so long that it could be easily disconnected by local powers - those who controlled the Silk Road[3] would control the profit of the trade. The history of the Silk Road[3] was one that different segments of it had been alternately conquered by different empires, which had defined different geopolitics on the Eurasian continent, and thus the rise and fall and revival of its trading cities. [New Silk Road: The Fate of the Old Continent, JIANG Jun, 2010]

《山海经》中描绘的以中国为中心的古地图
The ancient atlas shown in A Chinese Bestiary (Classics of Mountains and Seas) with China as the center

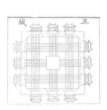 

左：宋人聂崇义根据《周礼·考工记》原则绘制的"王城图"，描绘出在帝制中国的王权礼制下城市的理想格局 / 右：按照里坊制规划而成唐长安城，当时是世界上最大的国际大都会
Left: "Map of the Imperial Capital of Zhou Dynasty" from the Illustrations on the Rites of Zhou, the book of Etiquette & Ceremonies, and Book of Rites, by NIE Chongyi; it is the model of the ideal city established by early Chinese society / Right: Chang'an in the Tang Dynasty. Planned according to and controlled by the Lifang street unit system, Chang'an in the Tang Dynasty was the largest international metropolis in the world at that time

宋朝取消了自发活动在时间和地点上的限制，街道的解禁和"里坊制"的突破令它的都城很快出现了如《清明上河图》中描绘的繁荣景象：市场从"坊市"走向"街市[5]"，时间上从"昼市"延伸到"夜市"，沿街
The Song Dynasty lifted restrictions on these informal activities in terms of time and place. The lifting of curfew in the streets and the disintegration of the Lifang system enabled the Song Dynasty's capital to develop thriving and prosperous scenes rapidly, as described in the Qingming Shanghetu (the "Upper River during Qingming Festival" scroll painting): the ma commercial center in the Song Dynasty. [Informal China: A History of Control & Out-of-control, JIANG Jun, 2006]

宵禁 Curfew[4]
街市 Free Market[5] >>> 7, 35

唐朝的统治者对自己的都城实行严格的"里坊制"，尽管唐长安城作为当时世界上最大、最持久的国际贸易"丝绸之路[3]"的起点，有着无比繁华的都市生活，但这些包括市场和红灯区在内的自发性内容，全部被分组布置在棋盘格状的指定空间中，同时也通过在入夜后实行宵禁的方式，强行指定了这些活动发生的时间。沿街立面不许开窗，街道在白天仅仅用于交通，入夜宵禁之后则连行走都会被判刑。

The rulers of the Tang Dynasty implemented the strict "Lifang (Alley and Lane) street unit system" in their capital. Although Chang'an was the starting point of the "Silk Road[3]" the world's largest and longest-lasting international trade route at that time, and enjoyed an extremely flourishing city life, the informal contents including the market and the "red light district" were all divided into groups and then assigned into specific checkerboard-like spaces. Moreover, through the enforcement of curfew[4] at night, the time for these activities to take place was also fixed in a strict manner. It was forbidden to open windows on the façade facing the street; the streets were only used for traffic during the daytime; at nightfall, when the curfew[4] was called, even walking in the streets would be punished.

宋朝政府的相对虚弱则使街道的管制主要诉诸于基层的邻里组织和保甲制度，通过实施上的权力下放和基层自治维持街道秩序。这种自治为多样性的社会阶层在街道公共空间中的鱼龙混杂和平等共生提供了基础，并形成了活跃的街头生活和街道文化。[《自发中国：一部控制与失控的历史》，姜珺，2006]

The relatively weak management of the street by the Song government mainly relied on basic-level neighborhood organizations and the "Baojia system" (translator' note: a system organized on the basis of households and used for guaranteeing local security), by maintaining order in the street through substantial decentralization and neighborhood self-government. This self-government provided the basis for the peaceful mixing and coexistence of different social classes in the public space of the street; it also helped develop lively street life and street culture. [Informal China: A History of Control & Out-of-Control, JIANG Jun, 2006]

城市中国 Urban China
自发中国 Informal China
目录 Contents

文明 Civilization[1]
蛮夷 Barbarism[2]
丝路 Silk Road[3]
宵禁 Curfew[4]
街市 Free Market[5]
围城之国 Walled-State[6]
门户城市 Gate-City[7]
宅院 Courtyard House[8]
园林 Private Garden[9]
土楼 Tulou (Clay House)[10]
华人大流散 Chinese Diaspora[11]
唐人街 China Town[12]
租界 Foreign Concession[13]
江湖 Jianghu (The Grassrooted Society)[14]
战时中国 Wartime City[15]
移民中国 Migrating China[16]
碉楼 Fortress Villa[17]
人民空间 People's Space[18]
核心家庭 Nuclear Families[19]
城乡二元 Urban-Rural Dual System[20]
单位大院 Danwei Dayuan[21]
大跃进 Great Leap Forward[22]
小高炉 Small Furnace[23]
人民公社 People's Commune[24]
三线城市 Third Frontier Cities[25]

下乡 Go to the Countryside[26]
计划经济 Planned Economy[27]
短缺经济 Shortage Economy[28]
消失的物品 Disappearing Objects[29]
越境偷渡 Escaping Borderline[30]
计生 One Child Policy[31]
超生 Chaosheng[32] (Illegal Child beyond the Plan)
特区 SEZ[33]
农民工 Migrant Workers[34]
村城 Rurbanization[35]
中国制造 Made in China[36]
一村一品 Branded Villages & Towns[37]
城中村 Urban Village[38]
打工杂志 Epidemic Mags[39]
艺术工业村 Fordism Art Village[40]
烂尾城世 Rotten-Tailed City[41]
时态城市 Tense City[42]
政商面相 Governmental Physiognomy[43]
文物转世 Reincarnated Context[44]
脏筑 Dirtitecture[45]
傍牌货 Copylefted Products[46]
房改 Housing System Reform[47]
私分 Privatized Communism Mansion[48]
中国回收 Recycled in China[49]
新农村 Socialist New Village[50]
新民居 New Village House[51]
冥品 Afterlife Fantasies[52]
唐人 Urban China[53]
潜意识发明 Subliminal Inventions[54]

策展编辑：姜珺+苏运升 / 团队：下划线工作室+上海易托邦建设发展公司
部分内容发表于《城市中国》杂志Vol. 1-40，本海报为统一主题下的再编排
所有图片未经特别标明作者或出处者为下划线工作室提供
Curating-Editors: Jiang Jun + Su Yunsheng / Team: Underline Office + eTopia
Part of the contents have been published in Urban China Magazine Vol.1-40, the wallpaper is a remix under a unified theme
All photos without special marked authors or sources are provided by Underline Office

四川西昌古城的南门瓮城，被商贩们自发占据
The urn-city at the south gate of Xichang City, Sichuan, is informally occupied by venders

1900年的《粤东省城图》，描绘了城墙中的正式城市和城墙外西关洪水般蔓延开来的自发城市
Map of the Provincial City of Canton (Guangzhou) in 1900, showing the formal city inside the wall and the informal city flooding beyond the wall outside the west gate

围城之国 Walled-State [6>>>8]
门户城市 Gate-City [7]

中国在其文明的大部分历史中垄断着黄河和长江，并在数千年中以之支持着其发达的农业。这种地理大一统在由海洋、沙漠和高原等天堑围合下的相对孤立状态中得以保存；作为人造天堑的长城[1]，则标示着农耕文明[1]在大陆中所能发展到的极限，并将中国同世界其它部分隔离开来。这种"围城之国[6]"如同一个宏观尺度的门禁社区，使得中国得以维持一种自给自足的经济和相对稳定的社会。[《新丝绸之路：旧大陆的宿命》，姜珺，2010]

Yellow River and Yangtze River, 2 of the top 10 great rivers in the world, have been monopolized by China in most of its history, and have supported China's agriculture for thousands of years. This geographical unification was preserved in a relatively isolated condition surrounded by natural blockades such as sea, desert and plateau. The Great Wall[1], as a man-made barrier, indicated the geographical extreme the agriculture civilization could develop to, and secluded China from the rest of the world. Within this "Walled State[6]," or a macro-scale gated community, China was able to sustain a self-sufficient economy and a relatively stable society. [New Silk Road: The Fate of the Old Continent, JIANG Jun, 2010]

十八世纪中叶之后，广州成为独享这一特惠政策的"特区[35]"。与后来邓小平设立的特区[35]相比，清廷的"广州特区[35]"多少出于一种被动的接受，而不是像后者那样主动的选择并带有实验性。尽管从全球贸易的角度看来，"一口通商"是一种束缚性政策，然而它却从客观上将广州造就成为一个炙手可热的国际贸易都市，以及一批富可敌国的封建买办商人；这些持有帝国特许经营执照的"官商"，在广州城西靠近珠江的"十三行"与来自世界各地的外商交易；他们的聚居和伴生的商业网络催发了一个关键的空前膨胀——西关[7]；这个因商而生的城市如潮水般从广州城的西翼门蔓延出去，它在机理上表现出的有机与活力与城墙[6]内那个正统方整的广州城形成鲜明的对比。无疑，这种自由贸易[9]尽管受到了来自中央的严格监管，但依然足以产生出一个自发城市。[《自发中国：一部控制与失控的历史》，姜珺，2006]

After the mid-eighteenth century, Guangzhou became the only "special zone[35]" to enjoy this preferential policy. In comparison with the special zones[35] set up by Deng Xiaoping in later times, the Qing government created the "Guangzhou Special Zone[35]" more or less out of passive acceptance, instead of Deng Xiaoping's active choice with experimental flair. From a perspective of world trade, "integrated customs clearance in one single port" was a constrained policy; however, it helped to build Guangzhou into a popular international city of trade, and to create a group of extremely wealthy feudal broker-merchants. These "official merchants", who held trading licenses issued by the government, traded with foreign merchants from all over the world in the "Shisanhang trading houses" in the western part of Guangzhou near the Pearl River. Their concentration in one area and the commercial network that developed in the wake of it triggered the unprecedented development of a gate-city: Xiguan(Western Gate). This city was created because of commerce, and quickly spread out from Guangzhou's Western Gate. Its organic and vital character was in sharp contrast to the classic and orderly town of Guangzhou located within the walls[6]. Undoubtedly, this "free" trade[9] was still under the central government's strict surveillance and control, but it enabled a city to emerge in an informal and spontaneous manner. [Informal China: A History of Control & Out-of-Control, JIANG Jun, 2006]

业界面，住宅区和商业区混为一体，城市的主要功能从唐朝的行政中心转型为宋朝的工商业中心。

its working hours were extended from the "daytime" to the "evening"; the facades on the street side were turned into attractive and showy commercial interfaces; residential and commercial districts blended into each other; the principal function of the city was also transformed from being an administrative center in the Tang Dynasty to being an industrial and

左：潮州的宅院[8]
右：苏州狮子林园林[9]中迷宫般的假山
Left: courtyard houses[8] in Chaozhou, Guangdong;
right: the labyrinth-like stone arts in Lion Forest Garden[9] in Suzhou, Jiangsu

土楼[10]是由客家移民[16]在中国东南山区为宗族村落兴建的碉楼[17]式的集体宅院[8]
Tulou, or Clay House[10], is a collective courtyard[8] house built by the Hakka immigrants in Southeast China for an entire clan, or a village, in the form of fortress

宅院 Courtyard House [8]
园林 Private Garden [9]
土楼 Tulou (Clay House) [10]

古代中国的"家国同构"关系被平行地反映在从国家到家庭、从城市到乡村的空间同构之中。它们几乎都可无一例外地被描述为"一组被墙围合起来的建筑群"：被长城围合的国家，被城墙围合的城市，被院墙围合的家庭。……这种"墙套墙"的同心圆模式之中，有关"天人关系"的风水理论和"人际关系"的礼法制度，被精确地映射到从城市规划到建筑营造、从方位尺度到形制色彩的日常空间之中。

The structural similarities between "nation" and "home" in the ancient China is reflected by the spatial similarities between nation, family, city and the countryside; almost all of them can be described as "a walled complex"—a nation enclosed by the Great Wall, the cities enclosed by the City's Wall, and the families enclosed by the courtyard wall. Operating under such a concentric circle-like "wall-within-wall" model, geomancy (which deals with the relationship between man and heaven) and ancient laws & order (which deals with the relationship between men) were accurately projected into all kinds of daily spaces. This is perceivable in urban planning and architecture, from the buildings' location and scale to their shape, type and color.

合院住宅以"宅[8]"合"院[8]"，"宅[8]"与"院[8]"有着虚实互补的辩证关系："宅[8]"可以是实的"院[8]"，"院[8]"可以是虚的"宅[8]"。"院[8]"不仅可以接纳天地之气，也为"宅[8]"提供了公共空间。而这种公共空间也因其所处的内外方位差异而有着全然不同的用途：道貌岸然的前庭被家长用于展现他在内政中的权威和外交中的地位，而隐秘的后院[8]则通常被开发成为风月无边的后花园[9]，这里，内外院落[8]展现了儒与道这两种本土价值观在空间上的辩证关系。[《中国住宅：集体价值的稀释与再造》，姜珺，2006]

In courtyard housing[8], the enclosed courtyard[8] and the enclosing houses maintain a complementary and dialectical relationship: the houses can be considered the indoor version of the courtyard[8], while the courtyard[8] can be seen as a form of open house. Taking in the essence of heaven and earth, the courtyard[8] provides the public space for the whole residence. The front and back yard, however, function differently: prim and proper, the former is used by the patriarch as a symbol of his superiority in the internal affairs and his status in external ones. The private back yard, on the other hand, is usually developed into a "rear garden[9]" for sexual activities. Here, the position of the two courtyards[8] exemplifies the dialectical spatial relationship between two local values: Confucianism and Taoism. Not only did they provide a way of shuttling between the courtyard houses[8] and the daily life, they also helped developing a dual personality of self-control and indulgence for the people living in the courtyard houses[8]. [China Housing: the Dilution and Reformation of Collectiveness, JIANG Jun, 2008]

租界 Foreign Concession[13]
江湖 Jianghu[14] (The Grassrooted Society)

对于上海租界[13]而言,这种结合不大可能像后来的浦东开发区那样通过政府渠道,在半封建半殖民地时代,它们本来就有着水火不容的对立性;租界[13]只能求助于本土的草根势力。随着二十世纪初全球性乱世的来临,租界[13]之外的世界变得日益混乱和充满仇恨,而租界[13]的相对稳定反而使之成为乱世之中的避难所,无政府沙漠中的绿洲;而这支被租界[13]扶持和纵容的草根势力,通过它在租界[13]、政府和底层社会之间的斡旋和中介作用,自发地结合成为势力庞大、盘根错节的黑帮;他们以租界[13]为靠山,以金钱、色相和暴力为手段,通过天罗地网的眼线和打手,控制了从每一家妓院、赌场、烟馆到每一条里弄的整个城市。政府的法律被地下势力的潜规则所渗透,它的决策可以被黑帮的意志所左右;到了上世纪20-30年代,这支源自社会底层的无形力量,已经成功地渗透到社会顶层的政界、军界和商界之中。[《自发中国:一部控制与失控的历史》,姜珺,2006]

For the foreign concessions[13] in Shanghai, it was unlikely for this combination to be realized through official governmental channels, as was true for the Pudong Developmental Zone later. In the half-feudal, half-colonial era, the Chinese government and the foreign concessions[13] held, by nature, irreconcilably opposed positions; therefore, the foreign concessions[13] could only turn to local grassroot forces. As conflicts and wars started to spread throughout the world in the early twentieth century, the world outside the foreign concessions[13] became increasingly chaotic and full of animosity; on the other hand, the relative stability in the foreign concessions[13] turned them into refuges in turbulent times, oasis in a desert of anarchy. Through its intervening and mediating role among foreign concessions[13], the Chinese government, and the bottom of society, these grassroot Jianghu[14], supported and indulged by foreign concessions[13], linked up to form a powerful and complex gang organization in an informal and spontaneous manner. [Informal China: A History of Control & Out-of-Control, JIANG Jun, 2006]

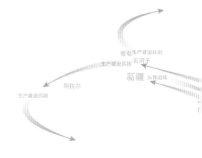

《拼贴全家福》,纽约美国华人博物馆收藏
Collaged Picture of a Family, Collected by Museum of Chinese in America (MOCA), NY

二战中被日军轰炸的上海
Shanghai being bombarded by Japanese in the WWII

华人大流散 Chinese Diaspora[11>>>16] 战时城市 Wartime City[15]

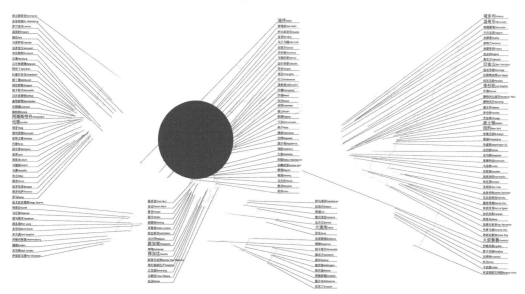

唐人街[12]世界地图,图解/下划线工作室,2007
Atlas of Chinatowns[12], diagram by Underline Office in 2007

"天堂银行"发行的纸币,欧洲唐人街的华人以欧元为母本设计成的冥品[54],用于在逝者的坟前焚烧悼念

Heaven Bank Note, an afterlife fantasy[34] object designed in the European Chinatowns[12] in the form of Eurocheque, is supposed to be burnt when commemorating the dead ancestors in front of their graves

唐人街 Chinatown[12>>>16]

是一块具有独立意志的"飞地",它将前现代中国的组织方式移植到异国他乡的土壤之中。当地社会的排斥不仅难以削弱它的能量,反而使之更为凝聚和强大,同时也更为封闭和排外。[《华人大流散中的隐蔽版图》,姜珺,2007]

is an enclave with independent will that transplants the organizational system of the pre-modern China into foreign outposts. The segregation of Chinatown[12] from the local community does not act to weaken the power of Chinatown[12] but instead increases its internal strength and cohesion, exclusivity, and "otherness". [Hidden Atlas in the Chinese Diaspora, JIANG Jun, 2007]

坐落于北京城心脏地带的天安门广场，是中国国家集体主义的心理中心[摄影/曾力]
Tian An Men Square, located at the heart of Beijing City, is the psychological center of China's national collectivism [Photo/by ZENG Li]

来自全国各地的红卫兵在"大串联"运动中齐聚天安门广场 [摄影/翁乃强]
Red Guard Soldiers from all over the country meeting together in front of Tian An Men in the collective movement of *Nationwide Travel* [Photo/WENG Naiqiang]

人民空间 People's Space [18 >>> 24]

在前现代中国的城市中，作坊、市场和会馆等地方集体空间只是局限在衙门、官道和城墙等国家集体空间的框架内；而在开埠之后的口岸城市中，它们在自由市场环境刺激下表现出对国家集体空间的突破性质。城市广场、公园、体育场、电影院等公共场所，车站、码头等交通枢纽，以及学校、医院等社会机构作为口岸城市的新型社会集体空间也同时发展起来，其中一些集会场所成为了后来民主革命的主战场；新中国建立后，它们又逐渐被新型的国家集体空间所取代，并全部以"人民"冠名——人民广场、人民公园、人民大道、人民公社[24]……国家集体主义以群众运动的方式将集体性渗透到从城市到农村的每一个个人空间，集体单位、集体学校、集体宿舍、集体农场……国家集体主义造就了无限责任的全能政府，而社会力量在国家事务中的长期缺场则导致了社会机能的萎缩。[《集体空间》，姜珺，2008]

In the pre-modern Chinese cities, the Local Collective Spaces such as workshops, markets and clan associations were just conformed within the framework of the State Collective Spaces such as governmental offices, official avenues and city walls[6]. Their nature of breaking through the State Collective Spaces had been amplified under the stimulation of the free market environment in the port cities after they were forced open during the colonial time. The public areas (city plazas, parks, stadiums, cinemas), transport intersections (stations, ports) and social institutes (schools, hospitals) were developed as new types of Social Collective Spaces, in which some meeting places became the main battleground of the Democratic Revolution later. When Communist China was founded, they were gradually replaced by new types of State Collective Spaces and entitled with People – People's Square, People's Park, People's Avenue[24], People's Commune[24]… by means of mass movement, the State Collectivism infiltrated into every individual space from the city to the countryside[24]: Collective Unit, Collective School, Collective Dormitory, and Collective Farm… The State Collectivism had made a mighty government with infinite responsibilities, while the absence of social power in the national affairs had resulted in the shrinkage of social functions. [*Collective Space*, JIANG Jun, 2008]

中国国内移民[16]地图
Chinese Domestic Migration[16] Map

移民中国 Migrating China [16]

在一系列行政命令、经济开发、战乱流亡的驱使下，人口的转移不仅伴随着资本、劳动力和生产的重组，同时也带来了语言、行为和文化上的流动和融合，从而直接或间接地强化了区域间的关联与合作，成为中国现代化背后的动力。移民本身，即是以人为载体的流动的现代性。[《中国移民地图》，《城市中国》Vol.16, 2007]

Driven by a series of political orders, economic development and war refuge, demographic shifting not only goes with reorganizing capital, labour force and production, but also induces the flowing and merging of language, behavior and culture, which consequently enhance the regional association and collaboration directly or indirectly. This has been a major force to push Chinese modernism. Migration[16] itself is a fluid modernism with human being as carrier. [*Chinese Migration Map*, Urban China Vol.16, 2007]

碉楼[17]的建筑语法即一个理想的洋屋顶和一个实用性楼体的相加
The architectural grammar of Fortress Villa[17] is basically an addition of an idealistic western top and a pragmatic defensive body

一栋独自矗立在广东开平乡间的碉楼[17]
A fortress villa[17] constructed isolatedly in the countryside of Kaiping, Guangdong

碉楼 Fortress Villa [17]

上世纪初，广东的四邑地区陆续出现由海外返乡的侨民[16]兴建的碉楼[17]。一方面，他们以西洋建筑为范本装饰住宅顶部，以彰显其从海外衣锦还乡的荣耀并获取尊重；另一方面，这些住宅被设计成碉楼[17]以防范客家和匪帮的入侵。碉楼[17]因而成为西洋屋顶与地方躯体的混血产物。[《城市回放》，《城市中国》No.3, 2005]

At the beginning of last century, Fortress Villa[17] were constructed by homecoming workers and emigrants[16] in the area of Kaiping, Guangdong. On the one hand, they collaged the decoration of western architecture on the top of the residential house to show glory of "returning from overseas" and to gain respect; on the other hand, the houses were designed into Fortress Villas[17] to resist invasion of Hakka and bandits. Thus Fortress Villas[17] became the hybrid offspring of western top and local body. [*Urban Review*, Urban China, Vol.3, 2005]

北京朝阳区金台里八号楼,在苏联影响下建于社会主义时期的单位大院[21][摄影/王迪]
Dayuan[2], or Big Courtyard[3], built in socialism time under the influence of Soviet Union. No.8 Building, Jintaili, Chaoyang District, Beijing [PHOTO/WANG DI]

单位大院 Danwei Dayuan [21>>>49]

"大院"是单位制度框架下产生的社区模式。由于各单位只对所属部门负责,围墙的功能在权力的再度垂直化中复活,用以圈围和管理一个个确定的基层社会单元;而在规模上,用于容纳"社会主义大家庭"的大院事实上已远远超越了家族概念,而成为了单位进行封闭式管理的社区。院落的集体主义价值在社会主义语境下被空前地放大。它被普及到从城市到农村的各级党政军机关,从一线城市到三线城市的各级企事业和厂矿单位,以及农村中的公社和大队。大院的"就地就业"模式,使得住宅被约简成单身个人或核心家庭的单位宿舍;而大院作为一个熟人社区,也使得这些同事们可以将更多的业余时间投入在"麻雀虽小、五脏俱全"的集体生活之中。[《中国住宅:集体价值的稀释与再造》,姜珺,2008]

The concept of compound (dayuan[3]) was a invented community model under the framework of danwei[2] system. Because each danwei[2] only reported to the department belongs to, the wall[4] had reclaimed its function of enclosing and managing the basic social units as power started to trickle down vertically again. As a container of the so-called "socialist big family[5]", the compound went far beyond the traditional concept of family; it was actually a community which had made possible the enclosed management style of danwei[2]. In the socialist context, the collectivist value of the old courtyard[6] was unprecedentedly maximized into ubiquity: its pervasiveness ran across all levels governmental and military offices, business and public entities, factories, communes[7] and production teams in the countryside[8]. The compound model encouraged—if no required—people to work in the area adjacent to their residence, therefore reducing housing into the dormitory of individual or family. As a community of closely-related people, the compound provided its residents a small-but-comprehensive collective way of life. As it turned out, a compound can be as sophisticated as a micro-city, with kinds of self-sufficient amenities and public spaces such as school, hospital and theater. [China Housing: the Dilution and Reformation of Collectiveness, JIANG Jun, 2008]

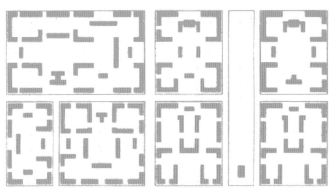

长春一汽厂某工人居住区平面图,在街区网络中由工人宿舍围合而成的大院[20]群
Plan of the residential community of FAW Motor Manufacture in Changchun, a cluster of Big Courtyards[2] enclosed by the workers' dormitories in the grid of street

核心家庭 Nuclear Families [19]

中国现代化的过程伴随着生产协作的横向化和社会化;行会、商会和公司等中间组织的崛起,则伴随着中国式"大家族"的萎缩。而在"大家族"的契约、利益和生产关系都交力击下的中国家谱》,姜珺,2007;摄影《家族图谱》/邵逸农+慕辰] The process of horizontalization and socialization of production cooperation coincided with the modernization of China, and while the rise of guilds, chambers of commerce, companies and other intermediate organizations witnessed the more closed and exclusive. [Chinese Family: Pedigree under the Impact of Modernization, JIANG Jun, 2007; Family Pedigree in Portraits, photograph by SHAO Yinong + Mu Chen]

1980年代后,大部分中国农民可以相对自由地在城里工作。然而由于户籍制度限制下形成的城乡二元结构,他们依旧被视为城市空间中的另类群体
After China's reform & open policy in 1980s, most of Chinese peasants were relatively free to move and work in the cities. However, because of the urban-rural dual system[5] confined by the household register system, they were still separated as an alternative group in the urban space

Eggs sold in the big co...

城乡二元 Urban-Rural Dual System [20]

"一五"时期所建立的重工业为主的国家工业体系,内在的具有"资本排斥劳动"的机制,而优先发展重工业也需要由农业未来为工业发展提供稳固充足的原始积累资本,这种经济建设体制使国家不得不将农民固定在农业生产之中,限制农民进入城市。1958年,全国人大常委会通过并公布了《中华人民共和国户口登记条例》,使户口登记有了全国统一完整的法律依据,城乡二元户籍制度由此确立。城乡分离的"二元经济模式"从此根深蒂固,直到改革开放后,户籍制度才逐渐有所松动。[《中国城市化一甲子》,《城市中国》Vol.40, 2010]

The national industrial system founded in the "First Five Year Plan" period was primarily based on heavy industry, with the mechanism of "capital repels the labor" inherently contained, at the same time the development priority on heavy industry also required the agriculture to provide a stable and sufficient primitive accumulation of capital for the industry, therefore such kind of economic development system determined the nation to fix the peasants in agricultural production and limited their entrance into city. In 1958, the Standing Committee of National People's Congress of China[9] passed and published the Regulations on Residence Registration of P.R.C., offering a complete and nation-widely effective legal reference to the household registration, which means the establishment of urban-rural dual system[6] of residence registration. The detached "urban-rural dual economic mode[10]" was deeply rooted since then, and not until the Reform and Opening-up, did the residence system started to get loose gradually. [60 Years of China's Urbanization, Urban China Vol.40, 2010]

1950年代的中国户口簿
Household register booklet in the 1950s

1974年河南辉县的"农业学大寨"运动中,人们被动员起来集体修建农田水利基础设施 [摄影/王世龙]
People working on water irrigation infrastructure in the movement of "Agriculture Learns from Dazhai" in Huixian County, Henan province, 1974 [Photo/WANG Shilong]

位于湖北省西部的三线城市[25]十堰是中国二汽所在的车城。它地处深山的区位为向外部市场输出产品造成了可想而知的障碍
Shiyan, the third frontier city[25] located in the west of Hubei province, is a motor-city for Erqi (the 2nd Motor Manufacturer in China), its location inside mountains brings considerable efforts in transporting its products to the market outside

人民公社 People's Commune [24 >>> 18]   # 三线城市 Third Frontier Cities[25]

"在这个新社会里,设立公共育儿院,公共蒙养院,公共学校,公共图书馆,公共银行,公共农场,公共工厂,公共剧院,公共病院,公园,博物馆等等;以后,把这些一个个的新社会连成一片,国家便可以逐渐地从根本上改造成一个大的理想的新村。" —— 毛泽东,《新村计划书》,1919

"In this new society, there are public nursery, public kindergarten, public school, public library, public bank, public farm, public factory, public theater, public hospital, park, and museum and so on; In the future, all these new societies will be connected into a whole, so gradually the entire country can be transformed fundamentally into a huge ideal new village[18]." —Mao Zedong, "New Village[24] Plan", 1919

三线城市[25]不仅在选址上参考了共产党的战时[15]经验,在造城方式上也沿袭了战时共产主义的传统。大规模资源调配的效率来自于军事化的、高度集中的行政指令体制;自给自足的微缩社会模式则来自于延安时期的"南泥湾精神"。山地、农村、军队、企业和城市在这里有着不可分割的复合性,并且由之产生了一种特殊的城市结构:其中心是一组沿着山谷展开的车间,车间之间的道路曾经是谷底的山涧,现在则变成了连接各工厂的流水线;它同时也作为城市的主干道连往另一端的宿舍区,沿途是工厂城市的公共建筑和社会服务设施:职工俱乐部、子弟学校、职工食堂、厂办医院……这一切被集中于"单位体制[21]"之下统一管理;由于所有的用户是同一流水线上的同事和家属,这些用于强化集体生活的机能在几乎与世隔绝的大山中,显得异常生动和饱满。[《三线的萎缩与再生》,姜珺,2008]

The Third Frontier Cities[25] adopted CPC's wartime[15] experience not only during site selection phase, but also maintained the wartime[15] communist tradition into city construction method. The efficiency of large-scaled resource distribution came from militarized, highly centralized governmental command system; while the self contained micro society mode had its origin in the "Nan Ni Wan Spirit" formed when CPC governed in Yan'an. The mountain, village, army, factory and city were inseparably compound here, gave birth to a unique type of urban structure: the center was a cluster of factory buildings winding along the valley, and the roads between plants used to be streams but now were assembly lines connecting various factories; it also functioned as a main road leading to the residence area on the other side of city, with public architectures and social service facilities of this "factory city" sitting on both sides — staff club, school for children of employees, staff dining hall, hospital run by factory — all of which were under the centralized management of "Danwei System[21]"; since all users were colleagues and family members on the same assembly line, those functions aiming at strengthening the collective lifestyle seemed extremely energetic and flourishing in the almost isolated mountains. [The Third Frontier in Shrinking and Reviving, JIANG Jun, 2008]

油画《地下长城》描绘了1960-70年代地下人防建设的场景 [图/许三多+伍振权]
The oil painting "Underground Great Wall" represented the underground civil defence construction in 1960s-70s [Painting/XU Sanduo + Wu Zhenquan]

为独立生产单位和利益共同体的意义便不复存在。1980年代之后,家庭的自然生育也被纳入计划体制,从而产生了一代家庭人口最小化的"标准家庭"[文《中国家庭:现代化冲

…milies' had submitted the entire contract, interests and relations of production to national planning[ ], they deceased in terms of independent production unit and community of interests. In 1980s it brought natural procreation of individual families into the planned system[ ] and resulted in a whole generation of 'standard family'[ ] with minimized domestic population more and…

《为1959年生产更多更好的钢而奋斗!》五十年代末一幅展现大跃进[22]雄心壮志的宣传画。
Strive to Produce More Qualified Steel in 1959! A propaganda poster showing the industrial ambition of Great Leap Forward[22] in late 1950s

大跃进 Great Leap Forward [22 >>> 55]
小高炉 Small Furnace[23]

在大跃进[22]运动中,农民们用他们在田间自发搭建的小高炉[23]进行"土法炼钢",但因设备低劣,质量难以过关 [摄影/晓庄]
In the Great Leap Forward[22] movement, product of the indigenous blast furnace[23] steel-making method is almost unqualified because of poor equipment condition [Photo/XIAO Zhuang]

以"全民炼钢"的方式实现国家在工业上的"大跃进[22]",天安门前林立的烟囱没有出现,取而代之的是举国上下遍地开花的"小高炉[23]",用分散取代集中,用人力取代机器。《自发中国:一部控制与失控的历史》,姜珺,2006

…le started China's industrial "Great Leap Forward[22]" by urging the whole country to takepart in steel production. In the end, chimneys did not stand in great numbers in front of Tiananmen Square; instead, backyard steel furnaces[23] were set up in every corner of the country - concentration was replaced by dispersion, and machines by manpower. [Informal China: A History of Control & Out-of Control, JIANG Jun, 2006]

1970年代初,知青们在黑龙江省的农村下乡插队,接受贫农的再教育[摄影/李振盛]
In the early 1970s, a group of educated youth went to work and live in the countryside[28] in Heilongjiang province, receiving re-education from the poor peasants [Photo/LI Zhensheng]

下乡 Go to the Countryside [28] >>> 37, 39, 40, 5_

1960年代初,由于大跃进[22]失败,城市人口就业途径变得空前狭窄,众多青年中学毕业后无法升学,面临就业困难。此种背景下,官方从1962年起在全国范围内有组织有计划地动员知识青年上山下乡[28]。1966年"文化大革命"兴起,一度打断上山下乡[28]运动的正常进程。第二年,北京市的红卫兵主动申请到边疆插队落户,拉开了"文革"10年大规模上山下乡[28]运动的序幕。"文革"期间,总共有1400万以上的城镇知识青年上山下乡[28],这意味着,十分之一以上的城镇人口在这种形式下被送往农村和边疆。这场运动一直延续到1979年才落下帷幕,一共席卷了1700万城镇知识青年。[《中国城市化一甲子》,《城市中国》,2010]

In early 1960s, the bankruptcy of "Great Leap Forward"[22] unprecedentedly narrowed the road to employment in the city, with large number of young people having no access to further education or job after graduating from middle school. In this background, since 1962 the government mobilized the educated youth to leave for the countryside[28] nation-widely, in planned order and organization. In 1966, the outbreak of "Cultural Revolution" once interrupted the regular process of sending educated youth down-to-the countryside[28] movement. However, in the next year, the Red Guards in Beijing volunteered to settle down in border areas, which began the large-scaled youth-sending campaign to rural areas[28] in 10-year-long "Cultural Revolution". In this decade, over 14 million educated young people from the city were sent to the countryside[28], which means around 10 percent of urban population were transported to rural and border areas[28]. This campaign lasted till 1979, with over 17 million educated youth from urban areas involved in. [60 Years of China's Urbanization, Urban China, 2010]

Weapon Dealer

火柴枪 Match Pistol

一种男孩们用自行车链条、橡皮筋和铁丝等回收[51]材料自制的玩具。其火药来自于火柴头,因而得名。它的射程与火力使之成为短缺经济[30]时期最普遍的男孩玩具。[图/下划线工作室,2005]

A DIY toy made by kids with ready found material such as bicycle chain, rubber band and iron wires. The name came from the powder inside, which was exploited from the head of matches. It's ability in shooting and firing made it into the most prevalent toy of boys in the era of Shortage Economy[30]. [Photo/Underline Office, 2005]

计划经济 Planned Economy [29]
短缺经济 Shortage Economy [30]

 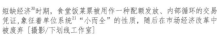

短缺经济[30]时期,食堂饭菜票被用作一种配额发放、内部循环的交易凭证,象征着单位系统[21]"小而全"的性质,随后在市场经济改革中被废弃 [摄影/下划线工作室]

Canteen coupons were used in the planned economy[29] time as self-circulated exchanging bills distributed in quota, showing the nature of being integrated and independant of Danwei[21] system, now mostly abandoned in the market economy [Photo/Underline Office]

消失的物品 Disappearing Objects [31]

假领 Fake White-Collar

当衬衣被穿在外套里面时,露出的只有领子。假领被设计成为一个完整衬衣的片段,其藏在外套里面的部分是空的。它在短缺经济[30]时期盛行中国,因为大部分工人必须在污染环境中工作、同时也买不起足够的衣服。假领使得"工作像蓝领,穿得像白领"成为可能。

When a shirt is worn inside a wool sweater, only the collar is exposed. A fake white-collar is designed as a segment of a complete shirt with its absent part hidden inside a wool sweater. …was widely used in the time of Shortage Economy[30] in China, when most of the workers had to work in dirty condition and couldn't pay for enough clothes. Fake white-collar made it possible for them to work as a blue-collar while dressing in a white-collar [Photo/Underline Office, 2005]

劳保口罩衣 Labor-Insurance-Mask Underwear

短缺经济[29]时代,在"发展经济、保障供给"的政策下,个人供需错位:劳动保险产品过剩而生活必需品不足。口罩是劳保产品,家庭主妇自发地用手工将过剩的产品制成家庭紧缺的、可以御寒的内衣裤,展现了在资源转换方面的民间智慧。[摄影/下划线工作室,2005]

In the time of planned economy[29], under the policy of "developing economy, ensuring supply", individual supply and demand is misplaced - labor insurance was superfluous and living necessities were insufficient. Masks were the product of labor insurance. Housewives spontaneously made the superfluous stuff into the items in shortage by handcrafts, which shows the folk wisdom in the resource conversion [Photo/Underline Office, 2005]

自制卫生巾 DIY Sanitary Pad

一条可以通过更换安装折叠手纸反复使用的自制卫生巾[摄影/下划线工作室,2005]

self-made pad in which folded toilet paper could be installed and replaced for women's period [Photo/Underline Office, 2005]

1970年代的深港边境线偷渡频频
The borderline[31] in-between Shenzhen and Hongkong in 1970s was one frequently broken through by stowaways [Photo/HE Huangyou]

越境偷渡 Escaping Borderline[32]

1970s珠三角偷渡重演[摄影/刘博智]
Reenactments of Illegal Immigration[19] in the Pearl River Delta in the 1970s [Photo/Pok Chi LAU]

计生 One Child Policy[33]
Provide guarantee with local registered residence certificate

超生 Chaosheng[34] (Illegal Child beyond the Plan)

Yating: Pls call home when you see this note. - Zhou Gang

1980年代之后,家庭的自然生育也被纳入计划体制[29],从而产生了一代家庭人口最小化的"标准家庭[19]"……传统中国家庭的金字塔型家谱被改组成"倒金字塔",家族内的血亲被缩减为直系血缘,而旁系血缘则逐步消亡。[《中国家庭:现代化冲击下的中国家谱》,姜珺,2007]

In 1980s One Child Policy[33] brought natural procreation of individual families into the planned economy system[35] and resulted in a whole generation of 'standard family'[19] with minimized domestic population… Traditional Chinese family'[19] pyramid was reversed to take the shape of 'inverted pyramid', which condensed kindred lineage of a family into a single lineal consanguinity with the collateral line gradually withered and disappear [Chinese Family: Pedigree under the Impact of Modernization, JIANG Jun, 2007]

人口在传统中国家庭[19]中不仅是一个"传宗接代"的遗传要素,更是一个"多子多福"的生产要素。这种观念根源于中国数千年的农耕文明[7]中占主导地位的劳动密集型农业。因而不难想象1980年代"计划生育[33]"在民间推广时所遭遇的抵制。大量没有合法户口的婴儿被超生,在"黑户"的阴影下秘密长大……[《中国家庭:现代化冲击下的中国家谱》,姜珺,2007]

In traditional Chinese family[19], population was not only an element of passing on lineage through generational procreation, but also a vital factor of labor production. This idea originated from labor-intensive agriculture that had dominating Chinese agricultural civilization[7] for thousand years. It was hence not difficult to imagine the resistance that One Child Policy[33] was meeting with when it was promoted in the countryside of China in 1980s. There were many babies born secretly beyond the national plan without an legal household registeration, who were living in a considerably secret life when they were growing up… [Chinese Family: Pedigree under the Impact of Modernization, JIANG Jun, 2007]

特区 SEZ ^{35 >>> 55}
(Special Economic Zone[35])

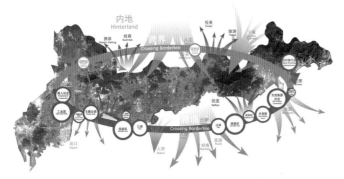

《深圳：虚界跨界[32]》：包括一线关和二线关的边界[32]是深圳城市化[55]背后的动力机制 [图解/下划线工作室，2007]

Shenzhen: Dissolving & Crossing Borderline[32], a diagram showing how the borderlines[32] in Shenzhen, including the 1st frontier and 2nd frontier, work as a mechanism behind Shenzhen's urbanization[55] [Diagram/Underline Office, 2007]

深圳被严格设计的空间形态体现了控制与开放相结合的结果：它是一个两道墙之间的城市——特区与香港之间的"一线"，和特区与内地之间的"二线"，它们分别建于毛泽东的50年代和邓小平的80年代；当年设立"一线"是为了与外界相互隔离，如今设立"二线"则是为了"一线"对外界的相对开放。深圳成为一个过渡空间，而内地与香港之间悬殊的经济势差，以及特区在关税方面的特惠政策，使之沿着深港边界延伸成一个狭长的、面向香港的线性城市，造就了一线和二线之间的繁荣。[《自发中国：一部控制与失控的历史》，姜珺，2006]

The strictly designed space of Shenzhen is an expression of the outcome of the combination of control and opening; it is a city located between two walls — "the first front between SEZ and Hong Kong and the "second front" between SEZ and the interior. These two "fronts" were set up by Mao Zedong in the 1950s and by Deng Xiaoping in the 1980s. In the early days the first front was set up to ensure China's separation from the outer world; the second, later front was set up to allow for a relative opening of the first front to the outside world. Shenzhen became a transitional space. The huge economic gap between the interior and Hong Kong, as well as a preferential policy on customs duties that SEZs ad, made Shenzhen spread into a long and narrow linear city extending along the boarder between that city and Hong Kong and resulted in prosperity between the two fronts. [*Informal China: A History of Control & Out-of-Control*, JIANG Jun, 2006]

耸立在深圳中心区的邓小平像显示了特区[35]在实验争议中得到的决策者支持
The propaganda billboard of Deng Xiaoping was erected in the center of Shenzhen to show the support from the decision-maker against the dissenters to the SEZ[35] experiment

农民工[36]宿舍：从内地迁徙到深圳谋生的农民工所处的高密度环境 [摄影/安哥]
Dorm of Migrant Workers[36], showing a highly dense living environment of those peasants who migrated[36] from the hinterland of China to the SEZ to make a better living [Photo/An Ge]

农民工 Migrant Workers ^{36 >>> 16}

……邓小平时代却承接了毛泽东时代的人口红利。中国农村剩余劳动力的海量库存，不仅令"中国制造[38]"保持着其它发展中国家难以企及的持久竞争力，也使得吸纳数亿农民进城的城市化[55]进程具有了举世无双的规模和持续性。对于城市中国[55]，这些非农化的新移民[16]不仅是实现工业化和城市化[55]的建设力量，也将在未来成为广泛的内需市场。[《从农耕中国到城市中国》，姜珺，2010]

... nevertheless, Deng Xiaoping's era received a bonus labor force thanks to Mao's population policy. The vast amount of surplus rural labor has not only sustained the competitiveness of 'Made-in-China[38]' products, unrivalled by any other developing country, but also added economies of scale and continuity to the urbanization[55] process characterized by the migration[36] of hundreds of millions of rural labor. For urban China[55], these new immigrants[16] are the labor force for industrialization and urbanization[55], as well as a vast domestic market holding great potentials for the future. [*From Agricultural China to Urban China*, JIANG Jun, 2010]

农民工们在进城的火车上眺望窗外，眼神中流露出对未来城市生活的向往 [摄影/王福春]
The peasant workers looking from the windows on a train going from the countryside to the city with the besh wishes looking forward to their future in the city [Photo/WANG Fuchun]

浙江嘉善的大舜镇被冠名为"纽扣镇"其纽扣产量在2003年达到600亿枚。大部分纽扣制造所需的贝壳都回收[31]自附近的另一个"珍珠镇"
Dashun Town in Jiashan, Zhejiang, is branded with "Button Town" as its button production has reached up to 60 billions in 2003. Most of its buttons are made of shells recycled[31] from another "Pearl Town" adjacent to it

一村一品 Branded Villages & Towns ^{39>>}

这些村镇以其为国内或国际市场制造的产品而冠名
named after the products manufactured for the national/global markets

[列表：制笔之都·山阴 Capital of Writing Brush·Shanlian；皮鞋之都·青田 Town of Leather Shoes·Qingtian；保暖内衣之乡·路桥 Town of Hot-water Heating·Luqiao；草帽之乡·长河 Town of Straw Hat·Changhe；针织之都·宁波 Capital of Knitting·Ningbo；消防器材之乡·黄岩 Town of Fire Apparatus·Huangyan；塑料打火机生产基地·章镇 Base of Plastic Casing Lighter·Zhangji；木地板之都·南浔 Capital of Wooden Floor Board·Nanxun；经编针织基地·海宁 Capital of Warp Knitting·Haining；长寿之乡·仙居 Capital of Elder·Xianju；花边之乡·萧山 Capital of Lace·Xiaoshan；袜子之乡大唐 Town of Socks·Datang；雨伞城·崧厦 City of Umbrella·Songxia；塑料城·余姚 City of Plastic·Yuyao；领带之乡·海宁 Town of Tie·Haining；泉水之乡·磐安 Town of Spring·Panan；水晶之都·浦江 Town of Crystal·Pujiang；气球生产基地·大溪镇 Product Base of Inflator·Yongkang；踏板车之乡·永康 Town of Scooter·Yongkang；袜业第一市·大唐 First County of Socks·Caota；伞城·崧厦 Town of Umbrella·Xiashang；童装之都·海宁 City of Children's Dress·Haining；棉被丝绸被丝棉被之都·湖州 City of Silk Quilt and Blanket·Tongji；化纤之乡·诸暨 County of Chemical Fiber·Zhuquan；电子王国·慈溪 Kingdom of Electronics·Cixi；椅子之乡·安吉 Capital of Chair·Anji；化妆品之都·义乌 Capital of Cosmetic·Yiwu；领带之乡·嵊州 Capital of Necktie·Shengzhou；电饭煲生产基地·临清 Capital of Electric Cooker·Linqing；钢笔之都·芬湖 City of Pen·Fenshu；蓄电池之乡·长兴 Town of Storage Battery·Changxing；球拍之乡·上虞 Town of Racket·Shangyu；塑料包装之乡·黄岩 Town of Plastic Package·Linhai；木梳之乡·天台 City of Filter Cloth·Tiantai；电器之乡·乐清 Capital of Electrical Equipment·Liushi；锡器之乡·黄岩 Capital of Tin·Huangyan；厨房用具之乡·永康 Capital of Kitchen Utensils and Applian；阀门之都·上海 Capital of Valve·Shanghai；磁性材料之乡·横店 Capital of Magnetism·Hengdian；汽车零部件产业基地·玉环 Base of Auto Parts·Yuhuan；摩托车之乡·台州 Capital of Motorcycles·Taizhou；皮鞋之乡·乐清 Town of (Leather) Shoes·Luchang；蜂蜜之乡·江山 Town of Honeybee·Jiangshan；水晶之乡·浦江 Capital of Xylitol·Puhua；石材之乡·瑞安 City of Pearl·Ruian；压力容器之乡·永嘉 Town of Pressure·Yongjia；丝织之乡·绍兴 Capital of Sartorius·Kaichan；环保产品之乡·永康 Town of Environmental Protection Filter；竹编之乡·嵊州 Capital of Bamboo Charcoal·Suchang；羽毛球之乡·杭州 Town of Badminton·Jiangshan；橡胶密封件之乡·慈溪 Town of Bamboo Weave·Changshu；皮革之乡·温州 Town of Tribute Silk·Sanda；皮革之乡·瑞安 Town of Leather·Wenzhou；模具之乡·黄岩 Town of Bront Shirt·Fengxiao；木雕之乡·东阳 Town of Modeu·Huangyan；指甲钳之都·温州 Town of Wooden Handicrafts·Dongyang；纽扣之都·桥头 Town of Nail Clipper·Wenzhou；Capital of Kincheng]

浙江省以产品冠名的村镇 [图解/下划线工作室，2006]
Towns & Villages in Zhejiang Province named after products [Diagram/Underline Office, 2006]

在温州奥康鞋业的一条皮鞋流水线上工作的打工妹们[36]
Migrant workers[36] working along an industrial line of leather shoes in Aokang Shoes Factory in Wenzhou, Zhejiang

渐进式改革中逐渐形成的"中国特色社会主义"理论，在工业化方面的实践体现为在苏联模式中务实地引入东亚模式，利用中国在海量流动人口上的比较优势，几乎无限量的廉价劳动力供给向劳动密集型的基础工业方向发展。中国由此从1950年代国际共产主义时期的"外援工业化"，到1960-70年代的"自主工业化"，过渡到1980年代全球资本主义的"外源工业化"，以价格优势跻身全球工商产业链条。[《城市中国：农耕文明的社会转型与动力机制》，姜珺，2010]

The theory of "Socialism with Chinese characteristics" formed step-by-step through the progressive reform, practices itself in the field of industrialization: pragmatically introduced the East Asian Model into the Soviet Model, utilized the comparative advantage of Chinese massive nomadic population, to march towards the labor-intensive fundamental industr with the seemingly unlimited supply of cheap labor force. China consequently moves from the "foreign aid industrialization" in 1950s of the International Communist period, to the "Independent industrialization" in 1960-70s, and then the "industrialization supported by external source" in the era of global capitalism in 1980s, ranking onto the internation industrial and trade chain predicated on her price advantage. [*Urban China: Social Transformation and Dynamic Mechanism of the Farming Civilization*, JIANG Jun 2010]

中国制造 Made i

1979 1985 1990 1995 1997 2002 2005

深圳宝安后亭村的村城化[37]进程
Process of rural urbanization[37] of Houting Village in Bao'an, Shenzhen

■ 老村 Old Villages
□ 城市化区域 Urbanized Territory

村城 Rurbanization [37 >>> 39, 52, 55]

在跨越珠江河口的东莞市东岸和深圳宝安和龙港的工厂区, 城市化[55]以分散的方式进行着, 不是以城市为中心作同心发展, 而是以根状将这个区域内每个乡村都作为扩张的立足点。就像牡蛎包裹着一个刺激物形成珍珠一样, 每个旧村落都成为它周围土地的发展核心。在这个过程中, 农民成了开发商和地主, 使用不同的工具来建设, 出租和租用住宅、工厂和宿舍区。这个去中心化的过程已成为创建中国最大的城市聚结的一个非常有效的方法, 并且速度举世无双。在这块介于广州和深圳之间的土地上, 城市化[55]的进程是微观城市的一种形式, 它并不依照城市规划进行, 而更多是作为一种创业机制, 在此前土地再分配所形成的农业框架中运作。[《厂区的细胞结构》, 艾德安·布莱威+徐建, 2006]

In the factory territory that spans the eastern side of the Pearl River estuary in Dongguan municipality and the Shenzhen districts of Bao'an and Longgang, urbanization[55] proceeds in a dispersed pattern, not moving concentrically away from city centers, but rhizomatically using every village within this zone as a beachhead for expansion. Like an irritant around which the pearl forms in an oyster, each village acts as a development nucleus for the land that surrounds it. In the process, farmers become developers and landlords using a diverse set of tools to build, lease and/or rent residential buildings and factory dormitory compounds. This decentralized process has been a remarkably effective method of creating one of the largest urban agglomerations in China, at a speed unprecedented anywhere in the world. The process of urbanization[55] in this terrain between Guangzhou and Shenzhen is a form of micro-urbanism, proceeding not according to urban planning but rather as an entrepreneurial system, operating within an earlier agricultural framework of land division. [The Cellular Structure of the Factory Territory, Adrian Blackwell + Xu Jian, 2006]

深圳龙岗大芬油画村鸟瞰
Bird's-eye view of Dafen oil-painting village in Longgang, Shenzhen

High-tech detective

城中村 Urban Village [40 >>> 44, 55]

城中村[40]是市场经济时期的快速城市化[55]与计划经济时期遗留的城乡二元结构[20]相碰撞的矛盾产物, 自发地成为爆炸式增长的城市新移民[76]的社会住宅[40]。[《社会主义新农村》,《城市中国》, 2006]

The Urban Village[40] is the contradictor product of the fast urbanization[55] in the market economy and the urban-rural dual system[20] in the planned economy[29], acting as an informal social housing program[40] for the booming immigrant[76] population. [Socialist New Village, Urban China, 2006]

Apartments to let

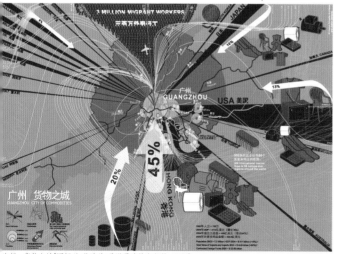

广州: 货物之城 [图解/细谷浩美+马谐霖建筑事务所, 2005]
Guangzhou, City of Commodities [Diagram/Hosaya Schaefer Architects + Joakim Dahlqvist, 2005]

Private detective,
conqueror of concubines

China [38 >>> 39, 48, 51]

城中村[40]的生与死 [图解+摄影/下划线工作室, 2005]
The process of the birth and death of an urban village[40] [Diagram & Photo/Underline Office, 2005]

Training of tricks in poker and Mahjong, victorious in every battle

大芬油画村出产的一幅《蒙娜丽莎》仿制品，售价约2000元
A copy of *Mona Lisa* made in Dafen Village in Shenzhen, sold for around 2000RMB

艺术工业村 Fordism Art Village ⁴² >>> 39, 40

大芬产业的非正式经济却有着"西洋美术"的正式形象，它制造的产品几乎涵盖了西洋美术史的所有断代，并将之打造成为可以在流水线上批量生产的趣味商品。大芬样本的奇特之处，正在于它将西方古典美学的光环，混合到东方农耕社会在全球工商产业链条的原始资本积累中，从而在朝拜性临摹与机械性复制之间，展现出崇高与卑微、梦想与现实之间的巨大张力。[《特区一村：中国城市化的大芬样本》，姜珺，2010]

The informal economy of Dafen is embodied by the formal image of "Western Fine Arts"; its products cover almost all of the genre generations in western art history, processing them into amusing goods massively assembled on the flow line. The amazing uniqueness of Dafen sample is its accomplishment to mix the aura of classical western aesthetics into the primitive capital accumulation process of an eastern agricultural country within the international industrial and business chain; therefore right amid the worship type of imitation and the mechanical reproduction... [*A Village by the SEZ: The Dafen Sample of China's Urbanization*, JIANG Jun, 2010]

海南省前财政金融中心信托大厦在1990年代的地产泡沫中烂尾⁴³
The former finance-office building of Hainan province abandoned as white elephant⁴⁴ in the collapse of bubble economy in early 1990s

烂尾城 Rotten-Tailed City ⁴³ >>> 35, 4

一个在泡沫经济中因资金链断裂而搁浅的造城运动，其中的半成品建筑作为不可消化的部分遗留下来，其内容与数量足以构成一个未完成的城市。[《烂尾城：现代巴别塔》，姜珺，2003]

A city-making movement that has been stranded because of the breakup of the financing in a bubble economy. The half-constructed buildings are left over as an indigestible part, whose content and quantity contribute to an uncompleted city. [*Rotten-Tail City: the Modern Babel*, JIANG Jun, 2003]

大芬油画村中的艺术生产 [摄影/下划线工作室+宋朝]
The art-manufacturing village-scape in Dafen Village [Photo/Underline Office + Song Chao]

Fake certificate

I will donate my liver voluntarily

Pay me my bloody money

打工杂志 Epidemic Mags ⁴¹ >>> 36

一个由少数正规品牌杂志衍生出非正式网络的垃圾杂志家族，常见于火车站等农民工的集散场所[《杂志城市》，《城市中国》Vol.1，2004]

A family of junk magazines derived literally from few formal brands into an informal network, mostly sold in train-stations from where migrant workers³⁹ are leaving for a long trip between the city and countryside³⁸ [*Magazine City*, Urban China Vol.1, 2004]

烂尾城⁴³各种被废弃的内容 [摄影/下划线工作室，2005]
Various abandoned programs of the Rotten-Tailed City [Photo/Underline Office, 2005]

一栋烂尾⁴³的高层通常被用于悬挂广告和标语，在盈利的同时掩饰自身的失败 [图/下划线工作室，2006]
A rotten-tailed⁴³ high-rise is always used for the carrier of billboards and slogans, making profit while covering its own failure [Illustration/Underline Office, 2006]

时态城市 Tense City[44]
一个由时间密度而不是空间密度定义的城市
A city defined by its TENSITY instead of DENSITY

   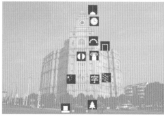
 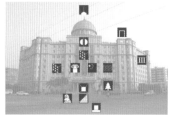  

政府建筑面相学[45] [图解/臧峰，摄影/下划线工作室，2005]
Physiognomy of Governmental Buildings[45] [Diagram/ZANG Feng, Photo/Underline Office, 2005]

政府面相 Governmental Physiognomy[45]

政府的权力集中和地方政府部门的竞争性相结合所产生的一套官方建筑语言，或称之为"政府面相学[45]"。[《中国式造城》，《城市中国》Vol.4, 2005]

The centralized system of country regime and competitive nature between its local governments and departments has generated a family of official facial structures, or the so-called governmental physiognomy[45]. [*We Make Cities*, Urban China, Vol.4 2005]

位于北京郊外的天子大酒店被设计成"福寿禄三星"雕塑造型——建筑面相学的极端类型[摄影/下划线工作室，2003]
Tianzi Hotel in the outskirts of Beijing, designed as a sculptural building - the the extreme typology of architectural physiognomy - with the images of three gods in China [Photo/Underline Office, 2003]

文脉转世 Reincarnated Context[46 >>> 44, 47]

在这些被迁徙的房子中，一种是那些被赋予特权的文物，它们被官方从原始基地移植到新址，"原封不变"地拆迁与重建；另一部分则是一种更为无意识的保护，来自老房子的材料——砖瓦和门窗——在大规模的地毯式拆迁中被重新释放成微粒，随着移民[16]一道迁徙到高处，并在一个更高的位置——新城的屋顶上——被加以重组。屋顶的平均高度为三层，这些被再利用的材料因而创造了第四层。一个戏剧性的叠加，一个意外的杂交，一个自发的城市在这里被无意识地创造出来。老村以"屋顶景观"的形态蔓延在新城之上，如同它的前世一样随意。文脉得以转世，覆在新城市主义的平面上；来自"水下白板"的幽灵，在一个新生湖面的上方，映射出似曾相识的天际线。[《屋顶上的幽灵》，姜珺，2004]

One type of "fugitive houses" are those that are privileged as antiques; officially dismantled transplanted from their original site and reassembled, unchanged in a new place. Another kind "fugitive houses" undergo a more piece-meal conservation: materials from old houses – roof tiles, bricks, doors, and window frames – were freed into particles from the massive carpet-bombing and taken with their owners to the higher levels they immigrated[16] to and then were re-constructed becoming the roofs of the brand new cities. While the average new houses are three floors, the reused antique materials become the fourth. A dramatic stack, an accidental hybrid, and an informal urbanism was created here, unconsciously. The old villages are now spread as the roofscape atop the new city, just as random as it was. Context is reincarnated, covering the plan of New Urbanism; ghosts form the underwater Tabula Rasa Underwater rises up to mirror the familiar skyline above the newborn LAKE. [*Ghost on the Roof*, JIANG Jun, 2004]

Belle House-Moving Company

三峡城市的之前之后[图+摄影/下划线工作室，2004]
Before and After of the Three Gorges Cities [Illustration & Photo/Underline Office, 2004]

窗宅 Window House

一个由各种窗框拼贴而成的违章扩建，窗框大都从拆迁工地回收而来。

A house extension made of diversified windows recycled[51] from demolition site.

浮宅 Floating House

一种由居住在海上的蛋家渔民建成的水上院宅。中央设有一个小屋的"浮院"事实上是渔业养殖的鱼网。

A courtyard house[9] on water built by the fishers living on the sea. The floating "yard" with a small house in the center is actually a meshwork used for raising fishes.

破土楼 Implosive Clay House[10]

土楼[10]是中国东南福建山区由客家人兴建的一种建筑类型。外观如城堡，以夯土墙作为防御性外围护，内部为木构建筑，以及或圆或方的内院，容纳了几乎一个村落的设施。随着其防御性在今天失去意义后，村民们由内向外地进行房屋扩建，夯土墙[6]不攻自破。

Clay house[10] was a typology in Fujian province in Southeast China, built by the Hakka (Guest People migrated[37] from mid-China). It's a castle-like building, with a clay wall[6] around as defence, as well as wooden houses and a round or square courtyard[9] inside, accommodating almost all the content of a village. As the function of defence had lost its meaning today, the villagers built extra houses from inside, which broke the unbreakable clay wall[6].

船宅 Boat House

广东阳江蛋民已长期远离大海。由于既无法在污染的海域捕鱼，也无力支付岸上的地租，他们选择将船只停泊在近城的河道中作为水上住宅以谋生，经营餐厅、茶馆、客栈等他生意。

The Dan-Fishers of Yangjiang in Guangdong Province have been kept from the sea for a long time. They berthed their boats in the narrow riverway near the city because they cannot go on with fishing in the polluted river, nor can they go on to the bank because the land price is too high. They built houses on their boats and made their living by operating restaurants, tea shops, inns and took other jobs on the land.

院宅厂 Courtyard[8] Factory

潮州一家由地主宅院[8]改建成的街道机械厂。由于宅内的墙板皆被拆除，先前相互隔断的房间被改造成了开放空间，从而使得房屋的木结构在机床上方一览无遗。房内墙上书着"打倒阎王，解放小鬼"，暗示着解放初年在这里发生过的地主与工农之间的阶级斗争。

This factory in Chaozhou used to be the private courtyard house[8] of a land lord, and has been reclaimed to be a machine factory. Since the walls[6] inside the yards had been completely demolished, the former divided room has been changed into an open space, and the Chinese traditional wood infrastructure of the house could be seen upon the machines. The old slogan "defeat the Pluto, liberate the little ghosts" which indicates "beat the land lords, liberate the workers" shows us the imagine of revolution at the early time of liberation.

大杂院 Big Mussy Courtyard[8]

北京四合院[8]原先用于一家人的居住。解放后被分给数家人使用；随着家庭人口的膨胀，各家在院内建起的连串小屋不断蚕食着公共空间，四合院[8]变成了大杂院。

Beijing courtyard houses[8], which had been originally used by one family, were re-divided into separated houses for different families, who built illegal small houses inside the courtyard[8] later and occupied the public space.

烂尾池 Swim Foot Groove

在海口的一栋大楼烂尾之后，它的基坑最终变成了一个雨水聚集而成的人工湖。因工程欠款继续居住在烂尾楼[43]中的农民工[16]将之用作游戏和洗浴的游泳池。

After the project of the skyscraper is abandoned, its base hole is eventually becoming an "artificial lake" with constantly storing water. The children of those migrant workers[16] residing in the rotten-tailed buildings[43] around the base hole take it as the swimming pool, playing and bathing.

风水长城 Fengshui Great Wall[6]

这面建于广州番禺的高墙事实上是一个巨大的中空构筑物，其尺度接近于古代城墙（30米高，5米厚，近300米长）。它的主要功能是保护一个富裕的工业村[37]的财富"不流外人田"。

The wall[6], located at Panyu, Guangzhou, is actually a huge hollow architecture, with the size near to ancient circumvallation (30 meters high, 5 meters thick and over 300 meters long). Its main function is to gather and protect wealth of a rich industrialized village[37] from another neighboring agriculture village, as well as keep the luck of wealth from leaking.

脏筑 Dirtitecture[47]

把建筑置入脏，你便得到了脏筑。
Put architecture into dirt, you get Dirtitecture[47].

脏筑与历史地理无关，它是建筑在某种失控情境下的突变产物，突如其来地有如天外来客；但它又与你息息相关：你寄居在它的体内，同时又改造着它，而这个过程如此简便易行，以至于你在厨房器皿的一个日常反应中就能得到它。[《脏筑》，姜珺，2003]

Dirtitecture is not related to history of geography. It is the mutation of architecture gone awry in an out-of-control condition, an unexpected alien invasion; however it is related to you in every way. You are living inside it and at the same time redefining it. This process is so everyday, that it can happen while reorganizing your kitchen utensils. [Dirtitecture, JIANG Jun, 2003]

傍牌货 Copylefted Products[48]

改革开放30年在外向型经济中成长起来的"中国制造[38]"尽管在产品本身的创新上乏善可陈，却在产业的规模、多样性和灵活程度上令人眼花缭乱，成为原始积累时期的"中国创造"形态……这些在"礼失而求诸野"的现代化过程中野蛮生长起来的企业建立自主Copyright的方式竟然是Copyleft[48]，这几乎是对知识产权制度的一个讽刺。在Copyright尊重创意、注重回报的另一面，是其对知识的条框化和先进技术的壁垒化，"中国制造[38]"的潜在创造性，则在于它以Copyleft[48]的"创意共享"和"活学活用"完成了它在既有的社会土壤中无法完成的自我教育……[《中国创造：大国产业升级的瓶颈与破局》，姜珺，2009]

Although "Made-in-China[38]", growing up in the export-oriented economy during the 30 years of reform and opening-up, has got little innovation in the product itself; however, it has also shown its vitality in the scale, diversity and flexibility of the industry, which made it a form of "Created-in-China" at the primary accumulation period. Paradoxically, "Made in China[38]" can be viewed retrospectively as potentially creative because its limits inspired instances of virtuous self-education precisely despite the unsupportive and frequently punitive social circumstances. With the help of "Creative Commons" and the "Learn and Practice" spirit of Copyleft[48] and open source movements… [Creative China: Bottleneck & Breakthrough of China in Industrial Transition, JIANG Jun, 2009]

篮球篮 Basketball Bucket

普通塑料桶容易在从河岸或水井中打水时被撞破。阳江农民将回收的破篮球改造成"颠扑不破"的水桶。（《即将消失的物品》，下划线工作室，2005）

The common plastic bucket is easily cracked by the collision with the riverbank or well wall when getting water from the river or well on a height. Peasants in Yangjiang recycle[51] the broken basketball into an unbreakable "basketball bucket". (Things in Disappear, Underline Office, 2005)

傍牌耐克系列
Pass-off NIKE Family

始于1998年的房改⁴⁹是中国城市化⁵⁵历史上的里程碑，它释放出爆炸性的内需并创造出一个庞大的有产阶层，启动了从生产驱动的城市化⁵⁵模式转向需求驱动转变。
The housing system reform⁴⁹ in 1998 was a milestone in China's urbanization⁵⁵, which released an explosive demand and created a large proprietary class, triggered the turn from a production driven urbanization⁵⁵ model to a demand driven one.

房改 Housing System Reform⁴⁹
私分 Privatized Communism Mansion⁵⁰

家具厨房 Furniture Kitchen

笼头锁 Lock of Bibcock

1950年代末的人们只能到公共大食堂用餐。在自家厨房开小灶是不合法的。"共产主义大厦"（或城市人民公社²⁴）曾经是一个现代化的样板建筑，其食堂被设在地下层，而居住间不设厨房。1980年代大锅饭制度被取消之后，住户们开始利用大楼宽阔的走道以家具形式建设自家的厨房。地下室的食堂则被改造成一家招待所。

By the end of 1950's, people went to dinner in the public dining hall. It was illegal to cook in private kitchens. Communist Mansion (or Urban People Community²⁴) used to be a paradigm building for moderization, with a dining hall in the basement but no kitchens in any cells. Since 1980's, the residents started to get use of the wide corridor and built private kitchens inside in the form of furniture. The basement dining-hall has been changed meanwhile into a cheap hostel.

在大厦的公共水房，为了防止自家水被盗用，每户家庭都不得不为各自的水笼头上锁。最便利的方式是铁丝在笼头开关上套装一个易拉罐。通过这种自制的方式，各家的用水安全得以保障。

In the public washing room, in order to prevent water from being stolen, each family has to add a lock to their respective tap. One of the cheapest and easiest ways is to use a pop can covering to the tap, locking through the iron wire buckle. In this self-made way, the safety of using water for each family is guaranteed.

Can't find a job, too hungry, pls help me with some food, thank you.

塑料皮鞋 Plastic Leather Shoes

Recycle⁵¹ mobilephone with high price

广东汕头的电子垃圾淘金镇⁵¹
Electronic Wastes Gold Rush Town⁵¹ in Shantou, Guangdong Province

左边农民工³⁶穿的塑料鞋，售价5元；右边为他制造的皮鞋，售价250元。
The plastic shoe on the left, sold for only 5RMB, worn by the migrant worker³⁶ as well as the leather one on the right made by him, sold for around 250RMB

中国回收 Recycled in China⁵¹ >>> 56

中国不仅出口中国制造³⁸产品，也进口被世界消耗过的废旧产品并加以反刍。电子垃圾在西方被处理时还不到一半折旧，在中国被再消费。无论是中国制造³⁸还是中国回收⁵¹，二者的"淘金热"都造成了大量污染。[《电子垃圾上的淘金客》，下划线工作室，2003]

China not only exports Made-in-China³⁸ products to the world, but also imports what have been digested from the world and ruminates. Electronic wastes are less than half consumed in the west and re-consumed in China. Either Made-in-China³⁸ or Recycled-in-China⁵¹ pollutes a lot in the gold rush. [Gold Diggers on the Electronic Wastes, Underline Office, 2003]

一个由回收⁵¹材料制成的门帘，隐约可见其前身的各种包装信息片段
A door-curtain made of recycled⁵¹ materials from tin cans, on which the segmental information from the previous package composes into a colorful collage

落实科学发展观　建设美好新农村
Carry out scientific method of development, build beautiful new village

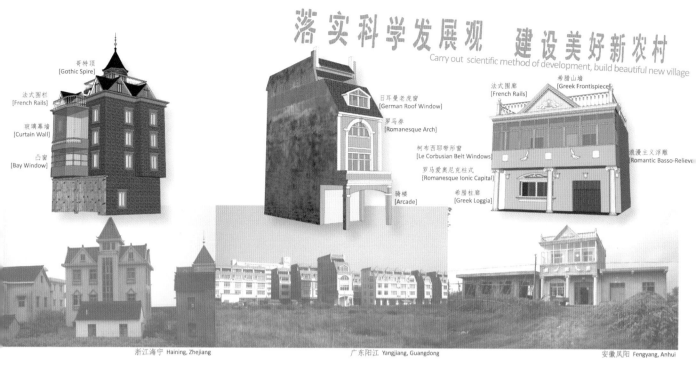

[Gothic Spire] 哥特顶
[French Rails] 法式围栏
[Curtain Wall] 玻璃幕墙
[Bay Window] 凸窗
[German Roof Window] 日耳曼老虎窗
[Romanesque Arch] 罗马券
[Le Corbusian Belt Windows] 柯布西耶带形窗
[Romanesque Ionic Capital] 罗马爱奥尼克柱式
[Arcade] 骑楼
[French Rails] 法式围廊
[Greek Frontispiece] 希腊山墙
[Romantic Basso-Relieve] 浪漫主义浮雕
[Greek Loggia] 希腊柱廊

浙江海宁 Haining, Zhejiang　　广东阳江 Yangjiang, Guangdong　　安徽凤阳 Fengyang, Anhui

18K奔驰金戒指
18K Ring of Benz

江苏省华西村建于两个时代的民居，左侧有着公共走廊的集体住宅和右侧分立、同时又高密度聚集的后集体独栋别墅
The village houses of two eras in Huaxi Village in Jiangsu province: collective houses on the left with public corridor, and post-collective houses on the right, seperated as independant villas while standing in a high-dense cluster.

新农村 Socialist New Village[52]
新民居 New Village House[53]

2005年10月，中央在"十一五"规划中，提出了建设社会主义新农村的重大历史任务。新中国在工业化的过程中，主要依靠农业提供原始资本积累，此时农民被固定在土地上，主要扮演生产者的角色。随着中国经济从外向型转向内向型，扩大内需成为当务之急，8亿农民所具有的巨大市场潜力将成为未来新的经济增长点，因此加快解决三农问题、提高农民收入，不只关系到社会稳定，更是未来经济发展的需要。[《中国城市化一甲子》，《城市中国》Vol. 40, 2010]

"Building Socialist New Village" was addressed as a major historical task in the Eleventh 5 Year Plan by the central government of China in Oct., 2005. The industrialization of new China after 1949 was mainly dependant on the primary capital accumulation provided by agriculture, when peasants were tied to their lands and roled as producers. As China's economy was turning from export-oriented to innerward-oriented, expanding the domestic demand becomes an urgent need. The massive potential of the market of 800 million peasants is going to be the new economic growth direction in the future. Accelerating the solution of rural problems, enhancing the income of peasants, is not only related to social stability, but also the demand for future economic development. [60 Years of China's Urbanization, Urban China Vol.40, 2010]

我想拥有一大堆的钱　钱
I wanna have a lot of money, money.

冥车 Mortuary Car
一辆纸质奔驰轿车，车牌号冠名"冥都"
A paper Mercedes-Benz with a registration number from the 'Capital of Hell'

寻求富婆
男：23岁，1.7米，未婚
端正，相貌体面，男子
性格开朗。从事服装，找一个有钱富婆做终身伴侣。要求
相貌端正，车敬不限。等你的来电 许生
电话：13570941259
诚心勿扰

Marriage Wanted
Looking for a Rich Lady,
as My Parnter for Life,
Waiting for Your Call

两包纸质套装，附有劳力士手表和诺基亚手机，以及一面印有美国国旗的钱包，仅售五元 [摄影/下划线工作室, 2006]
Two packages of suits, attached with Rolex watch and Nokia mobile, as well as a wallet with US national flag image, are sold for only 5RMB [Photo/Underline Office, 2006]

冥品 Afterlife Fantasies[54]
寄托对逝者来生美好祝愿的物品崇拜，折射他在世时的所欲所求
Fetish as best wishes to one's coming life, which parallelly reflects his desire in this one

冥宅 Mortuary House
一个为来生设计的标准农家宅，有着折衷主义的装饰、私家车、家庭主妇，甚至还配有一个保安。
A generic house for a peasant's coming life, with eclecticism decorations, private car, house wife and even a security guard.

城市中国 Urban China[55]

新中国建国60年以来，城市化[55]水平大幅提高，城市个数由建国前的135个增加到2008年的655个，城市化[55]水平由1949年的7.3%提高到2008年的45.68%；城市规模不断扩大，100万人口以上城市从1949年的9个，发展到2008年的122个；而伴随着建国后工业化在沿海和内地之间的几次来回迁移过程，中国的城市布局也由建国之初集中于沿海，而拓展到了广大内陆地区，城市空间分布日趋平衡；2000年之后，城市群崛起，据统计，2008年全国地级及以上城市（不包括市辖县）GDP占全国62%，城市化[55]在国民经济中日益扮演中心角色。[《中国城市化一甲子》，《城市中国》Vol.40, 2010]

In the 60 years since the founding of PRC, China's urbanization[55] process has progressed substantially, with city number increased from 135 before the 1949 to 655 in 2008, and the urbanization[55] rate from 7.3% in 1949 up to 45.68% in 2008, the urbanization[55] scale is continuously expanding, increasing the number of big city with population over 1 million from 9 in 1949 to 122 in 2008; following several times of industrial relocations between the hinterland and coastline, China's urban layout also changed from concentration around coastline region when PRC was newly founded, to the wide expansion in grand hinterland areas, and the urban space distribution has gradually reached a balance too; after 2000, city clusters started to emerge, according to the data, in 2008 the Prefecture-Level Cities' GDP accounted for 63% of China's national GDP, and the role of urbanization[55] in national economy is getting more and more important. [60 Years of China's Urbanization, Urban China Vol.40, 2010]

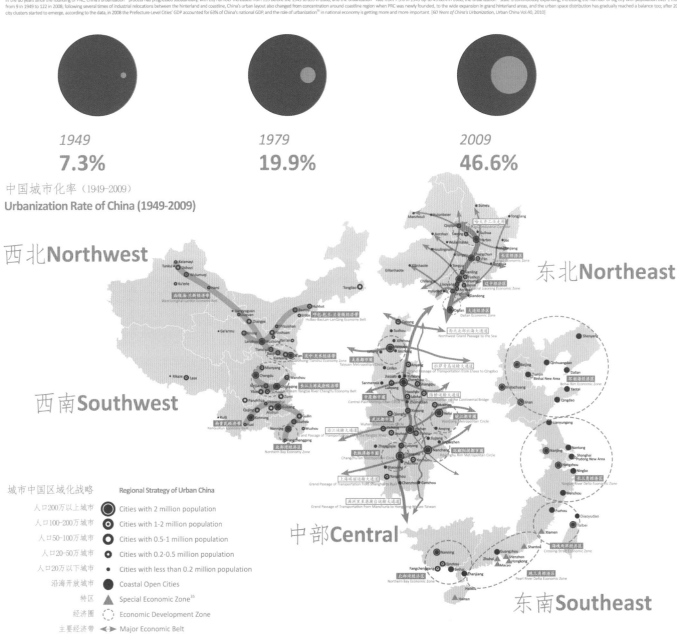

PR Wanted (for night club)
Male and female, with monthly salary above 30,000

VCD
32988 call
VCD of gays

避孕裤 Condom Shorts

由于艾滋病的威胁，为异性间避孕而设计的传统安全套已经不够安全。避孕裤的设计结合了避孕套和短裤，从而扩大了用户的保护区域。(《即将到来的物品》，下划线工作室，2005)

Because of the threat of AIDS, traditional condoms designed specifically for contraception are not safe enough. Condom shorts are designed as a combination of condom and shorts, which extend the protecting area for the users. (Things to Come, Underline Office, 2005)

收药
13719266655
Recycling medicines

绝情越无意
Your brutal rejection,
My nonsense of morality.

姜乐 Ginger-Cola

对于一些中国人而言，可口可乐不只是一种碳酸饮料。配上生姜，它就能煮成一种可以防治伤风感冒的中药。"姜乐"将这种民间配方正式化为一种中药饮料。(《即将到来的物品》，下划线工作室，2005)

For some Chinese, Coca-cola is more than a carbonate drink. Cooked with ginger, they can make them into a kind of Chinese medicine, which is able to cure the cold. Ginger-Cola formalizes this folk recipe into a Chinese-medicine-drink. (Things to Come, Underline Office, 2005)

伟克力 Viagrate

巧克力和情爱有关，伟哥和性爱有关。对有些恋人们而言，巧克力的口味太轻，而伟哥的药又太重。伟克力是一种被设计成伟哥形状的巧克力，起到了挑逗作用又不是太过份。它将可能成为情人节最畅销的礼物。(《即将到来的物品》，下划线工作室，2005)

Chocolate is about love, viagra is about sex. For some lovers, chocolate is too light while viagra is too strong. Viagrate is a kind of chocolate that is designed in the form of Viagra, which introduces the flirting effect of it without going too far. It could be a fast seller gift for the Valentine's Day. (Things to Come, Underline Office, 2005)

Hereditary medical skills
Cure your impotency

I love

两栖马桶 Amphi-Stool

中国人传统上习惯于蹲着入厕。当马桶于二十世纪初被引入中国时，它被视为一种现代化用品。然而，在公共卫生没有完全现代化的条件下，大部分安装在公共空间的马桶都达不到它们应有的洁净度，这迫使一些用户从现代的坐式入厕回到蹲式入厕。"两栖马桶"针对的就是这类用户：他们在坐或蹲之间可以有所选择。(《即将到来的物品》，下划线工作室，2005)

Chinese were used to squat in the toilet. When stool was introduced to China in early 20th century, it was respected as a modern object. However, given the unmodernized condition in public hygiene, most of the stools installed in the public space are not clean enough as they are supposed to be, which forces some users turn from sitters back to squatters. Amphi-stool is designed for these users: they can choose to either sit or squat. (Things to Come, Underline Office, 2005)

联防车 Mutual Defence Bike

一种加装了警报灯的改装车，后座上安有为警报灯供电的电池箱，作为群众性的自防、自治设施，为政府公安提供非正式的辅助力量。(《即将消失的物品》，下划线工作室，2005)

A refitted bicycle with a warning light powered by a battery installed to its back seat, as a local crisis management coping with the deteriorated public security informally. (Things in Disappear, Underline Office, 2005)

河北唐山的清洁车，车尾装有随车前进而转动的扫帚转轮 [摄影/下划线工作室，2009]
A cleaning tractor in Tangshan, Hebei province, installed with a rotating wheel of brooms [Photo/Underline Office, 2009]

潜意识发明 Subliminal Inventions

10 MILLION UNITS: HOUSING AN AFFORDABLE CITY
广厦千万·居者之城

Curator: Du Juan
策展人：杜鹃

| SHENZHEN AND CHINA / 深圳和中国 | 10 MILLION UNITS: HOUSING AN AFFORDABLE CITY |

China is launching a large scale social housing initiative to address the inadequacies of affordable housing provisions in the current market-based real estate development. The central government has completed a planning directive for 10 million units of affordable housing to be distributed throughout the country within the year 2011. Shenzhen has been given the mandate to construct 62,000 units, totaling 4 million square meters of floor area. The chain of operations to implement affordable housing is extensive, ranging from policy to planning, design, construction, distribution and eventually management. This task of rapid mass housing construction raises a critical question for the architectural profession: What is the role of DESIGN in this process?

'Housing an Affordable City' exhibition brings together government, enterprises, scholars, architects, planners, engineers, developers, and the public to examine the challenges and opportunities of providing low and mid-income housing in the dense environment of the contemporary city. The exhibition presents winning proposals of Shenzhen's "1 Unit—100 Families—10000 Residents Affordable Housing Design Competition". The exhibition also showcases innovative designs and research into the current issues of affordable housing, spanning different stages and scales of intervention ranging from construction details to national policy.

The default practice of affordable housing tends to be economized and downscaled versions of commercial real estate development concentrated in suburban areas of the city. As this has proven to be problematic from numerous examples around the world, designing innovative affordable housing necessitates the inventions of new concepts, value systems, technologies, and policies that lead to a more sustainable way of life. The Housing an Affordable City exhibition aims to establish design principles to be extended to all housing to promote responsible consumption of resources while improving qualities of urban life.

Exhibitors:

1. PEOPLE: I Live Here
   / Bai Xiaoci

2. IDEAS: 1 Unit.100 Families.10000 Residents
   / Affordable Housing Design Competition Entries

3. DETAIL: 10K House
   / Massachusetts Institute of Technology 10K House Design Studio

4. UNIT: WANKE Tu-Lou / URBANUS

5. BUILDING: Tree-Pot Tower / Standardarchitecture

6. CITY: Learning From the Informal City
   / Hong Kong University Urban Ecologies Design Studio

7. COUNTRY: 30 Years of Housing Policy Changes
   / Urban China Research Center

8. CONSTRUCTION: Exploring Standardization
   / Zhuoyue Group + Xiepeng Design

9. PREFABRICATION: BLOX
   / M3house + UAO Creations

Organizer:
Urban Planning, Land and Resources Commission
Shenzhen Center for Design
Co-organizers:
Human Settlements and Environment Commission
Housing and Construction Bureau
Zhubo Design Ltd

广厦千万·居者之城

2011年，住房和城乡建设部确定全国保障性住房建设任务为1000万套，标志着中国大规模保障性住房建设的全面展开。深圳同期建设任务是6.2万套，共计400万平方米。保障性住房实践是一个复杂的过程，涉及政策、规划、设计、建设到最后的管理和分配。这一过程和快速、大规模的住房建设，向建筑界提出了一个关键问题——设计在其中能做些什么呢？

"广厦千万·居者之城"保障性住房设计展广泛联合政府、企业、学者、建筑师、工程师、规划师、公众等各领域人士，共同探讨保障中低收入者居住在当代城市的混合高密度、可居性与宜居性带来的机遇和挑战。该设计展将展示"一户·百姓·万人家"深圳保障性住房设计竞赛的获奖及优秀作品。此外，设计展还将展览当今保障房问题的创新设计及研究，涵盖从建造细节到国家政策的不同尺度。

保障房建设往往是商业地产大楼盘模式在城市偏远地段的复制与简化，其弊端已被世界各地案例所证明。创新的保障房设计模式需要全新的思维、居住价值观以及技术和政策，来倡导可持续的生活方式。此次保障房设计展探索的居住文化及设计原则，可以推广至其他住宅建设，以促进合理的资源利用，同时保持都市生活的质量。

参展人：

1. 居民：深圳百面
   参展人：白小刺

2. 概念："一户·百姓·万人家"保障房设计竞赛
   参展人：保障房设计竞赛作品

3. 节点：一万元住宅
   参展人：美国麻省理工学院 10K House Studio

4. 单元：万科土楼
   参展人：都市实践

5. 房屋：树塔
   参展人：标准营造

6. 城市：自发城市
   参展人：香港大学 Urban Ecologies Studio

7. 国家：住房政策三十年
   参展人：城市中国研究中心

8. 建造：标准化研究
   参展人：卓越置业集团 + 协鹏设计

9. 预制：城市山林
   参展人：魔力方新型房屋 +UAO Creations

主办：深圳市规划和国土资源委员会
承办：深圳城市设计促进中心
支持：深圳市人居环境委员会　深圳市住房建设局
协办：筑博设计

Design by Sense Team

SHENZHEN AND CHINA / 深圳和中国

10 MILLION UNITS: HOUSING AN AFFORDABLE CITY

Bai Xiaoci / 白小刺

PEOPLE: I Live Here / 居民：深圳百面

In Shenzhen, 50% of the population lives in urban villages; 20% lives in factory dormitories; 25% lives in commercial housing; 5% lives in social housing. Bai Xiaoci makes use of photography to represent the geographical meanings and living details behind these four numbers, and uses documentary film to portray the distribution of different housing forms on the map. He tries to reflect the housing conditions of the 18 million population with statistical samples of 100 photos and at the same time points out the embarrassing situation of its current social housing.

深圳有 50% 的人住在城中村，20% 的人住在工厂宿舍，25% 的人住在商品房中，5% 的人住在保障房中，白小刺用照片来诠释这四个百分比数字背后的地理意义和生活细节，用纪实摄影的方法在地图上描绘了各种居住形态的分布情况，试图用统计学意义上的样本——100 张照片，来反映深圳 1800 万人口的居住现状，同时指出保障房在当下的尴尬境地。

Affordable Housing Design Competition Entries / 保障房设计竞赛作品

IDEAS: 1 Unit.100 Families.10000 Residents / 概念："一户·百姓·万人家"保障房设计竞赛

China's current 12th Five-Year-Plan has mandated the construction of 36 million units of affordable housing planned to house billions of people during the years of 2011-2015. According to this planning, Shenzhen government's assignment is to construct 240,000 units to house more than 800,000 people. This abrupt top-down task of rapid massive housing construction raises a series of problems such as planning, program, needs, policy, land resources, finance, design, construction, distribution and management. A critical question for the architectural profession has arisen: What is the role of DESIGN in this situation? 1—Unit 100—Families10000—Residents competition consists of three categories:

1- Unit
Category focuses on the design solutions of maximum spatial efficiency of a single habitation unit.

100-Families
Category encourages radical strategies to distribute affordable housing throughout the city's existing fabric with flexible clusters housing 100 families each.

10000-Residents
Category explores alternatives to China's typical large residential communities by encouraging the incorporation of mixed use and low-cost living environments.
The competition aims at promoting more systematic design solution for social housing through this event, exploring the position of utilizing design thinking to solve problems.

1. Kaisa "1-Unit" Gold Award —"Between Units" by Jiang Lin, Xun Changlei, Tu Quan
佳兆业 "一户" 设计奖金奖《户间》，蒋琳，郇昌磊，涂泉

2. Kaisa "Ao" Gold Award—"Habitation Daily" by Li Ying, Cao Taiming, Lin Yutao
佳兆业 "AO" 设计奖金奖《人居日报》，李颖，曹泰铭，林煜涛

SHENZHEN AND CHINA / 深圳和中国

10 MILLION UNITS: HOUSING AN AFFORDABLE CITY

Affordable Housing Design Competition Entries / 保障房设计竞赛作品

IDEAS: 1 Unit.100 Families.10000 Residents / 概念:"一户.百姓.万人家"保障房设计竞赛

中国第十二个"五年计划"确定 2011~2015 年全国建设 3600 万套保障性住房(可容纳上亿人)。由此自上而下分解,深圳的任务是同期建设 24 万套(容纳超过 80 万人)。这一自上而下、突如其来、大干快上的保障房建设任务,带来了计划/需求/政策/土地/资金/规划/设计/建设/分配/管理等方面的一系列困惑——设计在其中能做些什么呢?针对这个问题,"一户·百姓·万人家"设计竞赛策划方案包括了三个部分:"一户"题目针对的是居住空间内的效率设计;"百姓"题目引导关注以百户为邻里单元分解消化保障房任务的策略设计;"万人家"则侧重于探索在典型大社区中被忽略但又至关重要的低成本生活环境的规划设计。竞赛目的是通过这一创新活动,促进保障房问题得到更加理性与系统的设计解答,体现了设计中心"设计用来解决问题"的主张。

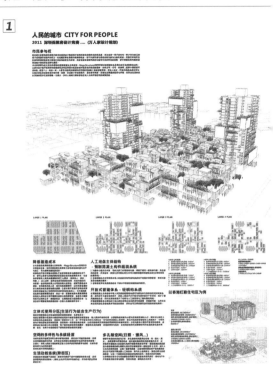

1

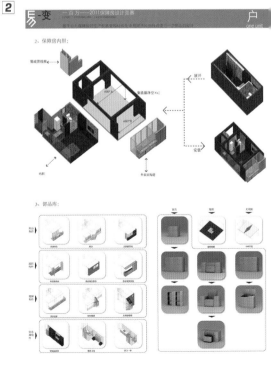

2

1. Kaisa "10000-Residents" Gold Award—"City for People" by Hsieh Ying-Chun Architects
佳兆业"万人"规划奖金奖《人民的城市》,谢英俊建筑师事务所

2. Kaisa Design Gold Award - "Flexible Change" by Shanghai Tongji Urban Planning & Design Institute, Shanghai ETOPIA Building Development Co., Dot Architects, ChengDu DODOV Design
佳兆业"综合设计"设计奖金奖《易变》,上海同济城市规划设计研究院、上海易托邦建设发展有限公司、度态建筑、成都多维设计事务所

DETAIL: 10K House / 节点：一万元住宅
Massachusetts Institute of Technology 10K House Design Studio / 美国麻省理工学院 10K House Studio

Northern Japan was stricken by a devastating earthquake and tsunami in March 2011. Thousands of homes were destroyed in the region. To help the dislocated population to resume a normal life as fast as they can, there is a tremendous need for dwellings that are inexpensive, easy to construct, and better designed than what they had before. We believe the 10K House approach is relevant in the post-disaster reconstruction in Northern Japan. Sustainability: Without an available infrastructure, 10K House has to harvest energy and treat waste in a self-sustained way. It should take advantage of latest green technology and products in building and energy industries in Japan as well as the Japanese housing traditions, especially in the area of pre-fabrication. With the awareness that mass production is an effective way to reduce cost, this studio will explore the use of digital fabrication to further transform a house into an industrial product and to have the budget under control. We collaborate with Muji, who has the wealth of market and production experience and the ambition of making Muji Houses. We see this undertaking as a way to extend the Muji furniture series outward and to develop a line of dwellings that would meet the needs of different customers. Although we will concentrate on the design of a detached house in this studio, we will be extremely interested in how an energy and waste independent house would impact the city. With minimum infrastructure, we are free then to envision a completely different urbanism, from the location of a community to the layout of streets.

2011年3月，日本北部受到强烈地震和海啸袭击，数千房屋被摧毁。短时间内恢复灾民的生活，需要大量廉价、容易建造、设计更合理的房屋。一万元住宅项目正是针对北日本的灾后重建项目。由于基础设施的破坏，一万元住宅能源和排污都要自我维持。达到这个目标需要结合日本建筑与能源工业的环保技术、产品和住宅传统，尤其是预制构件的使用。本课题组探索利用数字化制造方法将住宅转化为工业产品，大规模生产以降低成本。本项目与无印良品合作，凭借资金、生产经验及其建造"Muji House"的目标，无印良品将其家具系列拓展至符合不同顾客需求的住宅产品。我们的兴趣不限于该课题中的独立住宅设计，更关注能源排污自持住宅如何影响整个城市。对基础设施的极低依赖将造就非常不一样的社区和路网发展的城市。

3 & 4. 10K House Model
一万元住宅模型

SHENZHEN AND CHINA / 深圳和中国

URBANUS / 都市实践

UNIT: WANKE Tu-Lou / 单元: 万科土楼

Tulou is a dwelling type unique to the Hakka people. It is a communal residence between the city and the countryside, integrating living, storage, shopping, spiritual, and public entertainment into one single building entity. The circular shape of tulou facilitates its company with the existing buildings of any form. Such a design is different from the common practice of separating low-income communities from high-end urban communities. The concept of blurring the boundary between low-end and high-end communities can be conveniently materialized through the form of tulou. By reactivating the traditional living mode, the Tu-Lou Program originated a method for obtaining the land to build residential buildings for low-income families, and realized the co-existence of low-end communities with high-end ones in quite a decent manner, thereby contributing an architectural solution to social equality.

客家土楼民居是一种独有的建筑形式，它介于城市和乡村之间，以集合住宅的方式将居住、贮藏、商店、集市、祭祀、娱乐等功能集中于一个建筑体量，具有巨大的凝聚力。圆形的土楼可以使之方便地与任何形式的既成建筑群相伴，这也避免了将低收入社区从城市高端社区中分离出去的一般性做法。高、低端社区混杂的理念，通过土楼这一形式，能够很容易地实现。在重新激活传统的居住模式的同时，土楼计划在低收入住宅的土地来源问题上提出了创新性的方法，并且用一种非常体面的方式实现了低收入社区与高端社区的共存，从而为社会平等贡献了建筑学的解决方案。

10 MILLION UNITS: HOUSING AN AFFORDABLE CITY

广厦千万·居者之城

Standard Architecture / 标准营造
BUILDING: Tree-Pot Tower / 房屋：树塔

The goal of our design is to create a desirable, attractive community for young people, by designing dream apartments that occupy the smallest urban plot possible between 10 m² and 14 m² per floor. With this project we intend to explore a different way of social living. Our design process reinterprets the scale of a residential tower in order to transform it into a house scale building. This scaling process brings the perception of the building volumes and public spaces to a much more human size, balancing the relation between voids and solids. The towers are designed to be organized either together as a dense urban environment, or to stand alone in defined, enclosed plots, as dots in the existing urban fabric. This attitude makes the whole design flexible and adaptable to almost any kind of environmental situation. The proximity among the towers creates a feeling of a three-dimensional experience of the space, encouraging the social interaction between dwellers located in different floors. The irregular round shapes of the volumes emphases the perception of fluidity and permeability of the space.

本次设计意要营造一种对于年轻群体具有吸引力的超小型居住社区，其居住单元包括 10 平方米与 14 平方米两种，独层独户。在此案中，我们试图探索一种社会住宅的新的可能性。设计过程重释了住宅楼群的尺度，以使其更接近于"家"的尺度。这种重新定义使得建筑体量与公共空间更贴合人的感受。树屋的组织方式既可视作高密度城市环境本身，也可成为彼此独立的呈点状分布于城市织网中的小空间。这样的理解使得整个设计兼具柔韧性与对外在环境强大的适应性。树屋相间的状态所营造的三维空间体验激发了居住者间的互动。不规则的圆形体量强调了空间的流动性与彼此的渗透。

SHENZHEN AND CHINA / 深圳和中国

Hong Kong University Urban Ecologies Design Studio / 香港大学城市生态实验室

10 MILLION UNITS: HOUSING AN AFFORDABLE CITY

CITY: Learning From the Informal City / 城市: 自发城市

PROJECT PHASE I: DESIGN RESEARCH OF FORMAL/INFORMAL HOUSING
Mumbai; Caracas; Sao Paulo; Hong Kong; Singapore; Beijing

PROJECT PHASE II: DESIGN OF AFFORDABLE HOUSING IN SHENZHEN
Urban Planning; Housing Policy; Educational Programming; Infrastructure Occupation; Community Reconstruction; Building Renovation; Façade Intervention

Studio examine the role of design for affordable housing in dense urban environments. Rather than basing the design of affordable housing on existing commercial housing typologies and reducing unit sizes and other attributes, the studio establishes a design methodology based on the relationships between the spatial design along with policy, planning, social networks, informal economies etc. The research work examines the complexities of affordable housing by understanding the city as a synthetic ecosystem comprised of the Natural (resources and services), the Social (people and communities), the Economic (costs and affects) and the Constructed (buildings and infrastructure). Housing projects in selected cities, both formal and informal, are analyzed to establish parameters of comparison and evaluation. Lessons learned from this research will be applied towards generating new models of affordable housing in contemporary developing cities. Shenzhen is a unique site of experimentation in designing new typologies of affordable housing. The studio concludes with a series of design projects engaging in various scales and operations of interventions that address the potential roles of design in the topic of the affordability of housing and living in the city.

项目阶段一: 正式与非正式住房设计研究
孟买；卡拉卡斯；圣保罗；香港；新加坡；北京

项目阶段二: 深圳保障房设计
城市规划；住房政策；教育设计；基础建设使用；社区重建；楼房更新；立面介入

这些由工作室展现的作品评估了设计在密集城市环境中保障房的迫切问题中所扮演的角色。不同于把保障房设计建立在现有的商品房类型减少单位面积与其他设施的思路，这些作品通过理清空间设计与政策、规划、社会网络、非正式经济等方面的关系，提出一个新的设计方案。这项研究工作考察了保障房的复杂性，把城市看作一个由自然（资源与服务）、社会（人与社区）、经济（成本与效益）和建设（楼房与基建）组成的综合的生态系统。工作小组通过分析选定的几个城市中的正式与非正式住房项目来确立比较与评估的标准，并把在研究中学习到的知识应用到当代发展中城市的新型保障房设计中。深圳是实验研究新保障房类型设计的地方。最后，工作室通过一系列不同规模与操作方法介入的设计方案，来论述设计在保障房问题与城市生活中的潜在角色。

广厦千万·居者之城

267

Urban China Research Center / 城市中国研究中心
COUNTRY: 30 Years of Housing Policy Changes / 国家：住房政策三十年

Starting from 1978, the change of city housing supply of China from benefit housing system of the planned economy era to the comprehensive marketization of housing supply carries this aim to ensure that all resident have their own places to live and to develop social housing. Aiming at the economy growth of the local government contracts with the goals of society leading to the procrastination and replacement of social housing. This time the replacement of social housing policy provides not only the justice of distribution under the background of high speed economy growth, but also a outcome of reforming the city's management structure—also on outcome of financial relationship between central and local area.

从 1978 年到 2011 年，中国的城市住房供应从计划经济时期的单位福利系统，到住房供应全面市场化，主要解决的是居者有其屋的问题，再到保障房建设，可谓是中国房改的风雨三十年。伴随着住房商品化进程，城市住房的社会福利属性由垂直结构的单位体系转嫁到水平结构的地方政府之上。此次保障性住房体系的政策性重建，不仅是在经济高速发展的背景下对分配正义的一次伸张，也是城市治理结构转型的产物，是中央和地方之间的财政关系变迁的缩影。

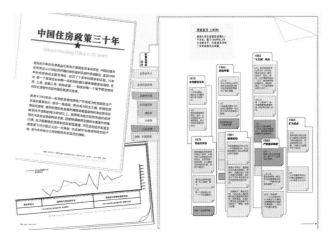

SHENZHEN AND CHINA / 深圳和中国

10 MILLION UNITS: HOUSING AN AFFORDABLE CITY

Zhuoyue Group + Xiepeng Design / 卓越置业集团 + 协鹏设计

CONSTRUCTION: Exploring Standardization / 建造: 标准化研究

Based on the outcome of the government's public housing standardization research, "the automated lifting platform construction method" was developed to achieve the cast-in-situ concrete shear wall structure system for the construction of the main body of high-rise housing block. "The automated lifting platform construction method" promotes the standardization of construction operation of the building, the industrialization and standardization of the production of the mechanical facilities, and hence enhances the overall energy saving performance. "The automated lifting platform construction method" significantly reduces the use of wood formwork and protects the forest resources, cuts down the production of construction waste and the needed amount of labor, and plays a positive role in upgrading the technology in the industry. "The automated lifting platform construction method" consists of seven parts including the mechanical transmission facility with automated lifting and linkage, equivalent story-height lifting frame, 3D steel-frame platform, the plastic framework and mold hung from the steel-frame platform, the power double-beam hoist, the crane fixed on the top of the lifting structure and the construction lift in the lifting structure. The lifting is controlled by computer, and the connection and disconnection between the mold and the frame is achieved pneumatically.

在政府保障性住房标准化研究成果的基础上研发的"自动升降作业平台工法",切合高层、超高层现浇钢筋混凝土剪力墙结构体系住宅建筑主体施工。"自动升降作业平台工法"对推进建筑主体施工作业走向标准化、机械设备制造标准化、产业化有积极作用,"工法"整体节能减排效果显著。"自动升降作业平台工法"对建筑行业大幅减少使用木材、保护绿色森林资源、减少建筑垃圾排放、减少劳务用工、促进行业技术升级具有正面作用。"自动升降作业平台工法"由具有自动升降、进出联动功能的机械传动机构,与层高等同升降架,空间钢结构平台,吊挂在钢结构平台下方的工程塑料模板与模架,电动双梁葫芦,固定在升降架顶部的塔吊及设置于升降架内的施工电梯七个部分构成。电脑控制机械上下动作,气动控制模板模架水平合拢与脱模。

M3house + UAO Creations / 魔力方新型房屋 +UAO Creations

PREFABRICATION: BLOX / 预制：城市山林

BLOX reflects the Chinese Ancient idea of self-sustained settlements embedded in a city. From the modern point of view, it could evoke more: how to reconcile the private space and public space; how to solve social problems with spatial means; how to re-build spatial qualities adapted to built densities; how to continue the tradition in the context of modern scale; how to adapt a micro-environment to an urban eco-infrastructure; and how to build up idealism through dirty reality.

BLOX is a form of "temporary architecture", which is allowed to be built on non-construction land. It could be implanted into one of the Shenzhen city urban public space, with the interior decorated with a family room layout. Thus, a tension will appear—between private living space and urban public space. It is an innovative and powerful method for inventing urbanity.

"城市山林"最初是用来形容镶嵌在高密度姑苏城内的苏州园林的。以现代城市学的角度去看，"城市山林"并不是一个简单的描述人居空间的形容词，它同时可以引发我们对于其他一些层面的思考。比如说，关于私人居住空间如何调和城市公共空间，关于建筑手段如何解决社会问题，关于空间密度如何影响空间质量，关于传统文脉如何应对现代尺度，关于人居微环境如何形成城市的生态基础设施以及关于理想主义如何实践于现实主义之中。

魔方组合屋（BLOX）是临时建筑，它可以存在于一些禁建的非建设用地之上。选择深圳市有特色的若干城市广场，将我们的设计魔方组合屋植入到城市公共空间中，内部按照一户人家的配置进行室内布置。于是，一种奇异的都市张力将出现——私人的居住空间 VS 城市公共空间。当一切建造成为现实，站在真实的"房子"前，你会明白，这给都市更新提供了一种新的可能性。

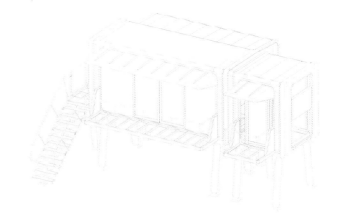

2011 REBIRTH BRICK DEVELOPMENT
再生砖进展 2011

Curator: Jiakun Liu
策展人：刘家琨

| SHENZHEN AND CHINA / 深圳和中国 | 2011 REBIRTH BRICK DEVELOPMENT / 再生砖进展 |

The "Rebirth brick Program" started in June 2008, following the Wenchuan earthquake on May 12th, 2008. The plan was originally designed to help local people conduct self-help production and reconstruction work. The basic idea of the rebirth brick is: taking the fractured ruins materials as aggregate, blending the cutting-off straw as fiber, adding cement, etc., and then making light bricks by local brick factories for the purpose of reconstruction work in the affected area. The rebirth brick is not only the 'regeneration' of waste materials, but also the mental and emotional 'regeneration' of post-disaster reconstruction.

As reconstruction work unfolded, there emerged problems, such as rearrangement and planning of lands, large-scale and rapid production of construction materials, and small-scale production by individual manual work can not satisfy the actual requirements. Therefore, factories were built in the affected areas for machinery production of rebirth bricks, which featured more stable quality and high-efficiency production. As a kind of cheap and fundamental construction materials, rebirth bricks entered formal and permanent construction material market, and were broadly used in reconstruction projects in villages.

As reconstruction work of the affected areas was gradually completed, rebirth brick is not limited to the concept of earthquake emergency relief work anymore. It enters the field of environmental protection— 'regenerated utilization of waste materials of demolished buildings'. Current products are becoming more diversified, with such developments as the permeable base, permeable floor tile, load-bearing standard brick, hollow fender brick, surface brick etc.. Now, these bricks are used in urban public buildings. It is believed that the establishment of rebirth brick factories and research and production of relative products will have a broader future.

"再生砖计划"始于2008年6月，即5.12汶川地震后一个月，当时的计划着重于民间个人生产自救，其基本原理是：用破碎的废墟材料作为骨料，掺合切断的秸秆，加入水泥等，由灾区当地原有的制砖厂制作成轻质砌块，用作灾后重建材料。它既是废弃材料在物质方面的"再生"，又是灾后重建在精神和情感方面的"再生"。

地震重建工作展开以后，面临土地重新整理规划以及大规模快速生产建筑材料的问题，个人的手工小规模生产不能满足现实状况。地震灾区开始办厂，再生砖进入机械生产，质量更加稳定且生产高效，作为一种便宜的基本建材，进入了正规的永久性建筑材料市场，使用于村落重建工程中。

随着地震重建工作的逐步完成，再生砖也逐渐脱离地震应急救灾概念，进入了"拆除建筑废墟旧材料再生利用"这一广泛的环保产品领域，现有产品逐渐多样化，如透水基层、透水地砖、承重标砖、空心围护砖、面层砖等，并已开始使用于城市公共建筑，再生砖工厂的建立和产品的研发制作可望有更广阔的前景。

| SHENZHEN AND CHINA / 深圳和中国 | 2011 REBIRTH BRICK DEVELOPMENT / 再生砖进展 2 |

| 1 | 3 |
| 2 | 4 |

1. After the earthquake, cleanup of the wreckage is an expensive and exhausting task of great difficulty. After sterilization, the rubble can be recycled.
地震之后，废墟清理是一个耗资费力的事，难以处置。防疫喷洒处理完毕后可以作为再利用材料。

2. The seasonal agricultural wasted straw is easy to obtain in summer in Sichuan suburban areas. As raw material, it reduces the cost and sufficiently avoids combustion pollution.
季节性农业废料的秸秆，在四川农村的夏季常见易得，作为原料加入再生砖，可降低成本且避免燃烧产生的空气污染。

3. Using debris from the ruins as the aggregates, along with the wheat branch pieces as the reinforcing fiber and finally mixing it all together with cement, this process can be turned into light-weight bricks for emergent service.
用破碎的废墟材料作为骨料，掺和切断的麦秸作纤维，加入水泥、沙等，制成轻质砌块，用作灾区重建应急材料。

4. "Rebirth-brick" has been produced by large-scaled machines, which are fast with high efficiency and stable quality, and has many varieties.
再生砖已进入大型机械生产，质量稳定且生产高效，品种多样。

SHENZHEN AND CHINA / 深圳和中国　　2011 REBIRTH BRICK DEVELOPMENT / 再生砖进展

1. Various Types of Rebirth-bricks
各种类型的再生砖

2. Houses built with "Rebirth Bricks" by local peasants after "5.12 Wenchuan Earthquake"
"5.12 汶川地震"后灾区农民用再生砖修建的房屋

3. Implementation of "Rebirth Brick" in Public Buildings
再生砖在公共建筑中的应用

4. "Rebirth Brick" exhibited in 2011 Shenzhen Biennale
再生砖在 2011 深圳双年展展出

GLOBAL
PERSPE
全球视野

CTIVES

6 UNDER 60
6 小于 60

Curator: Rochelle Steiner
策展人：Rochelle Steiner

| GLOBAL PERSPECTIVES / 全球视野 | 6 UNDER 60 / 6 小于 60 |

"6 Under 60" is a collaborative research endeavor and interactive, multi-media exhibition organized and presented by the University of Southern California (USC) School of Architecture, School of Cinematic Arts and Roski School of Fine Arts. An inter-disciplinary team of USC faculty, research associates, and students in architecture, design, curatorial practice, and interactive media have analyzed six cities that emerged or were transformed within the last 60 years—Chandigarh (India, 1953), Las Vegas (USA, 1959), Brasilia (Brazil, 1960), Gaborone (Botswana, 1964), Almere (The Netherlands, 1976), and Shenzhen (China, 1979).

With the rapidly increasing rate of international urbanization since the middle of the 20th century, the development of new towns has seen a steep incline in locations expected and unexpected ranging in Asia, the Middle East, and Europe, as well as remote locations in Africa and South America. The research investigates the original intentions, goals, catalysts, and master plans of these six cities; how their progress unfolded; and the developments that have made these cities what they are today. The presentation includes empirical data about the population growth, quality of life, industrial growth, cultural vibrancy, and economic success of each city. Geographic, natural, social, political, and economic conditions over the past 60 years of their development has also been considered, as well as how those aspects have affected the image of and position taken by each city, and both local and global outcomes.

Cities now house more than half the world's population, with many—including Shenzhen and others in China, India and various parts of the world—showing rapid acceleration and growth. Changes in the global political, economic, social, and cultural climate—as well as conditions that have unfolded in specific places and across the globe over this 60 year period—have made these cities as diverse as their geographic locations in the world. The research undertaken includes the collection and analysis of both qualitative and quantitative information about these six cities, and the formation of a lens through which to look at these locales both individually and collectively for points of distinction as well as commonality.

The "6 Under 60" exhibition is an immersive environment that includes infographics and visual imagery presented on interactive touch tables. The project invites visitors to explore the data about each city, while also drawing comparisons between and among them. Available information includes original planning and archival documents, architectural plans and maps, historic photographs and media representations, and graphic representations of data related to each of the six cities. The exhibition also includes moving images and sounds projected within the space that reflect the qualities of each city. A website (6under60.usc.edu) also makes a selection of information available to the wider public.

In planning the exhibition, the team considered questions about the nature and experience of architectural exhibitions—particularly formats and opportunities for viewing, reading, and interpreting information within this context, and more specifically within the context of the international biennale format. In an attempt to challenge and augment the typical presentation of plans, diagrams and documents, the team instead focused on creating an interactive experience, whereby visitors can immerse themselves in the cities while also exploring specific aspects of each of the six cities based on particular interests.

With the collection of vast amounts of data about these cities, an inevitable topic for exploration became the archiving of information. As such, we endeavor to create an urban archive that considers such questions as: what differentiates planned cities from those that develop more organically; what do these cities have in common despite their emergence and growth under differing geographic, political, and economic conditions; and how do these six cities provide a model for the future of new cities?

The USC School of Architecture, the USC School of Cinematic Arts, and the USC Roski School of Fine Arts plan to continue the research about Emergent Cities that began with "6 Under 60" as an ongoing interdisciplinary approach and project. Specifically, this initiative on the nature of Emergent Cities will explore contemporary issues that bring to light the changing landscape of cities, particularly those that are still in their initial century of existence, and how art, public art, design, architecture, landscape architecture, and the cinematic and media arts can contribute to and illuminate this process of emergence. We will also consider how the proliferation of new digital technologies and devices contributes to the way we understand and design for cities.

As cities become further networked, adaptable, responsive and interactive, the USC arts community is presented with an opportunity to collectively rethink our experience of and contribution to the evolution of our environments. By considering how individuals and societies might use networks, devices and embedded technologies to augment, capture and index the changing landscape of cities, we invite participating faculty and students to imagine and implement a series of location-specific public art interventions, interactive experiences and dynamic, real-time information visualizations that explore the dynamics and processes of Emergent Cities.

Led by Rochelle Steiner, curator and dean of the Roski School of Fine Arts, the Core Project Team members for "6 Under 60" are: Qingyun Ma, Della & Harry MacDondald Dean's Chair in Architecture, USC School of Architecture; Scott S. Fisher, professor & founding chair Interactive Media Division; associate dean of Research, director of Mobile & Environmental Media Lab, USC School of Cinematic Arts; Stefano di Martino, director, M. Arch Program, professor of Practice in Architecture, USC School of Architecture; Jennifer Stein, adjunct professor and research associate, Mobile & Environmental Media Lab, USC School of Cinematic Arts; and Steve Child, designer and adjunct faculty, Roski School of Fine Arts. Additional Principle Team members include: Sarah Loyer, Josh McVeigh-Schultz, Jon Rennie, Brettany Shannon and Kevin Tanaka. Dozens of additional USC students from these disciplines have also been involved in all aspects of this research project and exhibition.

| GLOBAL PERSPECTIVES / 全球视野 | 6 UNDER 60 / 6 小于 60 |

"6小于60"是由南加州大学（USC）建筑学院、电影艺术学院和Roski艺术学院共同策划主办的一个合作研究实验及互动多媒体展览。一个由南加州大学的科研教学人员、研究伙伴以及建筑设计、展览策划和交互媒体领域的学生组成的跨学科团队，研究分析了昌迪加尔（印度，1953）、拉斯维加斯（美国，1959）、巴西利亚（巴西，1960）、哈博罗内（博茨瓦纳，1964）、阿尔梅勒（荷兰，1976）及深圳（中国，1979）这6个在过去60年间出现并转变着的城市。

自20世纪中期起，国际性的城市化进程和速度都在急速增长。在亚洲、中东、欧洲以及遥远的非洲和南美的一些意料之内或之外地区，新城镇正以迅猛之势发展起来。这一研究的对象是这六个城市的最初原动力、目标、促进因素和总体规划，它们的进程是如何开展的以及这些城市的发展如何。这次双年展的展示包括了每个城市的经验主义的资料，例如人口增长、生活质量、工业增长、文化活力和经济成效。过去60年里，这些城市发展所经历的地理、自然、社会、政治和经济情况都被纳入了考量范围内。

当今，城市聚集了超过一半的世界人口，这正显示出了迅疾的加速与增长，包括深圳和中国的其他城市，印度和世界其他地区。全球政治、经济、社会和文化气候的变化以及过去60年间在特殊地区和全球范围内的情况，使得世界上各个地方的城市显示出了地区多样性。这一研究包括了这六座城市的一系列质与量的信息收集与分析以及个体和群体从不同角度对这些地方的观看。

展览"6小于60"创造出了一种沉浸式的环境，通过地理信息及视觉图

像互动触摸屏的方式得以呈现。这一项目邀请观众一同来探索每座城市的信息，同时进行对比。观众可以搜索原始规划和文献档案、建筑规划和地图、历史图片与媒体展示以及与六个城市相关的地理信息。空间中投影的图像与播放的声音都展示出了每座城市的特点。公众还可以通过网站（6under60.usc.edu）来浏览六座城市的相关信息。

在规划展示的过程中，策展团队考虑了建筑展的特点及经验，尤其是在语境中观看、阅读和阐释信息的形式与可能性，尤其是在国际双年展这一特殊语境中。尝试对典型的计划、图表和文本展示进行挑战，展示团队关注于创造出一种新的互动性体验，当观众根据各自的兴趣探索这六个城市的方方面面时，便让自己进入这些城市里。

此次展示包含了这六座城市的海量的信息，于是对这一特定话题的探讨成为了信息的收集。正是如此，我们试图创建一个探讨以下问题的都市文献库：规划型城市与有机成长的城市之间有什么差别；除了在不同地理、政治和经济条件下的出现与发展之外，它们之间有什么共通之处；这六个城市为未来的新都市提供了什么样的模式。

南加州大学建筑学院、电影艺术学院与 Roski 艺术学院计划以"6 小于 60"为出发点，将其作为一个进行中的跨学科方式与项目，继续开展新兴城市的研究。这一对新型都市的主动研究将着重于当代议题的探讨，为改变都市景观，尤其是那些仍然处在变革初期的地区，艺术、公共艺术、设计、建筑、景观建筑、电影和媒体艺术是如何作用于城市兴起的。我们还将考虑如何把新的数字技术和设备应用到理解和设计都市的方式中来。

当城市之间的关系越来越网络化，相互回应及互动愈加频繁，南加州大学的艺术社群一同来重新思考我们的经验和我们对身处的环境的进化的所为。通过思考个体与社会如何使用网络、设备和具体技术来为这些变化中的城市景观进行促进、捕捉和分类，我们邀请了参与的教师与学生来想象并实现一些在地的公共艺术介入、互动式体验以及动态的实时信息视觉化，以此来研究新兴都市的动态及过程。

项目策划：Rochelle Steiner（南加州大学艺术学院院长、策划人）；核心项目成员：马清运（南加州大学建筑学院院长），Scott S. Fisher（南加州大学电影艺术学院交互媒体系教授，移动与环境媒介实验室主任），Stefano di Martino（南加州大学建筑学院教授，研究生部负责人），Jennifer Stein（南加州大学电影艺术学院移动与环境媒介实验室研究助理），Steve Child（南加州大学艺术学院设计师）；其他主要项目成员：Sarah Loyer, Josh McVeigh-Schultz, Jon Rennie, Brettany Shannon, Kevin Tanaka。上述院系中的许多学生也全面地参与了此次项目的研究与展览工作。

| GLOBAL PERSPECTIVES / 全球视野 | 6 UNDER 60 / 6 小于 60 |

Satellite map and typical urban tissue
卫星地图和典型城市肌理

CHANDIGARH, INDIA / 昌迪加尔，印度

In 1947 Punjab Province was divided into East & West Punjab in the partitioning of the area between India and Pakistan. During this time Albert Meyer and Matthew Nowicki were working together to plan the new capital of Punjab. It was important that the city be placed in a new location to start fresh, not just removed from their post-colonial ties, but also divorced from past Indian ways in order to focus solely on the future. In 1948 the government of Pubjab in consultation with the Government of India approved the site at the foothills of the Shivaliks. Two years later Le Corbusier landed in India to plan the new city.

By 1952 the foundation of the dream city of India's first Prime Minister, Sh. Jawahar Lal Nehru, Chandigarh had been laid, the following year marking the formal opening of the city. Le Corbusier's master plan of Chandigarh was created to correspond to the human body, with the Capitol Complex located at the head, the City Center at the heart, the leisure space being the lungs, the cultural and educational institutions representing the intellect, the road system being the circulatory system, and industry the viscera. While Le Corbusier was convinced that people wouldn't drive as they would be able to walk everywhere, the plan failed because the concrete was far too hot. The government sector, made up mainly of concrete, lacks pedestrian traffic for this reason.

By 2001 the city's population reached 900,635. During the reorganization of the state, Chandigarh assumed the role of capital city of both Punjab and Haryana. One highlight in the contemporary city is the unplanned sculpture garden, a 25-acre sculptural maze made of recycled materials. In 2008 the Le Corbusier Center was opened, a space to cultivate an appreciation of Chandigarh's Modern heritage located in the Old Architects' Office. Finally, Chandigarh now promotes itself as an ideal site for Bollywood filming with its beautiful locations, clean environment, and accessible transport to Mumbai.

1947年，印度旁遮普邦（印度西北部）被分割成了分属印度和巴基斯坦的东旁遮普与西旁遮普。在此期间,阿尔伯特·迈耶和马修·诺维斯基一起为印属旁遮普邦设计新的首都。这个城市在全新的选址上开始了全新的城市建设，既从过去的殖民纽带中摆脱出来，也远离了印度过去的传统，全然地面向未来。1948年，旁遮普邦政府在与印度联邦政府协商后，被批准在西瓦利克山脉的丘陵地区选址。两年后，勒·柯布西耶进驻印度接手昌迪加尔的设计规划工作。

1952年，印度首任首相贾瓦哈拉尔·尼赫鲁心中的梦想之城举行了奠基仪式，昌迪加尔在接下来的时间里开始了自己的建设。勒·柯布西耶对于这座城市有着一套与人类身体相呼应的蓝图：首都综合设施是城市的"头"；市中心是城市的"心脏"；休闲区是城市的"肺"；各种文化和教育机构象征着城市的"智力"；道路交通体系是城市的"循环系统"；而各种产业是城市的"内脏"。当时，柯布西耶构想的是人们不需要驾车就可以游走城市的大部分地区，这个计划的失败是由于混凝土建筑温度过高，而大部分政府部门建筑由水泥构成，于是步行交通因此匮乏。

2001年，昌迪加尔的人口达到了900635人。印度各邦重组之后，昌迪加尔成为了哈里亚纳邦和旁遮普邦的共同首府。这个当代都市的一个亮点是未规划的雕塑公园，是一个由垃圾再生材料构造的占地25英亩的雕塑，或者说是迷宫。2008年，位于柯布西耶原办公地址的柯布西耶中心正式开放，这是为了表达对于这一现代遗产的敬意。终于，现在的昌迪加尔用自己美丽的城市、建筑，清洁的环境和与孟买之间便捷的交通，将自己塑造成了宝莱坞电影的理想拍摄地点。

GLOBAL PERSPECTIVES / 全球视野　　6 UNDER 60 / 6 小于 60

Vision plan
最初远景规划

1956

1976　　1986　　1996

2006　　2011　　2030

City Growth diagram
城市生长

CHANDIGARH, INDIA / 昌迪加尔，印度

| GLOBAL PERSPECTIVES / 全球视野 | 6 UNDER 60 / 6 小于 60 |

Satellite map and typical urban tissue
卫星地图和典型城市肌理

LAS VEGAS, UNITED STATES / 拉斯维加斯，美国

Las Vegas, located in the middle of the Nevada desert, was born in 1905 as a convenient railway hub and repair location between Los Angeles and Salt Lake City. However, in 1959, changes in gaming regulation led to a new birth for this American city. With the advent of the Nevada Gaming Commission created by the Gaming Control Act came new regulations to ensure honest, competitive gaming free of corrupt elements.

The new gaming regulations coupled with the city's flashy image—strongly supported by continuous celebrity visits throughout its history—shaped the face of Las Vegas with the help of a great deal of publicity. With a growing tourist economy, more and more hotel casinos were built. Unique to the city are the erections and implosions of numerous casinos beginning in the mid 1990s.

More recently, the strip has been fitted with skywalks at Tropicana Boulevard and a monorail began running between the MGM Grand and Bally's hotel-casinos. The Fremont Street Experience was also opened in the 1990s, changing the face of downtown with light shows, live entertainment, and activities for all ages all in a pedestrian friendly area.

By 2000, urban sprawl was increasing at twice the rate of urban population growth in the US; Las Vegas was the fastest growing metropolitan area. 2003 marked a new advertising campaign for the city, coining the tagline, "What Happens Here, Stays Here."

在 1905 年时，位于内华达沙漠里的拉斯维加斯，作为洛杉矶和盐湖城之间的铁路枢纽和补给地，应运而生。然而，在 1959 年，赌博法规的改变使得这个城市获得新生。随着《赌博控制法案》催生的内华达州赌博委员会的出现，新的法规确保了这里的赌博业的正当性，使赌博从此不再作为腐朽的活动而存在，这项活动变得实在且具有竞争力。

新的赌博法规以及整个城市历史上不曾间断的名人到访效应，都成为了拉斯维加斯最好的公共宣传，造就了这个城市现在的形象。随着旅游经济的增长，这里建成了越来越多的酒店式赌场，尤其是 20 世纪 90 年代中期起，越来越多的赌场在这里如雨后春笋般建立起来。

之后，热带大道(Tropicana Boulevard)与拉斯维加斯大道(Strip)交叉的地方建起了玻璃天桥，米高梅赌场大酒店和百利赌场酒店之间架起了单轨电车，90 年代弗里蒙特街大体验（The Fremont Street Experience）景观带的开放、户外声光秀、现场娱乐表演，还有步行街上那些老少咸宜的活动，都进一步改变了城市的面貌。

到了 2000 年，美国的都市扩张以两倍于城市人口增长率的速度进行着，而拉斯维加斯是其中发展最快的地区。2003 年，拉斯维加斯以"在这里发生的一切，都会留在这里"作为城市的标语开始了新一轮的城市广告宣传。

| GLOBAL PERSPECTIVES / 全球视野 | 6 UNDER 60 / 6 小于 60 |

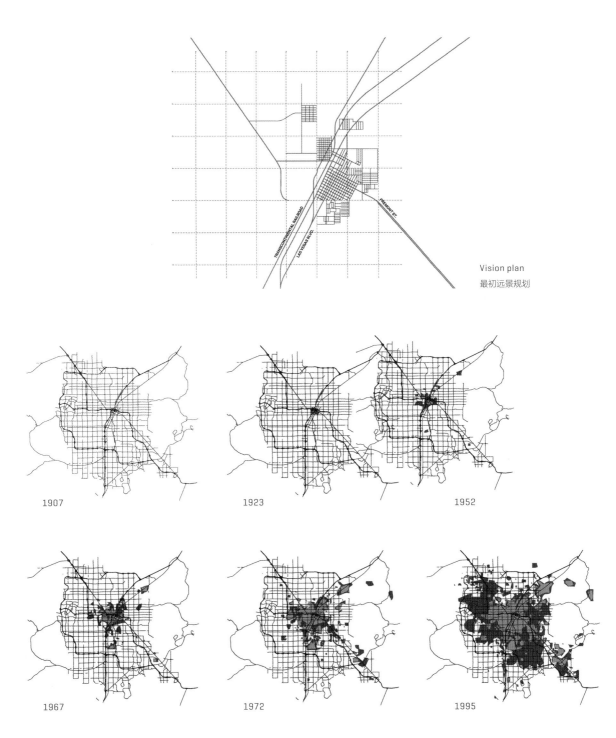

Vision plan
最初远景规划

1907   1923   1952

1967   1972   1995

City Growth diagram
城市生长

LAS VEGAS, UNITED STATES / 拉斯维加斯，美国

GLOBAL PERSPECTIVES / 全球视野　　6 UNDER 60 / 6 小于 60

Satellite map and typical urban tissue
卫星地图和典型城市肌理

BRASILIA, BRAZIL / 巴西利亚，巴西

Brasilia was designed in the late 1950s as the new administrative center of the federal government and planned to encourage settlement in the interior territories of Brazil. Oscar Niemeyer, chief architect for public buildings, organized a design competition in 1957 that was won by the planner Lucio Costa based on rough sketches. President Juscelino Kubitschek oversaw the building of the city—asking for fifty years of progress in five—and the city was officially inaugurated in 1960.

Brasilia was intended as an efficient urban environment built on two axes, one straight and the other curved in response to the topography. Yet it had virtually no sidewalks and few intersections, leading early settlers to experience a sense of anomie, alienation felt due to lack of daily interaction. The city's modern architecture—mainly designed by Niemeyer—such as the pillars of Alvadora Palace, the hemispheres and towers of the National Congress, and the crown-like cupola of the Metropolitan Cathedral makes it a visually iconic global city.

While Brasilia was envisioned to be a utopian, classless place, from its inception Brasilia had a secondary city made of construction workers' housing. This housing was built organically out of need, shifting the modern utopia vision almost immediately into a post-modern city. By 1980, 75% of Brasilia's population lived in the unplanned settlements of Brasilia, while the city proper had reached just half its target population of 557,000.

The city remains the capital and by 2009 the population has reached nearly 2.4 million. With museums, outdoor artworks, Parque da Cidade (the city's largest park), and sports (Brasilia hosts the basketball team, Universo BRB), the city is bustling.

作为巴西联邦政府规划的新行政中心，巴西利亚于20世纪50年代晚期开始进入设计流程，这也是巴西政府试图鼓励本国内地迁居移民的计划。公共建筑总设计师奥斯卡·尼迈耶在1957年组织了设计大赛，设计师卢西奥·科斯塔凭借草图赢得了比赛。当时的总统儒塞利诺·库比契克要求以5年的时间完成50年的建设进程，并最终使巴西利亚在1960年正式开城。

巴西利亚为了建设出更有效率的城市环境，有意地设计了两条不同的城市轴线，一条是直轴线，一条则是根据地势建成的曲轴线。然而，整个城市规划中没有人行道和十字路口，导致早期城市居民认为这里的城市生活是失范和混乱的，并由于缺乏日常的互动而让人觉得很疏离。这里的现代建筑大部分由尼迈耶设计，比如Alvadora Palace的立柱、国会大厦的半球体及高塔，还有巴西利亚大教堂（Metropolitan Cathedral）皇冠状的圆顶，都使得巴西利亚至少在视觉图形上成为了一座世界性的都市。

巴西利亚曾经被设想建成为一个没有阶级的乌托邦，在建城之初便在市内兴建工人住房。这些住房在还没有需求的时候便有机地建成了，这几乎使这个城市立刻把视野从现代乌托邦转移到了后现代城市上。到了1980年，75%的巴西利亚人口居住在城市里毫无规划的定居点中，而且整个城市的人口数量才刚刚达到它的原本目标人口数量557000的一半。

到了2009年，巴西利亚的城市人口已经到达240万。城市中遍布各种博物馆、美术馆、公共户外艺术品、Parque da Cidade 动植物园（巴西利亚最大的公园）以及各种体育设施（巴西利亚是Universo BRB 篮球队的主场），整个城市越来越繁华和越显熙攘。

GLOBAL PERSPECTIVES / 全球视野 | 6 UNDER 60 / 6 小于 60

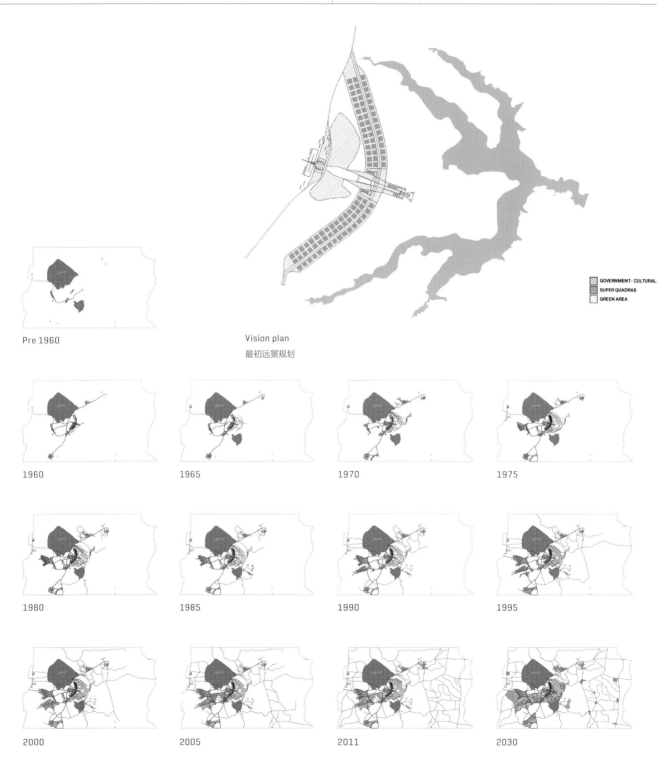

Vision plan
最初远景规划

Pre 1960

1960 1965 1970 1975

1980 1985 1990 1995

2000 2005 2011 2030

City Growth diagram
城市生长

BRASILIA, BRAZIL / 巴西利亚, 巴西

| GLOBAL PERSPECTIVES / 全球视野 | 6 UNDER 60 / 6 小于 60 |

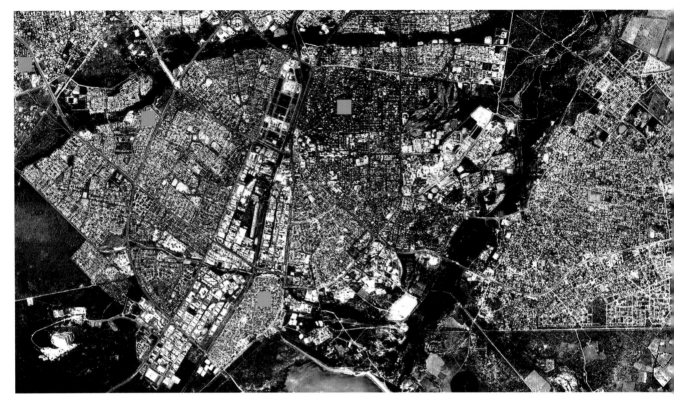

Satellite map and typical urban tissue
卫星地图和典型城市肌理

GABORONE, BOTSWANA / 哈博罗内，博茨瓦纳

Gaborone was named for Chief Gaborone who led his tribe from the Magaliesberg around 1880. The region was placed under British protectorate status in 1885, and remained so until Botswana gained independence in 1966. Gaborone, a post-colonial city, was planned in its location on state land allowing easier manipulation of the land near a water source. The first Master Plan was created by the Public Works Department in 1963, calling for a segregated town. The plan focused on the town functioning as a complete unit and on separating vehicles from pedestrian traffic.

Construction began in 1964 and by February of 1965 the first phase of the government buildings was complete. In a country with a $20 million Gross Domestic Product, the city cost $18 million in its first year. By 1966, three years after the start of the initiative, basic infrastructure was in place. The city was designed in the shape of a brandy glass, the large top part as the state seat and the thin stem as the economic engine. In this spatial design an economically segregated city was shaped. When the city was complete, it wanted to be compared to Brasilia and Chandigarh but lacked the utopian features that the others contained.

The Wilson-Womersley Master Plan was initiated in 1971 to extend the city northward attempting to eliminate economic segregation by mixing various cost housing areas so that school districts and centers contained varying categories of housing. The Broadhurst II Plan of 1979 focused on site and service development. Rural to urban migration contributed to 70% of the population increase in the 1980s. With these population increases came the Gaborone West Structure Plan, an expansion of Gaborone to the West over the Railway Line.

The Greater Gaborone Plan was established in 1994 and extends to the present. With a continuous slowing of population growth, the city expects all the empty State land to be inhabited by 2014. While the city was planned for 20,000 inhabitants, it has now reached 191,000, the vast population increase leading to the city's encroachment on tribal land. With stunning wildlife, the tourist economy is largely made up of safaris. The city now has four large shopping malls, a cultural center, and two golf courses.

哈博罗内，以哈博罗内酋长的名字命名，这名酋长在 1880 年将自己的部落从马加利斯堡山迁移至此。该地区自 1885 年起一直处于英联邦保护之下，直至 1966 年获得独立。哈博罗内开始进行规划设计时，力图使土地的改造和使用靠近水源。1963 年，市政工程部门提出了第一个隔离城市的规划，让城镇像一个独立的体系来运作，区分了步行和公共交通区域。

城市建设于 1964 年动工，第一批政府建筑在 1965 年 2 月便已经完工。第一年，这个年 GDP 仅为 2000 万美元的国家耗资 1800 万美元来建造哈博罗内。到了 1966 年，基础设施建设已完成。整个城市的形状被设计成一个白兰地酒杯，大的杯身部分是城市的所在，细窄的杯脚部分是城市的经济引擎。这种空间规划形成了一个经济隔离的城市。当城市建设完工时，曾被拿来与巴西利亚及昌迪加尔进行比较，但哈博罗内缺乏前两者那样的乌托邦式的特质。

Wilson-Womersley 事务所设计的蓝图于 1971 年开始，这个规划将城市向北扩展，试图混合各种档次和多种类型的居住区，让住房被包围在学校街区和市中心内，以消除城市的经济区隔。1979 年 Broadhurst II 计划则注重城市现状和服务的发展。20 世纪 80 年代增长的城市人口中，有 70% 来自于农村。随着人口的增长，哈博罗内的规划越过了铁路线，开始了向西递进规划的进程。

1994 年，更大的哈博罗内规划开始实施，城市随之扩张到了今天的规模。随着人口增速的持续放缓，哈博罗内希望 2014 年城市被居住使用。城市计划容纳 20000 名居民，现在已经达到 191000 名，大规模的人口增长促使城市开始侵蚀周边部族所闲置的国土用地。依靠丰富的野生动物资源，徒步野外旅行成为了这里的旅游产业的主要支柱。哈博罗内拥有四个大型购物中心、一个文化中心和两个高尔夫球场。

GLOBAL PERSPECTIVES / 全球视野

6 UNDER 60 / 6 小于 60

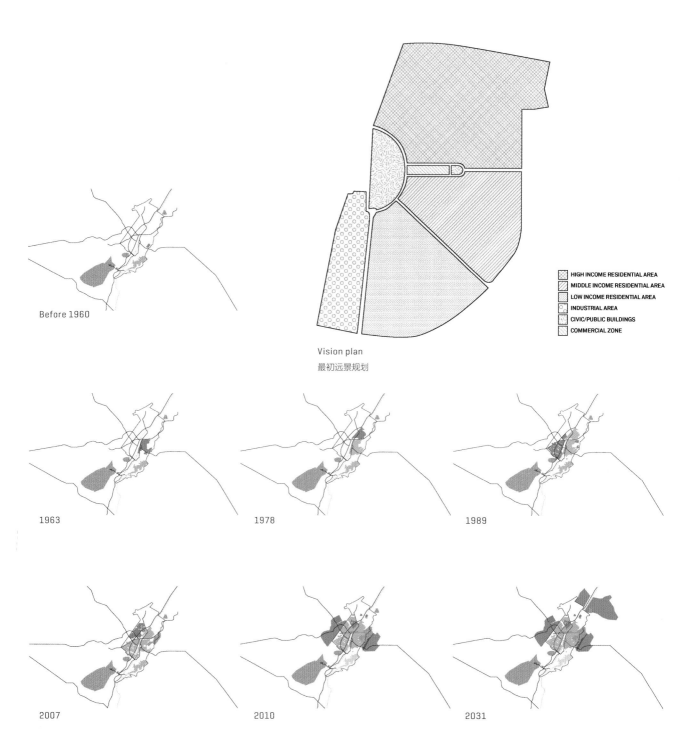

City Growth diagram
城市生长

GABORONE, BOTSWANA / 哈博罗内，博茨瓦纳

GLOBAL PERSPECTIVES / 全球视野　　6 UNDER 60 / 6 小于 60

Satellite map and typical urban tissue
卫星地图和典型城市肌理

ALMERE, THE NETHERLANDS / 阿尔梅勒，荷兰

Almere, first an overflow suburb of Amsterdam, underwent its first phase of construction in the southwest corner of Southern Flevoland in 1968. The city was conceived as a form of decentralized urbanism to solve regional problems in Amsterdam including rapid population growth, suburbanization, and traffic congestion. The goal was utopian: to build a city with a long-term flexible plan, allowing for social and technological changes to enable a healthy natural environment along with a diverse population and a stimulated urban culture. Since its inception, Almere has been one of the fastest growing cities in Europe.

The city was originally designed as a series of neighborhoods with their own facilities and identities, linked through a shared infrastructure. By 1973 "Almere 1985" was published, a polynuclear hierarchical plan for the city. In the 1970s the city established four main towns: Almere Haven (Port), Almere Stad (City), Almere Buiten (County), and Almere Hout (Wood). Reflecting urban development thinking of the time, Almere Haven is characterized by its maze of so-called "cauliflower neighbourhoods". Playful street patterns full of cul-de-sacs were designed to encourage social contact.

The 1980s were marked by competitions for free plots of land for people to build houses they had designed themselves. This approach to unique architecture was reintroduced in 2006, making Almere a 'do it yourself' city in its attempt to attract residents and developers. The mid 1990s marked a change in the city with an attempt to create a downtown space for economic reasons. This focus on economic development and boosterism has changed the city's earlier idea of providing homes for migrant people to offering homes to wealthier people.

By 2009 at least 134 nationalities are represented in Almere, while almost 17% of the total population was living at or under the poverty level; half had been that way for 3 years. Almere nonetheless continues to think toward the future with hopes to build two remaining nuclei: Almere Pampus (Out) and Almere Poort (Port) by 2025.

阿尔梅勒，最初是1968年开始在阿姆斯特丹的郊区——南弗莱福兰省西南角新建的小城。城市一开始的构想是以一种去都市主义中心化的形式来解决阿姆斯特丹本地人口膨胀、郊区化和交通拥挤的问题。整个目标是乌托邦式的：建造一个具有长期弹性规划的城市，持续的社会和技术的变革使此地具有健康的自然环境，多样化的人口结构和都市文化。从建市开始，阿尔梅勒就是欧洲发展最快的城市之一。

阿尔梅勒最初的设计是由公共基础设施连接而成的一个个社区，每个社区有自己的配备和定位。1975年，《阿尔梅勒1985》出版，为这个城市呈现了一个多核心递进式发展的规划。20世纪70年代，整个城市建成了四个主要的城镇：阿尔梅勒，Haven（港）；阿尔梅勒，Stad（城）；阿尔梅勒，Buiten（郡）；阿尔梅勒，Hout（林）。作为当时城市发展设想的反映，阿尔梅勒，Haven（港）被定位成迷宫一般的"花椰菜社区"。有趣的街道样式和遍布的死胡同的设计是为了激励社区的社交联系。

20世纪80年代的阿尔梅勒以在这片土地上发展竞赛为特征，人们被鼓励自己去设计和建造自己的住房。这种实现建筑的独特方式在2006年被再次引入，使阿尔梅勒成为一个DIY的城市，旨在吸引新的居民和开发商。20世纪90年代，阿尔梅勒因为某种经济原因而开始发展自己的城市中心区。过去着力于经济发展和城市振兴的举措让阿尔梅勒以为移民者提供家园为出发点，但逐渐演变成了为城市富裕人群提供新住宅的诉求。

到2009年，阿尔梅勒有134个不同的族群聚居，然而，其中17%的人口生活在贫困线及以下，这其中的一半人口已经在这样的情况下生活了三年。尽管如此，阿尔梅勒还是满心期待在2025年完成剩下的两个核心社区的建设计划：阿尔梅勒帕姆普斯（外）和阿尔梅勒港（港）。

GLOBAL PERSPECTIVES / 全球视野   6 UNDER 60 / 6 小于 60

Vision plan
最初远景规划

1965   1970   1975   1980

1985   1990   1995   2000

2005   2010   2030

City Growth diagram
城市生长

ALMERE, THE NETHERLANDS / 阿尔梅勒, 荷兰

| GLOBAL PERSPECTIVES / 全球视野 | 6 UNDER 60 / 6 小于 60 |

Satellite map and typical urban tissue
卫星地图和典型城市肌理

SHENZHEN, CHINA / 深圳, 中国

Shenzhen ["shen" means "deep" and "zhen" means "drains"] developed as a fishing village patterned with deep drains in the 17th century. The area remained a village until China's Open Door Policy was developed in 1978. Under the leadership of Deng Xiaoping Shenzhen became one of the first Special Economic Zones, a "window" for observing global trends in economic, scientific, technical, and managerial/market deviations and an experimental space for economic reforms. The city was selected due to its proximity to Hong Kong and ability for export as a coastal city.

Beginning in 1980 the Shenzhen Special Economic Zone was built from scratch. The former farmers became wealthy landowners with their valuable properties within the city. Government bureaus and executive units were transformed into economic entities by 1982, and by 1984 the economy become more export-oriented. From 1986 to the 1990s Shenzhen went under a restructuring phase. In 1988 the government lost control over the economic enterprises in the city. The mid-1990s were marked by a re-engineering phase in which strategies were deployed to develop the city into a city with modern environment and booming economy.

2007 marked the opening of the OCT-LOFT art park region created from renovated factory buildings. Shenzhen was named a member of UNESCO's Creative Cities Network and awarded the title of City of Design by UNESCO in 2008. In 2009, Shenzhen created plans to promote development of biotech, new energy, and Internet industries, which quickly became and continue to become vessels of economic growth. Shenzhen acts as the high-tech manufacturing hub in southern China.

In 1979 Shenzhen was a small fishing village consisting of 30,000 people. By 2003 that number reached 4.69 million and by 2010 the population reached over 10 million permanent residents. With attractions like "Windows of the World," a park consisting of reproductions of some of the most famous world tourist attractions, and "Minsk World," a military theme park, it is no wonder that Shenzhen is a tourist destination.

17世纪时，深圳（"圳"意为田边水沟）还是一个遍布很深的田边水沟的小渔村。直到1978年改革开放前，这里依然只是一个小村庄。在邓小平的带领下，深圳成为了第一个经济特区，是经济、科学、技术和管理/市场动向的"窗口"，也是经济改革的试验场。因为毗邻香港和位居海岸的出口能力，使得这座城市成为了这一政策的试点。

1980年起，深圳特区在平地中建立。农民们摇身一变成为了富裕的土地拥有者，在这个城市有着颇为可观的财富。政府与行政机构都在1982年转变为经济实体，到1984年则成为了一个以出口经济为导向的城市。从1986年到90年代，深圳都处在重建阶段。1988年，政府在城市中对企业的控制力已经失去了。90年代中期，采用了将城市建成一个拥有现代环境和经济激增的都市的策略，深圳由此进入了一个重新设计的阶段。

2007年，OCT-LOFT艺术园区的开幕迎来了一批重新改造的厂房。2008年，深圳被联合国教科文组织纳入了创意都市网络，并被誉为"设计之都"。2009年，深圳开始了一系列发展生物技术、新能源和互联网工业的计划，这迅速成为了经济继续增长的力量。深圳由此成为了南中国高科技生产的重镇。

1979年的小渔村深圳只有约30000人，但到了2003年，达到了469万人，2010年则激增到了1000万永久居民。有着世界著名旅游景点微缩景观的世界之窗和军事主题公园明斯克世界这样有吸引力的地方，毫无疑问的是，深圳还成为了一个重要的旅游站点。

| GLOBAL PERSPECTIVES / 全球视野 | 6 UNDER 60 / 6 小于 60 |

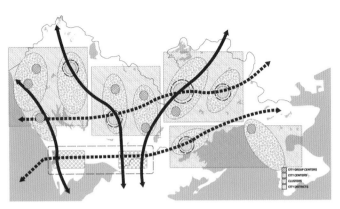
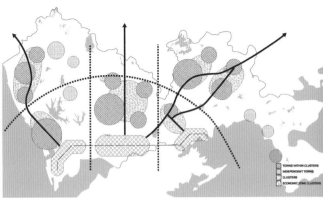

Vision plan
最初远景规划

1980-1990

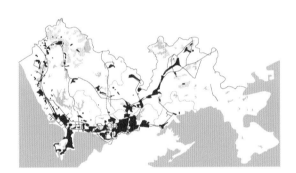

1990-2000

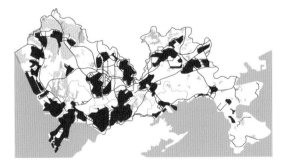

2000-2010

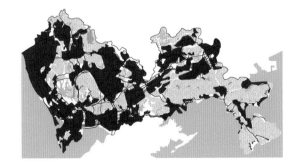

2010-2020

City Growth diagram
城市生长

SHENZHEN, CHINA / 深圳, 中国

AND THEN IT BECAME A CITY: SIX CITIES UNDER 60
然后，它成了一座城：未满 60 岁的 6 座城市

Curator: David van der Leer
Video works by: Surabhi Sharma, Cao Guimarães, Miki Redelinghuys, Sam Green, Astrid Bussink and Wang Gongxin
Workshops: Mary Ann O'Donnell and ATU
Exhibition Graphic Design: Masera Studio & Ana Rio
Research: Johanna van den Moortele

策展人：*David van der Leer*
录像作品：*Surabhi Sharma, Cao Guimarães, Miki Redelinghuys, Sam Green, Astrid Bussink*, 王功新
工作坊：马立安、观筑建筑发展交流中心
展览及平面设计制作：*Masera Studio & Ana Rio*
研究：*Johanna van den Moortele*

GLOBAL PERSPECTIVES / 全球视野   AND THEN, IT BECAME A CITY: SIX CITIES UNDER 60

and then it became a city:
Two Approaches to Urban Tipping Points

"While rare, sporadic case reports of AIDS and sero-archaeological studies have documented human infections with HIV prior to 1970, available data suggest that the current pandemic started in the mid-to late 1970s. By 1980, HIV had spread to at least five continents (North America, South America, Europe, Africa, and Australia). During this period of silence, spread was unchecked by awareness or any preventive action and approximately 100,000-300,000 persons may have been infected."

—Jonathan Mann

The development of many diseases such as HIV/AIDS is fairly gradual, until at a certain point, when the disease has affected a critical mass. A sudden shift takes place, and the disease rapidly develops into an epidemic that spreads like wildfire. This moment—when everything changes all at once—is commonly referred to as a tipping point. The term applies not only to epidemiology, but also to other scientific fields—for instance, it describes the dieback of the Amazon rainforest, or the decay of the Greenland ice sheet. The notion of these tipping points is not new. Friedrich Heinrich Hegel in 1812 addressed tipping points in his book Science of Logic: "There are no sudden changes in nature, and the common view has it that when we speak of a growth or a destruction, we always imagine a gradual growth or disappearance. Yet we have seen cases in which the alteration of existence involves not only transition from one property to another, but also a transition, by a sudden leap, into a . . . qualitatively different thing: an interruption of a gradual process, differing qualitatively from the preceding former state."

Since the 1970s the term tipping point has occasionally been applied to rapid changes in social systems—for instance, "the failure of one or more large financial institutions could represent a tipping point, destroying the whole system." Oddly enough, tipping points have been identified much less frequently in the development of cities, even with regards to population flux. This may be the result of our collective desire to see city change as constructive—we sometimes seem to think naively there is only improvement—and tipping points in science often are associated with going from bad to worse. With that in mind, it is not surprising that in the urban studies context, the Tipping Point Hypothesis of the 1970s is among the rare, and highly questionable, examples of the use of the terminology. This "hypothesis" was developed in the United States by real estate experts who believed that a tipping point was "the critical proportion of Negroes in a neighborhood that may trigger off a rapid White exodus." Making a statement like this these days would be considered outrageous of course—at least it shows how perception and culture can change for the better. I wonder how we may use the concept of tipping points to look at cities in a more upbeat manner than the 1970s hypothesis example? What could we learn from looking at these moments in the urban context? Are there even tipping points in our most rapidly developing landscapes? For instance, if we look at new towns: when do these turn from planning exercises into real places—into cities? Or more specifically, How does one apply the term, or identify the tipping point, of a city?

In most of the scientific uses of the term, researchers look for indicators that can identify a pattern early, before a rapid transition starts taking shape at a large scale. And being able to identify such a pattern for the growth or decline of cities is exactly what would be interesting for urban studies, and especially the development of new towns. What indicators for triumph and disaster can we identify early on, and what impacts would those have on the design of urban landscapes? Since antiquity, new towns have been developed throughout the world—often as safe havens for rapidly growing populations, as testimonials to political dreams, or simply as places for production and prosperity. These planned cities are among the most ambitious and costly human undertakings realized, and over the past 10 or 15 years we have started to visualize and analyze these developments through a seemingly endless barrage of maps, diagrams, plans, and statistics. Together these purport that we understand cities in general, and city growth and shrinkage in particular. But somehow I can't help but think that most of these analyses were generated by architects and planners for books that are so often written as self-promotional tools and employed on the hunt for new assignments. Paul Barker, former editor of the British New Society magazine, which he edited with Cedric Price and Peter Hall among others, recounts in the 1999 Non-Plan Revisited how planning is rarely analyzed thoroughly after completion: "So often, and this continues to be true, an urban plan was said to be fulfilled when it had only been completed. No one checked whether it did the job it set out to do." And indeed, a few maps and diagrams may not do the trick. Now having lived for more than half a century with some of the world's most problematic, and also some very successful, new towns, we need to experiment with modes of analysis that are more than PR stunts for designers, governments, and developers.

To learn when new planned towns stop being new and turn into actual cities, I would argue for two approaches that may help us identify what brings forth urban tipping points: first, one that is strongly analytical and currently highly criticized in both the design and scientific fields, and secondly, an approach that is much more related to the everyday experience of the proverbial "man in the street." Together, these different approaches to urban studies may help pinpoint possible tipping points—and could be a leap forward for the design fields.

Analytics-beyond-Analytics

The work of a multidisciplinary research team—with experts such as Luis Bettencourt, Jose Lobo, Dirk Helbing, Christian Kuhnert, and Geoffrey West, and the excellent researchers at the Cities Program at the London School of Economics (LSE) who tend to produce endless datasets that seem to show the epidemic growth of cities around the world—is among the best examples of the role that academia and science can and should play in the urban realm. This team has raised alarm bells on how despite much "historical evidence that cites are the principal engines of innovation and economic growth, a quantitative, predictive theory for understanding their dynamics and organization and estimating their future trajectory and stability remains elusive." They are developing dynamic models that are proving valid across urban systems—right now empirically proven only in the U.S., but possibly in other regions around the world in the future. The research and analysis of these teams is approaching a level of specificity whereby urban models can be summarized in equations such as Y(t) + YON(t), which analyzes the different elements to urban growth empirically. It shows how "most urban indicators scale with city size nontrivially, implying increases per capita in crime or innovation rates and decreases on the demand for certain infrastructure, is essential to set realistic targets for local policy." West and his team are close to drafting formulas for the development of urban supply systems such as they display in Growth, Innovation, Scaling and the Pace of Life in Cities (2007) where they can generate empirical evidence of when restaurants, car dealers, post offices, and pharmacies are needed and thriving in particular urban contexts. If we can create and produce data analyses like these—which go beyond the often beautiful, but less-than-analytical data visualizations of architects and designers—we may begin to understand the tipping points of cities better and anticipate potential development modes for the future.

Experience-beyond-Analytics-beyond-Analytics

Although it may sound tempting to focus solely on condensing complex city systems into pure analytical models such as aforementioned, re-reading Barker reminds us that "a city is not a computer program," and that "it has a life of its own." Although West and cohorts may argue differently, Barker believes—and in ways I tend to agree—that "no one is clever enough to know in advance, how cities will grow. You cannot tell which innovation will germinate and multiply a thousand-fold (like the mobile phone). Nor can we tell how people will decide to organize their lives, or how tastes in patterns of living will develop." Nor is it likely that we develop one coherent analytical model that can address all cities in all locations around the world. But if we believe—and it seems there is enough proof to do so—that at certain tipping points qualitative change is instituted in cities as a result of quantitative developments that can be captured in equations and analytical models, it may be good to look deeper at the quality of life in those cities, in parallel with the analytics-beyond-analytics approach. From its first issue in 1962 until it closed in the late 1980s New Society magazine advocated this position—that what ordinary people wanted, how they would lead their lives, rather than what experts said they ought to want, was key to understanding cities. And it is in this realm that the exhibition project and then it became a city plays a role. For this show and series of workshops I commissioned six video artists and documentary filmmakers from around the world, and a group of Shenzhen-based artists and architects to analyze the quotidian lives of ordinary people in six exceptional cities: namely six new towns under the age of sixty. None of the six artists working on the video pieces is using a checklist to codify the cities, nor is any of them an expert in planning issues, but in his or her own specific ways each speculates what it takes to turn these new towns into actual cities that feel like vibrant places to both inhabitants and visitors: Is it the number and the size of its trees? Is it the number or diversity of leisure activities? Is it in a multiplicity of architectural languages, or role of the arts and culture? Or is it its traffic and commuting times, pollution, or crime rates? Or is it as simple as in the passing of time? Surabhi Sharma traces personal bylanes to the Corbusier plans for Chandigarh (India, 1953), Cao Guimarães documents how a planned town such as Brasilia (Brazil, 1960) only becomes a city when its various elements gain autonomy, Miki Redelinghuys asks the inhabitants of Gaborone (Botswana, 1964) how to find the way to their home, Sam Green shows the roller coaster of emotions, good and bad, that many of us have for Las Vegas (U.S., 1960s), Astrid Bussink looks at Almere (Netherlands, 1976) and wonders how much its inhabitats are directed by the architecture they live in, and lastly Wang Gongxin highlights how the dynamic aesthetics and the life force of Shenzhen (China,

1979] have resulted in a "city of designers." Together their works celebrate a different, more ephemeral or poetic approach to the tipping points in urban life cycles, which is further elaborated on in the series of sociological and creative workshops hosted for and with the inhabitants of Shenzhen by Mary Ann O'Donnell and ATU/观筑 on a bus that drives around the city.

Intentionally both of the approaches—the analytical and the ephemeral—briefly outlined here in this text emphasize an analysis of the here and now, without yet venturing into design strategies or planning models for the future. By focusing on finding out more about these elements that may constitute urban tipping points, be it in science through analytical models, or through more experiential qualitative analyses such as in and then it became a city we may start to understand the rapid, almost epidemic, growth of cities around the world better. As we have learned from diseases such as HIV/AIDS ignorance rarely results in positive outcomes after a tipping point has been reached. If we get to know more about the parameters that go beyond mere design solutions we may be able to turn new towns and other urban planning modules more swiftly into vibrant cities.

GLOBAL PERSPECTIVES / 全球视野

AND THEN, IT BECAME A CITY: SIX CITIES UNDER 60

然后，它成了一座城：通向城市临界点的两种方式

稀有的、零星的病例报告和血清研究记录了人类在 1970 年之前就感染了艾滋病毒，但现有数据表明艾滋病流行趋势开始于 70 年代的中后期。到了 1980 年，艾滋病已经蔓延到至少五大洲，包括南美洲、北美洲、欧洲、非洲和大洋洲。在未暴发的时候，艾滋病的传播没有任何预防措施，以至于大约有 100000 到 300000 人先后在这期间被感染。

——乔纳森·曼恩

许多疾病如艾滋病的发生传播都是逐步的，当发展至某个时刻，疾病本身将会产生大范围的影响，直至突发性转变发生，疾病将会如洪水猛兽般地成为传染病。通常这种一切都发生根本性改变的时刻被称为临界点。临界点不仅常见于流行病学领域，也广泛应用在其他科学领域，比如在描述亚马逊原始森林的枯死和格陵兰冰川消融时。这并不是一个新的术语，弗里德里希·黑格尔早在他 1812 年的《逻辑学》（Science of Logic）一书中就用了临界点一词："本质不会产生突变。一般来说，当我们在说生长和消亡的时候，我们总是想象一个渐进的产生和消失的过程。存在的变化不仅包括一种属性到另一种属性的转变，也同时包括一种跃进式的转变……至本质上完全不同的事物：渐进过程的突然中断，形成区别于之前状态的质变。"

从 20 世纪 70 年代开始，临界点也会偶尔被使用于描述社会体系中的突变，比如"一个或更多的大型金融机构崩溃可以被视作整个金融体系崩溃的临界点"。但奇怪的是，临界点一词很少被用来描述城市发展的变化，即便是关于人口流动的也很少用到。这可能是因为在我们的集体意识中，天真地倾向于将城市的改变仅视为一种建设性的进步，而在科学领域中，临界点通常意味着从"坏"到"更坏"。在这种背景下，20 世纪 70 年代城市研究领域中的"临界点假设"那种经常受到质疑的术语就不足为奇了。这个"假设"是由美国的一些房地产专家发展出来的，他们相信"当社区中的黑人数量达到一定比例时，会导致社区白人的搬离。"这种表述在今天当然是不可以被接受的，但至少它说明了认知与文化是如何得以改善的。我很好奇今天我们在观察城市发展的时候，是否能以一种比 20 世纪 70 年代更好的方式来使用临界点这个概念。在城市发展的情境中，那些时刻可以让我们学习到什么？在我们的飞速发展的城市景观中是否也存在临界点？比如这些新兴城市，什么时候它们从蓝图演变成了现实空间，成为了真正的城市？更具体来说，对于一个城市，如何来应用和定义"临界点"呢？

在科学研究应用中，研究者们探寻在发生大规模突变之前的指标与变化模式。确定一个城市的成长与衰退模式正是城市研究的兴趣所在，特别是对于那些新兴城市的发展而言？哪些指标可以使我们在早期预见城市的繁荣和消亡？这对于城市景观的设计会有什么样的影响？自古以来，世界范围内一直在发展不同的新的城市。通常，它们会被建设成为激增人口的安全避难所，或作为某种政治理想的"证明"，或者仅仅是因生产和繁荣而形成的某些地点，这些城市矗立于各种极具野心和代价高昂的人类事业中。在过去的 10~15 年里，我们开始用看似大量的地图、图表、蓝图和统计数据来形象地分析城市的发展，这些材料意图说明我们是了解城市及其成长与衰退过程的。但是，我忍不住想到大部分的研究与分析是建筑师和城市规划者用来编著的工具，常常是作为一种自我推销的工具，用来获取新的项目委托。保罗·巴克与塞德里克·普赖斯和彼得·霍尔都是英国《新社会》杂志的前任编辑，在 1999 年发表的《没有被回顾的规划》中，叙述了城市规划在完成之后很少进行彻底分析的情况。"经常地，直到目前还普遍存在的状况是：当一个城市规划仅仅是完工时便声称它的目标实现了，没有人去检验这个规划是不是真的有效。"确实，几张图表和地图很难做到这一点。目前，世界上有一些最为成功同时也是最有问题的新兴城市已经存在了近半个世纪，我们的研究应当用一些分析模型去进行一些试验，而不是仅仅用来作为设计师、政府和开发商的公关材料。

在试图了解那些新的被规划的城市是在什么时候从一个新事物转变成当下的城市状况时，我建议可以用两种方法帮助我们认定是什么产生了城市的临界点。第一个方法是强有力的分析，但饱受设计和科学领域的批评；另外一个方法则是更多地与那些"张三李四"在城市中的日常经验有关。两种不同的方式一起使用，可能让我们的城市研究更加准确地了解临界点，并延伸到设计领域中。

超越分析学的分析学

由路易斯·贝当古、约瑟·罗伯、德克·黑尔宾、克里斯蒂安·库内特和杰弗里·韦斯特等专家学者以及伦敦政治经济学院 (LSE) 城市项目的优秀研究者们组成的多学科研究小组的工作，试图用海量的数据来呈现世界范围内城市的成长趋势，是学术界及科学界在城市研究领域中的角色的最好例证之一。他们的工作为我们敲响了警钟：尽管大量历史材料都证明城市是创新及经济增长的主要引擎，但用预测的理论来了解其动机机制、组织架构以及预估其未来轨迹和稳定性仍然是行不通的。研究小组发展出了一套动力模型来验证城市系统的有效性，目前仅在美国得到了实证经验证明，或许未来能在其他地区得到验证。他们的研究和分析一定程度地使城市模型可以用类似 Y (t) + Y0N(t)β 的公式来概括，以实证经验的方式来分析城市成长的各种要素，其中呈现了城市规模和各种指标的统一，意味着城市犯罪率及创新度的增长和对特定基础设施需求的减少成为了设定当地现实发展目标必不可少的数据。韦斯特和他的团队在 2007 年，发表的 "Growth, Innovation, Scaling and the Pace of Life in Cities" 中展示了他们近乎为城市供给系统的发展制定了某种公式，他们可以在研究中以实证经验证据说明在特定的城市情境中，餐馆、汽车经销商、邮局和药店会在什么时候被大量需要并呈现繁荣。如果我们可以创造类似这样的数据分析，而不是那些由建筑师或设计师提供的漂亮但缺乏实证经验分析依据的视觉说明，那么，我们或许可以开始更好地理解城市临界点本身，并在未来去预测有潜力的发展模式。

超越分析学的经历

尽管把聚焦于一个复杂的城市视作一个纯粹的分析模型听起来很诱人，但是当我们重新去阅读保罗·巴克的文章时，它还是会提醒我们 "一个城市不是一个电脑程序"，"城市有自己的生命"。也许韦斯特和他的团队会有不同的观点，但我更倾向于保罗·巴克所相信的 "没有人智慧到足以预测一个城市会如何成长，就像你无法预测一个创新和发明会产生多么复杂的效应和可能性一样，比如手机的出现。我们中没有人可以告诉人们如何决定和组织自己的生活，或是其对生存模式的体验会如何发展。" 也没有任何一个我们发展出的分析模式能适合世界上所有地区的所有城市。但是，如果我们相信（看起来也有足够的证据让我们相信）在一系列可以被公式及分析模型捕捉的量变到质变的临界点，使用上述 "超越分析学的分析学" 的方法的同时，我们可以更深度地来考察这些城市的生活质量。从 1962 年开始出版发行至 80 年代的《新社会》杂志一直主张：普通居民的需要以及他们的生活意愿才是了解一个城市的关键，而不是专家告诉人们应当如何生活。这也是 "然后，它成了一座城" 展览项目所扮演的角色。在这个项目和一系列的工作坊中，我委托了来自世界各地的六位录像艺术家和纪录片导演以及在深圳的艺术家和建筑师去分析六个特殊城市（六座都未满 60 年的新城市）中的普通人的日常生活。他们没有用一种清单式的方式去呈现一个城市，他们也都不是城市规划方面的专家，但他们用自己特定的方式去探寻：是什么把这些全新的 "城市空间" 转变成让此地的居民和访客感受到真实与生气的 "生活之地" ？是一个城市植被的数量和比例吗？是城市休闲活动的数量与品类的多寡吗？是城中建筑语言的多样性或文化艺术所扮演的角色吗？还是体现在一个城市的交通状况、上下班耗时、污染情况或犯罪率中？又或者答案只不过在它本身流逝的时光中？Surabhi Sharma（印度）从私人路径回溯柯布西耶所规划的昌迪加尔（印度，1953 年建城）；Cao Guimarães（巴西）记录了巴西利亚（巴西，1960 年建城）如何在众多城市要素获得自主性之后，才得以从一个规划真正成为真实的城市；Miki Redelinghuys（南非）记录了他询问哈博罗内（博茨瓦纳，1964 年建城）的居民如何找到回家之路的过程；Sam Green（美国）则为我们呈现了大家对于拉斯维加斯（美国，20 世纪 60 年代建城）的大相径庭、毁誉参半的各种情感；Astrid Bussink（荷兰）考察了阿尔梅勒（荷兰，1976 年建城）的居民是如何受到当地城市建筑的影响的；最后，王功新（中国）着力于描绘深圳（中国，1979 年建城）的城市生命力及它的动力美学机制是如何使之成为 "设计者之都" 的。与这些创作者各异的、感受性的及诗意的方式一道，马立安和黄静杰（观筑）合作的工作坊将在游弋于城市中的公共汽车上举办，他们用社会学的和富有创造力的方式进一步地详细讨论和说明关于城市生活圈临界点的观察与研究。

在此，我们用分析性的和感受性的两种方式，简要地概述了本文所强调的对于此时此刻的解析，但还尚未进入未来设计策略和城市规划模式。通过分析性模型的社会科学方式以及像 "然后，它成了一座城" 一样的实验性定性分析的方式，我们着力于探寻城市临界点可能的构成要素，或许可以开始更好地理解今天世界范围内高速的、风行的城市发展。正如我们在忽略艾滋病毒的教训中学到的，当事物到达它的临界点时，很可能产生不好的结果。如果我们能够知道更多的除设计解决方案以外的各种参数，或许才可以更顺利地将一个新城市的规划和模型真正变成一个有生气的城市。

GLOBAL PERSPECTIVES / 全球视野 | AND THEN, IT BECAME A CITY: SIX CITIES UNDER 60

City: Chandigare(India, 1953)
Title: Tracing Bylanes
Director: Surabhi Sharma(INDIA)

Surabhi Sharma, an independent filmmaker, was born in India in 1970. She studied film direction at the Film and Television Institute of India (FTII), Pune. Her films have been screened and awarded at various international and national festivals.

The beginning of the film switched between multiple static lens. Straight streets and town houses, a man standing in trance on an apartment balcony, leisurely walking people and patrolling guards, all the things were slowly moving or staying static. Huge spatial scale seemed to engulf human beings, as if only birds flying over showed a trace of vitality. The flipping side of the city was the individuals and their diverse lives. Immigrants gathered below the overpass looking for jobs; a barber opened up his own workplace besides a wall; a woman spread out a map and began the storytelling of entertainment time in the city. The female voice narrations as the background introduced audiences into a human scale of this city; expounded the philosophy and biology of living in Chandigarh from a more delicate perspective.

The bare concrete building facade is the tone of the city, the huge sound of the wind through the concrete construct parts seems to tell the plain sentiments of modernism in the 1950s. Two contrasting shots give more rhetoric meaning to an unsaid fact underlying this tone. The first shot captured a welcome slogan written "City Beautiful" at a highway entrance. The second shot belied a critical review from a citizen: "a shortcut does not exist in the city, because all the roads are straight square". Obviously, the city's beauty and shortcuts do not exist in the original plan. It is people that open up the bypass leading to those qualities. If it is said that French architect Le Corbusier planned an ideal form for Chandigarh, what this film recorded is how people give more meaning to this form and how they remedy the defects of the dehumanizing side of large-scale space.

城市：昌迪加尔（印度，1953）
片名：《发现旁道》
导演：Surabhi Sharma（印度）

Surabhi Sharma，独立女导演，1970年出生于印度拉贾斯坦，毕业于印度电影与电视学院导演专业，她所制作的影片已在世界各大电影节巡演并获奖。

影片的开始在多个静止的镜头间切换：笔直的街道和联排住宅，站在楼房阳台上沉思的男子，悠闲的行人和无事而巡逻的保安，一切都缓慢地进行，巨大的城市空间尺度似乎要将人吞没，仿佛只有鸟飞过时才闪现一丝活力。城市的另一面则是微观的个体和多样的生活。移民聚集在立交桥下面寻找工作机会，墙根儿的理发师开辟自己的工作场所，女人摊开地图讲述记忆中的饮食与舞蹈。影片背景中不同的女声旁白则为观众带入人性尺度，娓娓讲述昌迪加尔这座城市的生活哲学和个人体验。

裸露的混凝土建筑立面是这座城市的基调，穿过混凝土构件的巨大风声仿佛诉说着20世纪50年代朴素的现代主义情怀。影片中充满寓意的两个镜头呈现出强烈对比：进入城市的高速公路上有一座迎宾的大门，上书"城市美丽"，与此同时，居住者对它的评价却是"在这座城市不存在捷径，因为这些道路都是直方型"。很显然，城市的美丽与捷径并不存在于规划，而是人们开辟的"旁道"中。如果说法国建筑师勒·柯布西耶赋予了昌迪加尔一个理想的形态，这座影片所记录的则是人如何赋予空间形态更多意义，并修补大尺度空间的非人性要素。

| GLOBAL PERSPECTIVES / 全球视野 | AND THEN, IT BECAME A CITY: SIX CITIES UNDER 60 |

City: Brasilia (Brazil, 1960)
Title: Brasilia
Director: Cao Guimarães (Brazil)

Cao Guimarães, filmmaker and visual artist, born in 1965 in Belo Horizonte, Brazil. His work has been screened at the art museums around the world from the late 1980s, and he has participated in various film festivals after 2000.

Film "Brasilia" took a whole day as time clue, showing life images of residential and public spaces in Brasilia, a young modernist new town. In the morning, the parade ceremony on a large square was faring. At the far end of the square two lonely vendors stood. At noon, public buildings under the shadows of tree, graffiti on the red brick wall, garbage littering over green space, everything seemed static. The latter half of the movie applied a large number of images to reflect the crossover of various modes of transportation with daily life. The horse-drawn vehicles ran on the road, a converted bicycle sold goods, subway passengers seating and standing worn on numb looking. In the night, the bus station was crowded. Traffic flows coming across up and down the overpasses disappeared at the early hours of the morning, leaving the giant structure standing below the clouds of the sky.

Brasilia, a new town born in the socialist period, almost did not consider the setting of sidewalks and intersections in the initial design. The film starts from a flashing figure of cross as the metaphor of confusion and alienation caused by this lack. This is just a start. The film showed an anachronism in the space through a series of collage of images and mingled background music. For example, formalized apartment blocks in socialist workers' village were set to contrast with makeshift tents and mobile public toilets on the giant square. This contrast belies the dislocation between idealistic plan and real life. With the background sound of metal instrument and aircraft engine, the interaction between utopian dream imprinted on the urban landscape and rich urban life exuded a bizarre post-modernism.

城市：巴西利亚（巴西，1960）
片名：《巴西利亚》
导演：Cao Guimarães（巴西）

Cao Guimarães，导演和视觉艺术家，1965 年出生于巴西的贝洛奥里藏特市，其艺术作品在 20 世纪 80 年代末期于世界各地美术馆展映，2000 年后多次参加各个电影节。

影片《巴西利亚》以一天的时间为线索，展示了巴西利亚这座现代主义新城的居住空间和公共空间中的生活影像。早上，在蓝天白云下，广场上进行着稀稀拉拉的列队仪式，偌大的广场上只站着两个孤单的小贩。中午，阴翳之下的城市建筑，涂鸦的红色砖墙，城市绿地上的垃圾和吃草的马，一切仿佛都是静止的。影片后半段则运用大量镜头反映了城市中各种交通方式与日常生活的交割：马拉车在公路上奔跑，改装后的自行车，地铁上麻木的乘客，夜晚的公交站台人流交织，上下交叉的立交桥车流不息，到清晨时分则空荡荡，只有巨构在天空的流云下矗立。

巴西利亚，一座诞生于社会主义时期的新城，在设计之初几乎没有考虑人行道和十字路口的设置，影片一开头通过十字交叉的图底关系隐喻了这种缺乏所导致的混乱和疏离。这仅仅是一个开始，影片通过一系列的影像拼贴和背景音乐的交杂，于空间中展示时代错位，比如社会主义工人新村式的公寓住宅与巨型广场上临时搭建的帐篷和移动公共厕所形成的时代错位，体现出了人为规划与现实生活的距离。与此同时，乌托邦梦想在城市景观上的烙印与丰富的城市生活交割，在刺耳的金属乐器声和飞机引擎声的背景中，表现出一种光怪陆离的后现代气息。

GLOBAL PERSPECTIVES / 全球视野 | AND THEN, IT BECAME A CITY: SIX CITIES UNDER 60

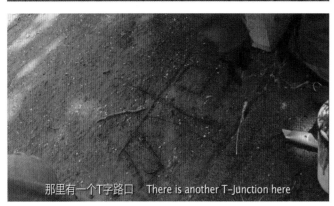

然后，它成了一座城：未满 60 岁的 6 座城市

City: Gaborone (Botswana, 1964)
Title: When a village grows up
Director: Miki Redelinghuys (Brazil)

Miki Redelinghuys, female documentary filmmaker, have long been active in the African continent. She began to settle in Cape Town in 2000 and start her career as documentary filmmaker thereafter. She has filmed several documentaries for broadcastings in South Africa.

In the film "When a village to grow up", Miki Redelinghuys tried to record the mixed identity of those who came from rural to urban in the rapid development of an immigrant city, Gaborone. The idea of interviewing part was inspired by a Gaborone tour guide's words: "Here in Botswana, in Gaborone. If you want to go somewhere…It's much easier to get directions through objects. Not by street." Be it real estate developer, shopkeeper or shoemaker under a mulberry tree, to find the way home was always a combination of story which was told through a complex object, rather than a series of street names. A T-junction, a roadside robot, a small shop, or a broken truck, any identifiable details of the daily life precisely constituted the memory of this city and its people.

In the film, the African countryside landscape overlapped with modern elements of the city. People thronging to this city from surrounding villages still bore the characteristics of village life. The coexisting of their rural and urban identity is like the juxtaposition between cow herd slow moving forward along highways and the giant billboards standing next to them. The noisy street music and informal stalls, the cranes in the city center and ongoing construction, these scenes highlight a vibrant and open possibility featuring the city's future modernity. The gist of this transition process has been concluded in a quote in the beginning of the film: "You have that transition from the village aspect to the modern aspect, but there is continuity. It is actually something you can see and feel."

城市：哈博罗内（博茨瓦纳，1964）
片名：《当一座村庄长大》
导演：Miki Redelinghuys（巴西）

Miki Redelinghuys，女纪录片导演，长期活跃于非洲大陆，2000 年开始定居开普敦，开始了纪录片导演生涯，为南非的电视台摄制了多部纪录片。

在影片《当一座村庄长大》中，导演 Miki Redelinghuys 尝试通过访谈的方式记录哈博罗内这座快速发展的移民城市中混杂的乡村和城市认同。导演从哈博罗内导游的一句话中寻找到灵感："在哈博罗内，如果你想要到什么地方去，通过物件识别方向会更容易，而不是通过街道。"无论是地产开发商、小卖部店主还是桑树下的一个鞋匠，找到回家的路永远都是通过一个复杂的物件组合所讲述的故事，而不是一系列路名的组合。一个丁字路口、一个路边机器人、一家小商店或一辆破卡车，日常生活中的细节事物恰恰构成了城市的可识别性和人们的记忆。

影片中，非洲乡村的景观与现代城市的要素相交叠，从周边村庄来到这座城市的人们带着对乡村生活的记忆开始城市生活，就像郊区公路上慢吞吞前行的牛群和旁边的巨型广告牌一样和谐共处。嘈杂的街头音乐和非正规的小摊小贩，城市中心处处可见的吊车和尚未完工的建筑，一切都在彰显一座从乡土迈向现代的城市所具有的活力和开放的可能性。这正应了影片开始时导演所引用的一句话："你看到的是一个从村庄到现代城市的转变，但是，这中间有持续性，实际上这就是你可以看到和感受到的一切。"

| GLOBAL PERSPECTIVES / 全球视野 | AND THEN, IT BECAME A CITY: SIX CITIES UNDER 60 |

很惊人的图像。

我们今天所创造的历史。"

然后，它成了一座城：未满 60 岁的 6 座城市

City: Las Vegas(U.S.A, 1960s)
Title: Las Vegas Portait
Director: Sam Green (U.S.A)

Sam Green, documentary filmmaker, was born in Detroit in 1966. He studied journalism from University of California Berkeley. His most recent projects are the "live documentaries" Utopia in Four Movements (2010).

The film is a visual portrait of Las Vegas. The director used some first-person voice-over musings to explain the particularity of Las Vegas in the United States and relate its historical time to its current development. The film edited many interesting street interviews. They were immigrant old couple, exciting tourists, honeymooners, dancing mobile parts seller, etc. From their words, we learned the city's unusual parts, and various reasons driving people here. In addition, the director subtly implanted archival image and footages of the 1950s and 1960s to describe many featuring points, like, the atomic bomb test in the desert, the demolition of the old buildings, the trainer injured by a lion. Accompanying its birth and development, these seemingly crazy but actually usual things exposed the characteristics of Las Vegas.

The director deliberately referenced Henry Ford to comment on the portrait of Las Vegas: "We want to live in the present and the only history that is worth a tinker's dam is the history we made today". This is a gambling city, an ever-changing city, a growing-on-the-desert city, in general, "destructive creation" perhaps is the best footnote of this city. The old building was destined to be demolished and reconstructed; fast-food-style wedding could not promise forever; only the openness and possibilities could ensure the vitality of the city. Through the whole film, a variety of commercial billboards and suburban apartment were used to epitomize the American dream. However, at the end of the film, a scene of half-baked buildings brought us back to reality. Director meant to prompt audience: their fate either will like previous buildings, being blasted and rebuilt, or left there as a sign of the end of the American Dream.

城市：拉斯维加斯（美国，20 世纪 60 年代）
片名：《拉斯维加斯肖像》
导演：Sam Green（美国）

Sam Green，纪录片导演，1966 年生于底特律，毕业于加州大学伯克利分校新闻学专业。于 2010 年与 Carrie Lozano 一同拍摄《乌托邦的四个运动》。

这部影片是拉斯维加斯的视觉影像志。导演运用了大量的第一人称旁白来解释这座城市在美国的特殊性，通过讲故事的口吻将拉斯维加斯的历史与现在结合在一起。影片剪辑了许多有趣的街头采访，如移居至此的老夫妻，兴奋的游客，蜜月旅行者，汽车零部件推销员，从他们的口中我们得知了这座城市的不同寻常，得知了人们来到这里的各种理由。除此之外，导演甚为精妙地将 20 世纪 50、60 年代的影像档案植入影片，如沙漠中的原子弹爆炸、旧楼房的爆破、驯兽师被狮子咬，伴随着这座城市的诞生、发展，这些既疯狂又平常的事情暴露了它的特点。

导演在影片中着意引用亨利·福特的金句以总结拉斯维加斯肖像的特点："惟一值得我们关注的历史是此刻我们正在创造的历史。"这是一座赌城，一座喜新厌旧的城市，一座沙漠上生长出来的城市，"毁灭性创造"也许是它的最佳注脚。旧的楼房要炸掉重建，快餐式的婚礼并不能承诺永远，一切皆有可能的开放性才能保证这座城市的活力。贯穿影片始终的各种商业广告牌和郊区式住宅是美国梦的缩影。然而，影片在结尾处所拍摄的 2008 年金融危机中造成的烂尾楼将我们拉回现实。导演提示观众：它的命运要么和以前其他的建筑一样，爆破重建，要么就成为美国梦终结的标志。

GLOBAL PERSPECTIVES / 全球视野   AND THEN, IT BECAME A CITY: SIX CITIES UNDER 60

然后，它成了一座城：未满 60 岁的 6 座城市

City: Almere (Holland, 1976)
Title: Creating Almere
Director: Astrid Bussink (Holland)

The Astrid Bussink, female documentary filmmaker, was born in 1975 in Eibergen, Netherlands. She obtained a master's degree at Edinburgh College of Art, and now lives in Amsterdam. Several documentaries made by her have been screened all around the world.

This short film meant to explore how Almere people build and use the space. The lens was locked up at a serial of city scenes, like runway on playground, parking lot in residential areas, waterfront trails, playgrounds, roundabouts, etc. By taking special image and sound footages, this film created an effect that behavior and activities of the people in these spaces seemingly followed the incessant guidance from the voice-over. The commands from the voice-over and the on-site feedback from actors constitute connections between people and space use. Whether the driver could adapt to the roundabout, whether the color of the glass wall is beautiful, whether pedestrians find street trash convenient to use, to what extent ground pavement could guide people behaviors in public space, all of above questions besetting designers could only be answered by putting design to real use.

When architects and planners design cities, they will keep thinking about the relationship between human and space, either on a sketch or in minds. Meanwhile, they will continue to modify the design to achieve the desired effect. The way this film was made provides another possibility of design. Every individual's behavior and interactions between people can test the rationality of the original design. High placement of camera in the film is supposed to create the same angle where designers overlooking the whole site. Almere, at the beginning of city building, emphasized on flexible planning; in the 1980s it promoted DIY building; and after 2006 it resumed this idea. This film can be seen as the footage of how Almere was designed and built. In this process, people participate in the design, and they better experienced urban life in designed spaces. This idealistic method of design does work in this city.

城市：阿尔梅勒（荷兰，1976）
片名：《创造阿尔梅勒》
导演：Astrid Bussink（荷兰）

Astrid Bussink，纪录片导演，1975 年出生于荷兰 Eibergen，在爱丁堡艺术学院取得硕士学位，现居于阿姆斯特丹，完成过多部纪录片并在世界各地参展。

这是一部探讨阿尔梅勒人如何建造并使用空间的短片。整部影片将镜头定格在一系列特定的城市空间场景上，比如运动场跑道、居住区的停车场、滨水步道、游乐场、环形交叉口等，并采取特殊的图像和声音剪辑，用不间断的画外音指导人们在这些空间中的行为与活动。声音指令和画面中行为的反馈构成了对空间使用的思考。驾驶者是否适应环形交叉口的行车方式，玻璃幕墙的色彩是否好看，步行街上的垃圾桶是否方便，不同的地面铺装对公共活动空间中的行为具有怎样的引导作用，这些设计师需要考虑的问题只有在真实的使用中才能得到解答。

建筑师和规划师在进行城市设计时，会不停地在草图纸上和头脑中琢磨人与空间的关系，并且不断修改设计方案以达到预期的效果。这部影片则提供了另外一种可能性，每一个人在空间中的行为及其与他人的互动都是对空间设计的测试。影片中，镜头始终被设置在一个高机位以模拟设计师俯瞰的视角，从而更适合看清全局与细部的关系。阿尔梅勒在建城之初便强调弹性规划，又曾在 20 世纪 80 年代和 2006 年后提倡 DIY 式的城市营造方式。影片所传达的是阿尔梅勒城市设计和建造的特殊理念：人们在参与中进行设计，在设计中更好地体验城市生活，这原本是一种非常理想的设计方式。

| GLOBAL PERSPECTIVES / 全球视野 | AND THEN, IT BECAME A CITY: SIX CITIES UNDER 60 |

City: Shenzhen (China, 1979)
Title: A Circle is drawn, A city is born
Director: Wang Gongxin (China)

Wang Gongxin, video artist, was born in 1960 in Beijing, China. He was dedicated to the exploration of video art in the 1990s, and is the first artists who create video works by fully using digital editing.

This is not a conventional documentary, but a collection of montages by using digital editing and animation. This film set Deng Xiaoping's action of "drawing a circle" in Shenzhen as the main metaphor which helped to deconstruct the scratching development process in Shenzhen. In addition to the use of footages of crowding street, other montage effect was composed of low pixel picture collage. Six Chinese characters (literally meaning "time is money") were set as masks. Then the director put picture of the high-rise buildings, people, traffic, and money flows in the foreground or background of these masks. He also used a fast flip animation, accompanied by continuous explosion of sound and the clock ticking, to create a tense rhythm.

By using the art of film editing, this film represented the symbolic landscape in deep political and economic implications, and deconstructs the history of China's reform and opening up, and the history of Shenzhen's city making. At the very beginning of this film, it implied the direct reason of Shenzhen's city making, which was the image as "an old man drew a circle". In addition to the direct instructions from Deng (in a simple stroke), the film also highlights the remarkable spatial development strategy, the coarse reality caused by the influx of all kinds of capital and human resources into a single city. This was a vivid manifestation of tremendous impetus of state-dominated economic reform brought to the city. The tension effect created by using montage represents the anxiety faced by individuals who experiencing this process. Slogans on red bands advocating economic efficiency form the unique landscape of the city, and omnipresent statues and photos of Deng Xiaoping constitute the deepest symbolic memory of Shenzhen.

城市：深圳（中国，1979）
片名：《可圈可点的城市深圳》
导演：王功新（中国）

王功新，录像艺术家，1960年生于中国北京，20世纪90年代开始致力于录像艺术的探索，是中国第一位完全以数码编辑创作录像作品的艺术家。

这并不是一部常规的纪录片，而是通过数码编辑和动画效果创造的蒙太奇。影片以1992年邓小平在深圳南巡时"画了一个圈"为核心意象，通过画圈的动画解构深圳从无到有的发展过程。影片运用了少量街道人群的影像，其他的蒙太奇效果使用的是低像素图片的拼贴。在拼贴的效果上，导演以"时间就是金钱"6个字为遮罩，将高楼大厦、人流、车流、金钱流等图片放入前景或背景，使用快速翻页的动画效果，并配以连续的爆炸声和时钟的嘀答声，制造出紧张的节奏。

影片运用剪辑艺术挖掘符号性景观背后的政治经济含义，对中国的改革开放和深圳的造城史进行解构。影片一开始就介入深圳造城的直接原因，即"一个老人画一个圈"。除了领导人的直接指示，影片也突出了可圈可点的空间发展战略，各种资本、人力集中涌入一地所造成的粗粝现实感，生动地体现出了国家主导的经济改革对这座城市所形成的巨大推动力。通过蒙太奇制造的紧张效果对应了发展中个体所面临的焦虑感。效率至上的标语式宣传形成了这座城市独特的景观，而无处不在的领袖雕像、照片则构成了深圳最深处的符号性记忆。

GLOBAL PERSPECTIVES / 全球视野

AND THEN, IT BECAME A CITY: SIX CITIES UNDER 60

David van der Leer commissioned six video artists and documentary filmmakers from around the world, and a group of Shenzhen-based artists and architects to analyze the quotidian lives of ordinary people in six exceptional cities: namely six new towns under the age of sixty.

None of the six artists working on the video pieces is using a checklist to codify the cities, nor is any of them an expert in planning issues, but in his or her own specific ways each speculates what it takes to turn these new towns into actual cities that feel like vibrant places to both inhabitants and visitors: Is it the number and the size of its trees? Is it the number or diversity of leisure activities? Is it in a multiplicity of architectural languages, or role of the arts and culture? Or is it its traffic and commuting times, pollution, or crime rates? Or is it as simple as in the passing of time?

Surabhi Sharma traces personal by lanes to the Corbusier plans for Chandigarh (India, 1953), Cao Guimarães documents how a planned town such as Brasilia (Brazil, 1960) only becomes a city when its various elements gain autonomy, Miki Redelinghuys asks the inhabitants of Gaborone (Botswana, 1964) how to find the way to their home, Sam Green shows the roller coaster of emotions, good and bad, that many of us have for Las Vegas (U.S., 1960s), Astrid Bussink looks at Almere (Netherlands, 1976) and wonders how much its inhabitants are directed by the architecture they live in, and lastly Wang Gongxin highlights how the dynamic aesthetics and the life force of Shenzhen (China, 1979) have resulted in a "city of designers."

These works celebrate a different, more ephemeral or poetic approach to the tipping points in urban life cycles. In order to deepen the audience's experience of the films, Mary Ann O'Donnell and ATU have organized a series of sociological and creative workshops that bring the films to the inhabitants of Shenzhen. The workshop venue is two-fold.

First, the films are shown on a bus that drives around the city, allowing visceral comparison between the city onscreen and the city outside our window.

Second, the bus stops at representational sites within the city – commercial, public, and educational, for example – to open the conversation about tipping points to a wider audience. Indeed, the workshops aim to enrich local discussions about the ends and means of planned urbanization by bringing the films to the general public and providing a forum for discussing when and how a city becomes a city (and not "just a place to work" as many have described early Shenzhen).

The bus drives around the city
在城市中行驶的公共汽车

Citizen are invited to watch 6 videos one the bus
经过设计师改装的大巴邀请市民同行，在城市中穿梭的同时观赏 6 个影片

然后，它成了一座城：未满60岁的6座城市

David van der Leer 委托了6位分别来自世界各地的视觉艺术家和纪录片导演以及深圳本土艺术家和建筑师，剖析6个城市年龄小于60岁的年轻城市的普通老百姓的生活。

做这些影片的艺术家们没有像罗列清单一样去编纂这些城市，他们也都不是问题规划专家，他们却用各自的方法推测如何把这些新城镇变成使城镇居民和游客都感到充满活力的地方：是树木的数量和大小？是休闲活动的数目还是种类的多样化？是建筑语言的变化丛生抑或艺术和文化所扮演的角色？是它的交通和通勤时间、污染或犯罪率呢，还是像时间推移那样简单？

Surabhi Sharma 通过车道追踪勒·柯布西耶为昌迪加尔（印度，1953年）制定的规划；Cao Guimarães 用文献证明一个规划好的城市，如巴西利亚（巴西，1960年），只有在各个方面都获得自主权时才能成为城市；Miki Redelinghuys 询问哈博罗内的居民（博茨瓦纳，1964年）如何找到回家的路；Sam Green 觉得我们大多数人对拉斯维加斯（美国，20世纪60年代）抱着跌宕起伏、或好或坏的情绪；Astrid Bussink 关注阿尔梅勒（荷兰，1976年）并好奇当地居民如何受他们居住的房屋建筑的影响；最后，王功新强调深圳（中国，1979年）的动态美学和生命力如何使之成为"设计者之都"。

为了使观众加深对影片的体验，马立安与ATU观筑举办了一系列社会学创意研讨会，把影片带给深圳居民。研讨会有着双重意义：首先，公共汽车一边行驶于城市各个角落一边播放这些影片，它使我们可以对屏幕内和车窗外的城市有真实的比较。第二，公共汽车将会停在城市里有代表性的地方，例如商业、公共、教育地带，使之向更多受众打开有关临界点的对话。事实上，研讨会的目的是通过影片加深大众对规划的城市化进程的目的和方式的讨论并提供一个平台来讨论一个城市何时和怎样成为一个城市（不是大家所描述的早期的深圳——"只是一个工作的地方"）。

Volunteers prepare for the workshop
志愿者在进行工作坊搭建准备工作

Children are encourged to draw their dream city at the workshop
青少年参与到工作坊中，绘制心中理想的城市

GLOBAL PERSPECTIVES / 全球视角

Roundtable: Cities Under 60 (Forum Transcript)

DEC 10th 2011 Shenzhen Biennale Roundtable@ HUA Club, OCT Art and Design Gallery, Nanshan District, Shenzhen

Moderator:
Terence Riley [Chief Curator, Shenzhen Biennale]

Roundtable discussion on the Shenzhen Biennale special exhibition on new cities across the globe for Grouped exhibits in house 2 [G Under 60: And Then It Became A City, Six Cities, Hua Gang, curators and videographers and theorist].

SPEAKERS

Surabhi Sharma, Director, Title: Tracing Bylanes, City: Chandigare [India, 1953], India
Rochelle Steiner, curator and dean of the Roski School of Fine Arts
David van der leer, Curator, Assistant Curator, Architecture and Urban Studies, Guggenheim Museum
Astrid Bussink, Director, Title: Creating Almere, City: Almere [Holland, 1976], Holland
Stefano di Martino, Director, M. Arch Program, professor of Practice in Architecture, USC School of Architecture
Qingyun Ma, Della & Harry MacDondald Dean's Chair in Architecture, USC School of Architecture
Michelle Provost, International New Town Institute [INTI], Holland

圆桌会议：城市未满60（论坛纪录稿）

2011年12月10日深圳双年展，深圳南山区华·美术馆（华·沙龙）

主持人：
泰伦斯·瑞莱 (Terence Riley)

圆桌会议60在深圳双年展特别展览中举行，全球新城市展览在2号展馆举行[G小于60：于是它成为了一个城市，六个城市，华港，策展人和录像者和理论家]。

演讲者：
Surabhi Sharma（项目录像作品导演，《发现旁道》，印度）
Rochelle Steiner（项目策划人，美国南加州大学艺术学院院长、策划人）
David van der Leer（项目策划人，古根海姆博物馆建筑和城市研究所助理策展人）
Astrid Bussink（项目录像作品导演，《创造阿尔梅勒》，荷兰）
Stefano di Martino（美国南加州大学建筑学院教授、研究生部负责人）
马清运（美国南加州大学建筑学院院长）
Michelle Provoost（国际新城研究中心 (INTI)，荷兰）

Terence Riley: When I was preparing for this biennial, I totally understood that Shenzhen government would like to apply Shenzhen as the theme of the biennial, so I curate a themed exhibition called Shenzhen Build. I also positioned Shenzhen in the concept of New Cities for discussion. Like other emerging cities, Shenzhen is very young and rising abruptly. The capital, energy and time that contributed to the planning and construction, can be explained that new cities present the great achievement of people. So I went through several cities emerged during 20th century and compared them to each other, such as Chandigarh in India, Gaborone in Botswana, Las Vegas in US, Almere in the Netherland and Shenzhen. The comparison and assessment leads to a summary of whether people's endeavor brings benefit, whether people enjoy such urban life, and whether the city is built according to the urban planer's schedule. So far, all the guests here are very interested in city, even though they're not urban planers, however, they are all urban researchers, planners, and residents. So, here's my first question: is urban planning a science? If not, what is it?

Note: The 6 cities in the project are Chandigarh (India, 1953), Las Vegas (USA, 1959), Brasilia (Brazil, 1960), Gaborone (Botswana, 1964), Almere (The Netherlands, 1976) and Shenzhen (China, 1979).

David van der Leer: First of all, we have to analyze the importance of city, but over-analysis also brings problems, for example, taking the cities as mathematics formula, urban planning as science. I think, science does not contain many aspects of cities. So when making the videos, we put in a lot of subjective views. The exhibition in an interactive way presents the analysis and comparison of 6 cities under 60 years old, by video-graphing them and collecting various data. Besides, the videos are showed in a bus which traveling around Shenzhen. It is very interesting that the videos about the city can be watched during the urban trip.

Ma Qingyun: If a city has a development model, it means the city complies with certain kind of rules, such as density and vitality, so we can say the city has certain rationality. But does it mean that urban planning is science? My answer would be NO. Urban planning is not only a science. In China, we understand China from the scientific perspective, maybe because we are more concerned about the hardware in the process of researching and executing urban planning.

Michelle Provoost: Science requires hypothesis. From this aspect, urban planning is a science, especially for the newly planned cities. Urban planners often search for a model, a method that is suitable for political, economic and social environment. Regardless of modernism in 20th century or various styles in 1960s, urban planners always hoped their planning could be treated as scientific works. However, those new cities are proved as failure 20 and 30 years later. Because they do not satisfy the expectation and have flexibilities, or are already out of style, we can call it as a science suffered continuous failure.

Ma Qingyun: Can science nject an un-doubtable, un-challenged power in it? In such case, does science really work? If the design fails, is it the problem of the designer or decision maker's original design?

Surabhi Sharma: From certain angles, new cities are not very successful. But there is another problem: some people tend to preserve all the past of a city, and turn it into a museum. If all the Indian cities are the places for historical preservation rather than residences, it would be lost in the time. Some cities are designed in a poetic way rather than in a scientific way, but after some years, it is found out that the city could not reflect the Indian social and economic reality. For example, Chandigarh has not progressed much since 60 years ago. This is a problem itself.

Stefano Di Martion: The cities in our research have many similarities. The conclusion is: urban planning is not the exactly science. City is a cultural presentation about the residents. No matter what you do, everyone gathers to inhabit. It is the nature of human being. Indeed these cities have their weaknesses, but like other cities, these new cities are continuously developing and suitable for retirement. For example, Las Vegas has good weather. this may match the demands from a specific group who want to spend the life there. So we need to discuss how to evaluate these cities. Certainly, our work so far is very limited.

Astrid Bussink: I also want to respond to the question of whether the cities are failure. So many people in the Netherlands reckon Almere is a failure, because it becomes an accessory of Amsterdam. Originally people wished that Amsterdam residents could move to Almere, but they did not, because they thought they belonged to Amsterdam. People who live in Almere love their city, but the outsiders think it is a failure.

Stefano Di Martion: Some cities may never develop further over certain phase. Now Paris has become a failed city. We also think New York is declining because there is little space for development. I think, a successful city must continuously create new spaces, which are not only the physical spaces, but also cultural and imaginary spaces. This is how a city preserves the vitality.

Terence Riley: It is very interesting to talk about failures of a city. We can discuss the success and failure is subjective or objective. If a city grows irreconcilably with the original plan, it might be a failure, or an unexpected success. The influence from residences makes it different from the original plan. I would also refer to the political factor. We firstly brought out the question of whether urban planning is science or not, because we found that urban planners and architects nowadays take urban planning as a political action, a still and quite formal practice. They think any urban issue is a technical issue, and design can resolve these issues. When you try to improve urban planning methodologies, you have to take into account of what kind of people should be involved in urban planning along with technical experts.

Michelle Provoost: In my point of view, all the new cities are planned by top-down political decisions. Architects have a general plan, but seldom include the future residences into the design and development. So I totally agree your opinion. If we say those new cities are failure, we would say the urban planning is totally failed. We could not ignore the power from the residences, and the historical cities are built by multiple layers of architecture, rather than a single vision. So if one person can be added in the plan, He or she should work from bottom-up, and bridge the gap between the top and bottom. Shenzhen can do some more interesting work. I guess the urban planning of Shenzhen is different from the anticipation 30 years ago.

Terence Riley: Let's see more concepts of failure. In 1970s, when I was still studying at an architecture school, the post-modernism began to emerge. British critic Charles Jencks wrote a book on it, and described Brasilia in a quite negative way, far behind Paris and London. I thought I would see a lot of wired space in Brasilia. However, I found it very inhabitable. And the architecture design was excellent. The size of the city is only 1/10 of Washington, and full of residents. People adopt pitched roofs due to the hot weather. I also found the ground floor is quite useful, so columns are added to raise up the building, underneath which children can play and trees can grow. It is a smart design to create better living qualities, which is different from my original expectation of a giant object. When Brasilia is compared to London and Paris again, I feel the critic is not correct. We may have to look into if the cities with planning are better than cities without planning at a similar age. Are there any advantages of the planned new cities? For instance, three cities in Texas became homogenized during the growth. Chinese cities also lose their uniquenesses. Shenzhen, Guangzhou and the whole PRD, t become a mega-city with no differences from each other. Is it a success brought from planning?

Stefanodi Martion: It is difficult to judge good or bad without the factor of time. Because the issues we mentioned are beyond the city itself. We're only talking about the newly planned city. Urban planners may face more extensive issues: where the resource omes from? How the resource affects the city? Then our discussion is not only limited in the planning of new cities, but also the planning of the relationship between new cities and natural environment.

David van der Leer: (to further discuss whether planning or non-planning is better) one measurement for planning is to see of the city becomes sustainable, low carbon and green.

Terence Riley: We just started comparing these cities several months ago, so it is difficult to answer right now. I saw that the whole world deviates from the general plan. Chandigarh, Brasilia and Shenzhen require to be the motor cities. If it is planned to be a motor city, how can it be an energy-saving city? If we really want to save energy, less planning is necessary.

Michelle Provoost: We do not have a definitive ecological standard, so people have various understandings on ecology-friendly city. We saw a lot of such slogans in China, even in many urban planning proposals. If the carbon-neutral Masdar City is built up from scratch in a desert, how much is it ecology-friendly? We may say that the inner circle of Paris is friendlier because it has higher density, and encourages pedestrian rather than private cars. So I think urban planning could not help us to answer how a city can become more ecology-friendly. Some people would say that there's very limited space for constructing new cities, but there is now a new wave of building new cities. The scales are even much larger than before. The original urban planning of Almere has a large number of public transport. But it is pitty that the city is not ecology-friendly enough.

Terence Riley: Let's go back to the issue of failure. We could not say that all the 6 cities are failures. For example, Brasilia with few population is only 50 years old, where the government is located. After work around noon or afternoon, people just take a flight back to home in San Paolo. If a city is only a place for work rather than social interaction, we may say that it does not have sociality.

San Paolo is a spontaneous city without urban planning. The mass transport was recently deployed. The traffic causes a cruel situation: the low-income families take 3 hours on commuting, so that they could not enjoy the family time. Although Shenzhen has less problems than San Paolo, is the city is always chasing for vitality. However, is the vitality enough? Is aesthetics still necessary?

David van der Leer: I think Shenzhen is a young city. I don't know exactly how large it is, but I feel that half of its population or even 60% are 20 to 30 years old. This city is full of energy and passion. You can tell from the audience here.

Ma Qingyun: I want to remind all that the history of Shenzhen might not be shorter than Beijing. Beijing might be even younger. The changes made Beijing become a younger city than Shenzhen. But now only Forbidden City is saved from demolition; Hu Tong, and old towns are actually fake. As time changes, Beijing will be a new city, and Shenzhen will no longer be a new city. All of these may not be caused by urban planning.

Surabhi Sharma: Time is a factor that makes city more interesting. Beijing can be called a new city in 20th century based on the proportion of new buildings. Certainly the "quantity" sometimes is not most important. The cultural memory often is defined by the old part of a city. The city may have different layers, which make Shenzhen very interesting. After many years, these 6 cities we mentioned today would not be called 'new towns'.

Stefanodi Martion: Some cities are definitely new. New York used to be a radical new town, which forms a basis for the stat quo. In fact the city is less than 200 years old. Urban planning is a catalyst, which can successfully make a city. For example, both of Chandigarh and Brasilia are the formally designed cities through regional planning. Brasilia is very interesting. The original plan was to have 500,000 residents. In 1980s, the figure was only 60% reached. The city has expanded extensively, and the earliest plan

GLOBAL PERSPECTIVES / 全球视野 ROUNDTABLE: CITIES UNDER 60

is only a small part of the urban area now. A city can not be planned from the beginning to end. In Brasilia only the architecture was built in the early stage, but they are only part of the city. Urban planning can be the impetus for further development: bring energy to the city to ensure the sustainable development.

Astrid Bussink: Now we shall talk about Almere. I think this city is still under development. You can see some change in years. In 1970s, the city's aim was to build houses for working class, but later, this area became a plan with high criminal rate. In 1990s, people were looking for a new life style that everyone could build their own houses. Almere has the ever-changing urban concepts at the different stages. I totally agree with Michelle that urban growth needs time. Only after decades it becomes interesting. Some cities grow according to the plan, but Almere changes the plan all the time. We may say, Almere is a series of new cities.

Surabhi Sharma: Chandigarh has two parts, the main planned area and new suburbs. Both of them are developing and already become messy. I always say that some general plans are failure: they could not respond to the reality. Without doubt Chandigarh is a very beautiful city with pleasant environment and expands speedily. However, it becomes e ugly and chaotic after man-made technologies are added. Energy is absent in those so-called suburbs, and they are not a living city. The city is a still place, where only officials go, but young people or diverse age group would not go.

Michelle Provoost: I think Shenzhen is exceptional lucky. It takes advantage of weather, geography and economic policies. 14 years ago when I arrived in Shenzhen first time, the city was very uncomfortable: no trees, nothing, and only ugly buildings. After 14 years, the city is totally changed. From this perspective, Shenzhen is not only a very beautiful city, but also full of energy. It is much better than other new cities. Some new towns are pretty, but have no energy at all.

Ma Qingyun: I would like to talk about the concept of time or city memory. In Shenzhen, I guess less than 1/4 people grow up here, and 3/4 population stay partly in Shenzhen, and partly in their hometown. The migration affects the city growth. Some people I met in Shenzhen 15 years ago have already moved back to their hometown or other cities, but they all have memories about this city. Shenzhen is attractive to all the migrants. I grew up in a new city too. In 1950s, almost all the new cities were constructed based on American model. Many of them were built to satisfy the production requirements, just like industrial cities in America. The second type of new city is to serve the army. My family lived in a military new town. There were approximate 14 families in the community. I lived there until 18 when I went to university. The trees were big there, under which we often read. The children usually hang out together rather than staying at home. The memory about that place is the most precious for me. I never mix it with that of Rome, Paris or Manhattan. When I went back to the town, I found it is still there, but trees were smaller than I felt, because I grew up. My memory has a sense of romantics, without limitation and regulation. When you were a child, you can do anything you want. Likewise for the first generation who immigrated to Shenzhen, they might have the same thoughts. They thought Shenzhen might be the best place for them, and some young people might agree with it soon.

圆桌会议：6 小于 60

Terence Riley： 在我准备策划本次双年展的时候，我理解深圳市政府希望这次双年展能够把深圳作为一个主题，比如说举办一个"深圳建造"的主题展，或者把深圳放在一个新城的概念当中进行探讨。深圳和其他的新城一样，非常的年轻，是真正拔地而起的新城。通过投入到这个城市的建造规划的资金、精力和时间，可以看到新城代表了人类的成就。因此，我在策展的时候，就考虑在20世纪挑出几个城市进行比较，比如印度的昌迪加尔、巴西的巴西利亚、博茨瓦纳的哈博罗内、美国的拉斯维加斯、荷兰的阿尔梅勒及中国的深圳。通过对比评估，总结人类所投入的努力是否带来了回报，是否能够让人们在城市中享受到生活，而且城市的发展是否符合规划师的计划。看来各位在座的嘉宾对城市非常感兴趣，他们可能不是专业的城市规划师，或者他们都在研究城市、规划城市，而且也是城市的居民。**因此我向大家提出第一个问题：城市规划究竟是不是一个科学，如果不是科学，是什么呢？**

注：《城市小于60》项目分析的六个城市分别是昌迪加尔（印度，1953）、巴西利亚（巴西，1960）、哈博罗内（博茨瓦纳，1964）、拉斯维加斯（美国，20世纪60年代）、阿尔梅勒（荷兰，1976）及深圳（中国，1979）。

David van der Leer： 我们首先考虑到了分析城市的重要性，但是对城市过分分析也会带来一些问题，比如说有人最后会把城市全部简化成数学等式，把城市规划当成科学。但是我觉得科学没有办法涵盖城市的方方面面，因此，我们在给这6个城市制作影片时，加入了一些主观的想法。我们的展览将6个小于60年的城市的分析成果展出，并比较了这6个城市。我们对6个城市进行了摄像，收集了各种数据，并在展览中通过互动的方式对比这些数字。另外，我们还制作了录像，把录像放在一个大巴中，大巴在深圳的各个地方转一周，播放6个城市的音像。能在城市当中看城市的录像，这非常有趣。

马清运： 如果一个城市有发展模型，就表明了这个城市遵守了某些规律，比如说密度、活力，也就是说这个城市有一定的理性。但是，这就等于城市规划是一门科学吗？我的答案是否定的，城市的规划并不仅仅是科学。在中国，我们想从科技的角度理解城市，可能因为我们在研究规划和执行城市规划的过程中，更关心的是硬件。

Michelle Provoost： 科学需要假设。从这个角度上来说，城市规划是一门科学，尤其是新城市的规划。城市规划师总是需要寻找一种模型、一种方式适应城市所在的政治经济和社会环境。不管是20世纪的现代主义还是60年代的一些风格，城市规划师一直希望把自己的作品当成是一件科学作品，但是大部分的新城在20~30年之后都被认为是失败例子，因为没有能够满足人们的预期，或已经过时了，或没有灵活性，所以我们可以称之为不断失败的科学。

马清运： 科学是不是能够注入一种不可置疑、不可挑战的权威在里面？在这个案例里面，科学到底是不是真的有用，如果设计失败了，是否是由于主导者或者管理者最初设计时的问题？

Surabhi Sharma： 从某些角度来看，新城确实做得不是很好。但也有这样的问题：有些人欲将城市的过去全部保留，将城市变为一个博物馆。如果印度的城市都是保存历史的场所，而不是住的地方，就等于已经迷失在时间当中。有些城市设计得并不是很科学，而是以一种非常诗意的方式来设计。但一段时间以后发现，这不能够反映出印度的社会和经济现实，昌迪加尔市和60年前的城市差不多，这本身就是一个问题。

Stefano Di Martion： 我们研究的城市有很多相似的地方。最后得出的结论是：城市设计并不一定存在真正意义上的科学。城市是一个文化的表现，是住在这里的居民文化的表现。不管你的职业是什么，大家都会到一个地方居住，这是人的本性。确实，这些城市有自己的起伏，但是与很多其他城市一样，这些新城市都在不断地发展，而且这些城市都很适合退休的人居住。比如说拉斯维加斯，气候很好，可能符合某些人群的需求，有些人说不定就想在那里养老。所以，我们需要仔细讨论用哪些指标衡量这些城市。当然，我们这个工作本身是很有限的。

Astrid Bussink： 我也想回答一下这些新城市到底是不是失败的。很多荷兰人都认为阿尔梅勒是一个失败的城市，因为它成为了阿姆斯特丹的附属品。本来人们希望阿姆斯特丹的人能够搬过去，但是阿姆斯特丹人对自己的城市很有归属感，不愿意搬到阿尔梅勒。住在里面的人喜欢这个城市，而住在城外的人认为阿尔梅勒是一个失败的城市。

Stefano Di Martion： 城市有可能发展到一定阶段就不再发展。巴黎现在已经成为了一个失败的城市。纽约，我们也觉得它开始走向颓败，因为它现在已经没有什么发展空间了。我认为一个成功的城市一定要能够不断创造出新的空间，不仅是物体空间，还有文化和思想空间。我觉得通过这样的扩张才能够保持这个城市的生命力。

Terence Riley： 关于失败的城市的问题，谈论起来肯定很有意思。我

GLOBAL PERSPECTIVES / 全球视野　　　　ROUNDTABLE: CITIES UNDER 60

们可以讨论成功和失败是客观还是主观。比如一个城市的发展和城市规划的预期不一样，它既可能是一个失败的案例，也可能是一个出其不意的成功，因为是当地居民对城市的影响让这个城市和原有规划产生不同的。我还要谈一下政治的元素。我们一开始问大家"城市规划是不是一个科学"是因为我们发现现代城市规划师或建筑师，无论是有意还是无意，把城市设计变成了一种政治的行为，一个非常静止、正式的行为。他们以为任何城市问题都是技术方面的问题，而且设计师可以完全解决这些问题。如果你想要去改善城市规划的方式，你需要去想除了技术专家以外，把什么样的人纳入到城市规划团队里面。

Michelle Provoost： 我觉得所有的新城市都是自上而下的规划，都是一种政治的决策。建筑师有一个总体的规划，但是很少有城市在设计和开发的时候是和未来的潜在居民一起设计的。所以我完全同意你的说法。如果我们称这些年轻的城市是失败的话，应该称为城市规划整体的失败。我们不能忽略居民的力量，有历史的城市都是由一层一层的建筑组成的，而不是基于某一个单一的理想。所以，如果加入一个人的话，可以找一个可以做自下而上的工作的人，能够帮我们在上下之间起一个桥梁作用。所以，我想在这个方面，深圳也可以做更有意思的工作。我估计深圳的规划可能跟30年前的预期并不一样。

Terence Riley： 我们再来看看失败的概念。70年代的时候我还在读建筑学校，当时后现代主义开始出现，英国的评论家查尔斯·詹克斯（Charles Jencks）写了一本书介绍后现代主义。把巴西利亚描述得很负面，远不如巴黎和伦敦，说巴西利亚问题很多。我以为到巴西利亚会看到一些很奇怪的空间，后来发现这个城市的可居住性很强，而且设计的建筑都非常好。城市的大小大概只有华盛顿的1/10，居民区已经住得满满的。当地人根据天气炎热的现状采用了斜屋顶，还发现楼下的空间比较有用，就专门在建筑下面加了很多柱子，将楼抬起，小孩在下面玩，也可以保证树不断地生长。本来以为会看到一个庞然大物，结果发现了非常聪明的设计，人们也有很好的生活质量。因此，再把巴西利亚和伦敦、巴黎进行比较的时候，我觉得评论家书里写的是错误的。我们在比较的时候，要看同一年龄的城市规划之后是不是要比不规划的城市好，这些规划过的新城市是不是有一些优点。比如说美国得克萨斯州的三个城市，他们在成长过程中好像完全同质化了。像中国很多的城市，基本上没有了自己的个性，深圳、广州，整个珠三角区域看上去都一样，成为了一个大城市。一个区域内的城市群发展成为一个统一的城市，这是规划所带来的成功吗？

Stefanodi Martion： 没有时间这个因素，可能很难判断一个城市的好坏，因为我们所说的这些问题已经超越了城市本身。我们现在研究的只是规划的新城市，还有很多其他城市。城市规划师面临的问题会更加宽泛。比如说这个城市的资源从哪里来？这些资源对城市产生怎样的影响？这样我们的讨论就不限于新城市的规划，还包括规划新城市和自然环境之间的关系。

David van der Leer：（继续讨论城市究竟是有规划好还是没有规划好）一个标准可以是看规划是否使城市变得可持续、低碳绿色。
Terence Riley： 几个月前我们才开始对比这些城市，这个问题确实很难回答。我看到这个世界似乎越来越偏离总体规划。昌迪加尔、巴西利亚、深圳的规划都要求他们成为汽车城市。如果总体规划就要建成汽车城市，城市真的能够节省能源吗？如果真的想要省能源，我们就需要少一些规划。

Michelle Provoost： 我们所说的生态的标准还没有确定，因此人们对什么是生态友好的城市有不同的理解。中国有很多城市都说"打造成生态友好的城市"，在城市规划方案当中都有这样的表述。但如果一个碳中和城市马斯达尔城（Masdar City）在沙漠当中拔地而起的话，究竟有多少生态友好性呢？可能有人会讲巴黎的内城更加生态友好，因为它的密度更高，而且不鼓励私人轿车，鼓励人行。因此，我觉得规划可能没有办法帮助我们回答怎样的城市更加生态友好。有人会说现在没有再造很多新城市的余地，但现在又有了造新城市的浪潮，这些新城市可能规模比以前的更大。阿尔梅勒最早的规划中就有很多公路交通，没有形成生态友好的城市，这也是非常遗憾的事情。

Terence Riley： 回到失败的问题，不能说这六个城市都是失败的城市。比如说巴西利亚只有50年的历史，人口非常少，而且是政府所在地，中午、下午下班后大家都乘飞机回到圣保罗。如果一个城市仅仅是一个办公场所，不是一个社交场所，它可能就没有我们所说的城市的社会性。圣保罗是自发形成的城市，没有什么规划，才刚刚部署大众交通。车辆造成了一个很残酷的情形：工资非常低的穷人上下班需要花三个小时，基本上没有时间享受家庭生活。深圳的问题要比圣保罗少很多，但是它的问题是想追求活力。活力就够了吗？不要美丽吗？

David van der Leer： 我们不能这么讲，我觉得深圳对我来说是很年轻的一个城市。我不知道它究竟多大，但我感觉人口中有一半到60%

可能都是20~30岁的年轻人。这个城市充满了活力和热情。大家看看我们在座的各位多么年轻，还可以看到他们充满了热情。

马清运： 我想提醒大家，实际上深圳的历史不比北京短，北京可能更加年轻一些。北京的变化可能会使北京成为比深圳更加年轻的城市。可能北京现在只剩下紫禁城没被拆，你们喜爱的胡同、老城，实际上都是假的。因此，我们可以看到随着时代的变迁，北京成了一个新城市，深圳不再是新城市。这可能并不是规划所造成的。

Surabhi Sharma： 时间是使城市更加有趣的一个因素。看看北京新建筑的比例，可以说它是20世纪的新城市。当然，"量"有时候不是最重要的。一个城市的文化记忆往往是由老城部分定义的，城市可能有不同的层次，这就使深圳也非常有趣，这六个新城市随着时间的积累，以后也不会被叫做新城市。

Stefanodi Martion： 有些城市肯定是新城。纽约最早的时候是一个非常激进的新城，因此形成了现在的纽约，但是实际上也不到200岁。城市规划可以成为一个催化剂，成功打造一个城市。比如说昌迪加尔和巴西利亚，完全是经过设计的城市，经历了区域规划等正式的规划。巴西利亚非常有趣，原来规划是50万居民。在80年代，居民只达到计划人口的60%。当时执行规划的时候，这个城市铺得非常开，而最早规划的区域只占现在城市中很小的一部分。其实这个规划是一种催化剂，我们没有办法将一个城市从一开始到消亡都规划好。巴西利亚早期通过规划将所有的楼都建好，这是可以的，但是这仅仅占整个城市的一部分。规划可以成为继续发展的一个动力：给城市注入活力，保证环境的可持续发展。

Astrid Bussink： 现在谈谈阿尔梅勒。我觉得这个城市还在发展中，每几年都会有一些改变。70年代的时候，阿尔梅勒的目标非常明确，就是让工薪阶级的人有房子住，但后来这里成为了犯罪率很高的地方。20世纪90年代的时候，人们希望有一种独特的居住方式，就是能够按照自己的要求去建房子。其实阿尔梅勒在不同的时间段有着不同城市的概念，而且一直在变化。我同意Michelle的观点，很多城市需要时间才能成长，有了几十年的历史才变得很有意思。有些地方按照总体规划建设，但是阿尔梅勒一直在改规划。可以说，阿尔梅勒是一系列的新城市。

Surabhi Sharma： 昌迪加尔现有的主规划区和新郊区都在发展，而且已经变得非常混乱。我一直在说这些总体规划是失败的：它没有能够根据实际情况进行变更。昌迪加尔毫无疑问是一个非常漂亮的城市，在以非常快的速度扩张。昌迪加尔的核心是环境宜人，但是在加入新技术以后有些失败，这是人为的失败，让它看起来很丑陋、混乱，而且这些所谓的郊区本身没有任何的能量，不像是一个活生生的城市。只有官员才去，不会有年轻人或是不同年龄层次的人，是一个静止的城市。

Michelle Provoost： 我觉得深圳真的是例外的幸运。不仅仅是气候得天独厚，还有经济方面，这个地方有点像压力锅。14年前我第一次来的时候，这个地方让人很不爽：没有任何树，什么都没有，都是很难看的大楼。但现在这个压力锅放了14年之后，城市有了翻天覆地的大变化。从这个角度来看，现在深圳不仅仅是一个非常漂亮的城市，而且有很多的能量，比很多新城市都做得好。很多新城市漂亮但是没有能量。

马清运： 我想谈谈时间的概念或是城市的记忆。在深圳，我猜大概不到1/4的人口长居于深圳，3/4的居民有一段时间在深圳，有一段时间住在老家。人口流动其实会影响到城市的发展。我15年前在深圳认识的人有一半已经回到自己的老家或者是去了其他的地方，但是他们都对深圳有非常好的记忆。这个其实和深圳的吸引力有关。深圳一直对进来的居民有无尽的吸引力。我是在一个真正的新城市中长大的。50年代的时候，一种新城是按照美国的方式来建的，很多的新城就是为了满足生产的需求，像美国的工业新城一样，第二种是为了服务军队的军区新城。我就住在军区新城，大概14家人住在一个小区，我住到18岁后去读大学。那个地方的树都很大，我们会在树下读书，小朋友们从来不在家里面，一起在外面做坏事或是一起做好事。这些对我来说是最好的记忆，我不会将这些记忆与在罗马、巴黎或是曼哈顿的记忆混在一起。我对于巴黎的印象不一定会有这么好，我回到小镇以后，发现小镇还在，但是树变得越来越小，因为我长大了。我小时候看这个树像山一样高，长大了回去看树，其实那么矮小。我的记忆带有一种浪漫主义的感觉，没有任何的限制和规则。那时候，你是一个小孩，可以在里面为所欲为。我觉得对深圳第一代移民来说，说不定会有同样的想法，觉得深圳是世界上最好的地方，估计有很多年轻的居民很快也会认同这个观点。

GO WEST PROJECT: ALLMETRO POLIS
西游项目: 所有都市

Curators: Michiel Hulshof Daan Roggeveen
策展人: Michiel Hulshof, Daan Roggeveen

GLOBAL PERSPECTIVES / 全球视野

GO WEST PROJECT: ALLMETROPOLIS

It's 2030.

In the heart of Western Europe lies a vibrant city that people refer to as the 'Dutch Miracle'. In twenty years, the population has increased from 250,000 to 3 million inhabitants - and the end of the explosion is not yet in sight. In and around the city a thriving economy has developed: a combination of green industries, high tech and biotechnology. The city has a mixed character: high rise buildings interspersed with a lot of nature. The public transport network is one of the best in the world. The three universities in the city are top-notch. With the 2028 Olympics, the city has also presented itself in a sportive way. The population is highly educated and is coming from all over the world: Russia, North Africa, Central America, China and India. The official language is English.

Welcome to the Flevopolder.

Welcome to the Allmetropolis.

The similarities between the Chinese city of Shenzhen and the Dutch Flevoland region with the city of Almere are stunning. Both areas started their urban development in the late 1970s: Shenzhen as a promising Special Economic Zone (SEZ) and Almere as promising New Town on Flevoland, the world's largest man-made island. Both are located in a river delta, adjacent to major international metropolises: Shenzhen near Hong Kong and Flevoland next to Amsterdam. In spite of these similarities, both urban areas developed in a completely different manner.

The Go West Project put into perspective the economic, social, spatial and ecological developments of Shenzhen and Flevoland. This forms the starting point of a research that shows the possibilities and opportunities of using the Shenzhen model in a European context.

In September 2011, Michiel Hulshof and Daan Roggeveen (Go West Project) published How the City Moved to Mr Sun—China's New Megacities. With Allmetropolis they continue their research into China's urban development. Allmetropolis is an initiative by the Go West Project in partnership with the International Newtown Institute (INTI).

The exhibition has been made possible with the support of the City of Almere.

www.gowestproject.com

西游项目: 所有都市

现在是 2030 年。

在西欧的中心地区有一座充满活力的城市,它被人们称为"荷兰的奇迹"。在过去的 20 年中,人口数量由 25 万增长到 300 万,然而爆发之势还远远看不到尽头。城市和周边地区也出现了繁荣的经济发展:由绿色产业、高新技术和生物技术组成的产业链。这座城市具有丰富的个性:美丽的自然景观簇拥着高耸入云的摩天大楼,城市的交通运输网络居世界领先地位,市内还有三所世界顶级大学。在 2028 年举办了奥林匹克运动会后,这座城市更能展现自己的体育优势。城市居民都受过高等教育,他们从世界各地云集而来:俄罗斯、北非、中美洲、中国和印度。官方语言定为英语。

欢迎来到弗雷沃庞德尔。

欢迎来到"所有都市"(Allmetropolis)。

中国的城市深圳和荷兰弗莱福兰省的阿尔梅勒市之间的相似之处多得令人吃惊。两者都于 20 世纪 70 年代开始了它们的城市发展进程:深圳作为特许的经济特区,阿尔梅勒作为弗莱福兰省的新兴城市,也是世界上最大的人工岛屿。两者都位于河流三角洲处,毗邻主要的世界级港口:深圳毗邻香港,而阿尔梅勒毗邻阿姆斯特丹。虽然拥有这些共性,深圳和阿尔梅勒却有着不同的发展状态。

西游项目着眼于深圳及弗莱福兰省的经济、社会、空间和生态发展。这些因素作为研究的出发点,展示了在欧洲的语境下运用深圳模式的可能性。

于 2011 年 9 月,Michiel Hulshof 和 Daan Roggeveen(西游项目)出版了《城市如何移向孙先生——新生的中国大城市》(How the City Moved to Mr Sun-China's New Megacities)一书。通过"所有都市",他们继续对中国城市发展进行研究。"所有都市"是西游项目与国际新城研究所(INTI)合作的一个项目。

展览由阿尔梅勒市提供赞助。

www.gowestproject.com

GLOBAL PERSPECTIVES / 全球视野　　GO WEST PROJECT: ALLMETROPOLIS

ALMERE, FLEVOLAND

Flevoland and Shenzhen have some striking similarities: they both started their rapid urbanisation at the end of the 1970s.
Before that time, Flevoland was a mostly empty piece of land, reclaimed from the South Sea in the Netherlands.

弗莱福兰和深圳有着惊人的相似之处，即两地的城市化均从 20 世纪 70 年代开始。
在那之前，弗莱福兰由荷兰南部海域填海造陆而来，大部分地区荒凉空旷。

FLEVOLAND in the Rhine River Delta
FLEVOLAND
Area / 地区　　　　　　　1,400 sq km
Population/ 人口　　　　　400,000
Density/ 密度　　　　　　 286 / sq km
GDP　　　　　　　　　　 $ 11 billion
GDP/capita/GDP/ 每人　　$ 29,000

RANDSTAD
Area (Urban)/ 地区　　　 4,300 sq km
Population/ 人口　　　　 7,1 million
Density/ 密度　　　　　　1,650 /sq km
GDP　　　　　　　　　　 $ 307 billion
GDP/capita/GDP/ 每人　　$ 46,700

SHENZHEN

PHOTO: HE HUANGYOU
摄影：何煌友

Shenzhen was a small and poor fishermen's town in the South of China. Like the rest of China, the economy was state-led.

同时期的深圳还只是一个中国南部的小渔村。
国有计划体制统治着这里的经济，与其他中国地区相同。

SHENZHEN IN THE PEARL RIVER DELTA
SHENZHEN
Area / 地区　　　　　　　1,950 km^2
Population/ 人口　　　　　10 million
Density/ 密度　　　　　　 7,700 inh/sq km
GDP　　　　　　　　　　 $ 146 billion
GDP/capita/GDP/ 每人　　$ 14,500

HONG KONG
Area (Urban)/ 地区　　　 1,104 km^2
Population/ 人口　　　　 7 million
Density/ 密度　　　　　　6,480 inh/sq km
GDP　　　　　　　　　　 $ 325 billion
GDP/capita/GDP/ 每人　　$ 45,736

西游项目：所有都市

Both New Towns were planned next to an existing international international metropolis :
Flevoland is next to the Randstad (Amsterdam, Rotterdam, Utrecht and The Hague)...
..while Shenzhen is next to Hong Kong

两座新城均位于已有的国际化都市附近：
弗莱福兰临近 Randstad（阿姆斯特丹、鹿特丹、乌特勒支和海牙）

CONCEPT: SUBURBAN IDYLL
The News Towns in Flevoland were meant to become a paradise for the middle class: a safe, healthy living environment where every family had their own house with their own garden.
Advertisement:
"In Zeewolde, you find forests, water for surfing, golf courses and your company all next door"

概念：田园式的栖居
位于弗莱福兰的新城最早被规划成中产阶级的居住天堂，即一套别墅、一个花园、一个家庭和一个安全健康的生活环境。
广告：在 Zeewolde 尽享森林、冲浪、高球……
而你的公司，就在隔壁！

···while Shenzhen is next to Hong Kong
……深圳，临近香港

CONCEPT: BUSINESS IDYLL
Shenzhen was created to be an experimental ground for the practice of market capitalism. It was meant to become one of the regions in China that would 'get rich first' and 'open up' to the international world.
Propaganda poster:
"Special Economic Zones - China's great open door"

概念：商业天堂
深圳成立之初被定义为市场经济的试验田，它将"先富起来"并且成为"对外开放"的前沿阵地。
宣传海报：
经济特区——中国对外开放的大门

| GLOBAL PERSPECTIVES / 全球视野 | GO WEST PROJECT: ALLMETROPOLIS |

ALMERE, FLEVOLAND

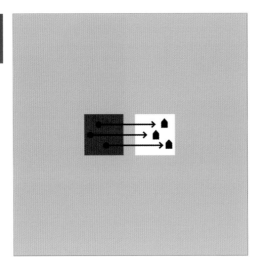

POLICIES FOR A GOOD LIVING ENVIRONMENT
Flevoland doubles the size of the Randstad. Policies were directed at creating a good living environment:
- Low density spatial development
- Creation of pedestrian zones and bicycle paths
- Advanced system of public transport with bus-only-lanes
- Experiments with new residential concepts
- Focus on green zones and playgrounds for children
- Development of leisure areas

政策：创造良好的居住环境
Flevoland 的尺度为 Randstad 的两倍。
政策的制定直接关系到创造一个良好的居住环境。
- 低密度发展
- 修建步行道和自行车道
- 先进的公交车专用道系统
- 实验新型居住理念
- 关注儿童活动空间
- 发展休闲空间

SHENZHEN

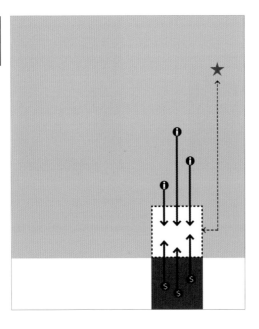

POLICIES FOR ECONOMIC DEVELOPMENT
Shenzhen is a buffer between China and the outside world. It attracts money from Hong Kong, Taiwan and oversees and workers from China and combines it with cheap land. Policies were directed at creation a good business environment:
- Tax breaks for companies
- Posibilities for foreign investment
- Loose migration laws, loose labor rights
- Good infrastructure
- Deregulation
- Land availability

政策：创造良好的商业环境
深圳是中国与世界的缓冲区域。它吸引来自港澳台和海外的资金和来自内陆的劳动力，与廉价土地结合。政策的制定直接关系到良好商业环境的建立。
- 公司免税政策
- 鼓励外资投资
- 宽松的移民政策，宽松的劳动力政策
- 良好的基础设施建设
- 放宽管制
- 宽松的土地政策

NEW LAND
Flevoland is the world's largest manmade island, reclaimed from the South Sea in the Netherlands between 1939 and 1968.
Originally, this New Land was meant to serve agricultural purposes, making the Netherlands self-sufficient in food production.

新大陆
弗莱福兰是世界上最大的人造岛屿，于 1939 年至 1968 年在荷兰南部海域填海造陆而成。岛屿营建之初是为了缓解荷兰的农业用地不足的局面，使荷兰在农产品方面达到自给自足。

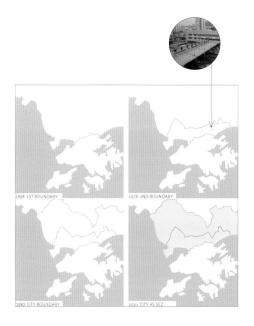

BORDER TOWN
In 1898 the first boundary marked the current border between Hong Kong and the mainland China. By introducing the so called '2nd boundary' in 1978, the Chinese government turned the border town of Shenzhen in a buffer zone. This buffer was appointed a Special Economic Zone in 1980. The 2nd boundary had its own border control system. However, the actual city of Shenzhen was much bigger than the SEZ. Only in 2010, the whole city was integrated. The physical border of the 2nd boundary still remains, although it is not in use at this moment.

边城
1898 年，一道边界线将中国香港与中国大陆分割开来，1978 年中国政府将深圳变为一个缓冲区域，称为"第二边界"，这一缓冲区在 1980 年被指定为经济特区。第二边界拥有其自身的控制系统。然而，深圳的范围远比所谓的特区要大。直到 2010 年，整座城市才转变为经济特区。
尽管第二边界的物理存在已停用，但依然还存在。

| GLOBAL PERSPECTIVES / 全球视野 | GO WEST PROJECT: ALLMETROPOLIS |

ALMERE, FLEVOLAND

1976: FIRST INHABITANTS: PIONEERS
The very first 25 inhabitants of Almere, the largest New Town on Flevoland, came from Amsterdam. They fled the poor living conditions in the big city and were called 'The Pioneers'. According to a newsreport on the National Evening News 'the keys they got open to them a new world that they only have to fill in according to their own wishes'.

1976 年，第一批居住者：拓荒人
阿尔梅勒是弗莱福兰省中最大的一座新城，第一批入住阿尔梅勒的居住者有 25 人，来自阿姆斯特丹。他们逃离了大城市喧嚣的居住环境，被视为"拓荒者"。当时国家晚间新闻报道提到"新房钥匙为他们（第一批居民）打开的全新生活已经十分圆满，现在他们只需要在其中添置个人喜好的物品"。

SHENZHEN

1978: FIRST INHABITANTS: WORKERS & OFFCIALS
Before 1978 Shenzhen was inhabited by farmers and fishermen. The first new arrivals were officials from Beijing who came to execute policies aimed at the creation of a special zone. After that, migrants arrived from all over China: thousands of contruction workers who built roads and factories, and workers who populated the new factories, founded by State Owned Enterprises or businessmen from Hong Kong and oversees.

1978 年，第一批居民：工人与管理者
在 1978 年之前，深圳的居民大多是农民和渔夫。第一批外来居民是来自北京的官员，他们来践行创建深圳特区的政策。自此，来自中国各地的城市移民开始充斥深圳。成千上万的建筑工人盖房修路，国有或外资生产线上的工人日夜开工。

西游项目: 所有都市

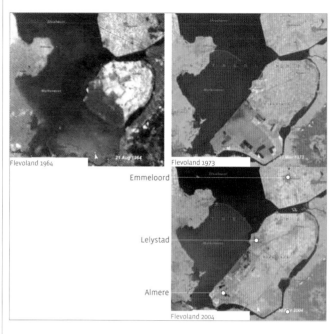

FLEVOSPEED

The Dutch government planned the first New Towns in Flevoland far away from the Randstad urban region. This emphasised the rural character they had planned for the New Land.
First the New Town of Emmeloord was planned, then the new provincial capital Lelystad. But urbanisation only really took off when the government started the development of Almere, close to Amsterdam and the rest of the Randstad urban region.

FLEVO 速度

荷兰政府计划在远离 Randstad 城区的弗莱福兰省建立第一个新城， 在新的土地上准备着重农村化的规划。
首先是埃默洛尔德，其次，新的省会城市莱利斯塔德也在计划中，但是直到政府开始对临近阿姆斯特丹的阿尔梅勒和 Randstad 的剩余城区投入开发，城市化才真正起飞。

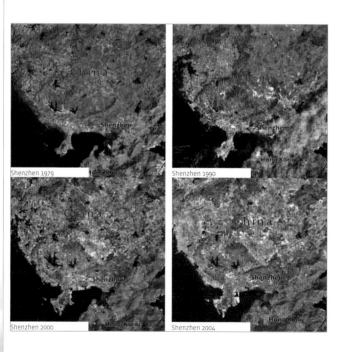

SHENZHEN SPEED

In less than 30 years time Shenzhen developed a unique urban landscape in which thousands of highrise buildings and factories (grey) have replaced previous agricultural and vegetated areas (green) at a speed never seen before on this planet.

深圳速度

在少于 30 年的时间里， 深圳以前所未见的速度发展成为一个独一无二的城市， 成千上万的高层建筑和工厂代替了之前的农田，灰色代替了之前的绿色。

GLOBAL PERSPECTIVES / 全球视野 — GO WEST PROJECT: ALLMETROPOLIS

ALMERE, FLEVOLAND

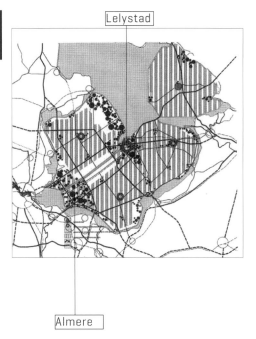

MULTINODAL

In the first regional plans, the New Towns in Flevoland were planned as a multimodal network, separated by large green areas. In the centre the provincial capital Lelystad would arise. However, in reality the most western town Almere would soon outgrow Lelystad in size.

在初始的区域性计划中，新城弗莱福兰被计划为一个多节的网络，由大面积的绿色区域隔开，中间的省会城市莱利斯塔德将冉冉升起。然而，事实上，最西边的城市阿尔梅勒却很快在面积上超过了莱利斯塔德。

SHENZHEN

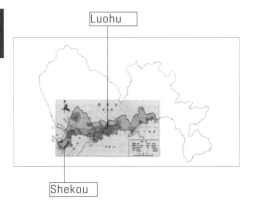

MULTICORE

Planning of Shenzen followed the logic of a border town. Every time a new border crossing with Hong Kong was opened, a new city district emerged around it. Starting point were the border crossing at Luohu and the ferry harbour at Shekou. Luohu became Shenzhen's first commercial centre, and Shekou its first industrial area.

当初规划深圳是作为一个关口城市，每次一个通往香港的新的关口开放，一个新的城区就出现了。最开始是罗湖和蛇口码头，罗湖是深圳的第一个商业中心，蛇口则成为了深圳的第一个工业区。

西游项目：所有都市

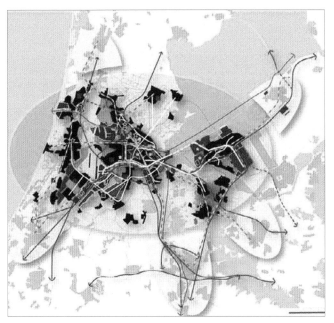

PART OF RANDSTAD CONURBATION
Since the beginning of the 2000s, Flevoland doesn't see itself as a separate entity, but as an integral part of the Randstad conurbation. In this view, Flevoland can supply the Randstad with the much-needed space for economic and residential expansion.

大都市的一部分
从 2000 年初开始，弗莱福兰并不是作为一个个体存在，而是作为构成 Randstad 都市的一个必要的部分存在。这样看是因为弗莱福兰可以为 Randstad 的经济和居住扩展提供必需的空间。

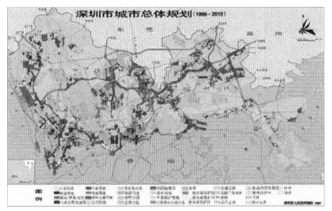

Statutory planning
From Hong Kong, Shenzhen copied the concept of Statutory planning. However, in these dynamic conditions the planning mechanisms were hard to maintain. In this map urban plan for Shenzhen in 1996, the prediction for 2010 was that Shenzhen would have 4.3 million inhabitants. Right now, the city has an estimated 15 million inhabitants and even 20 million mobile phone connections. Thus, the green areas of this 1996 urban plan are allready filled in. Not statutory planned, but allowed by lower level governments. Or just illegally built.

法定计划
深圳效仿香港法定计划的概念，然而，在不断变化的条件下计划机制难以维持。在地图上，1996 年深圳的城市化发展预计在 2010 年将有 430 万居民。目前，这里估计已有 1500 万居民，2000 万部移动电话。这样一来，1996 年规划的绿地已经被填满，没有法定的计划可循，是被低层政府允许，或者是被不法地建造的。

GLOBAL PERSPECTIVES / 全球视野　　GO WEST PROJECT: ALLMETROPOLIS

ALMERE, FLEVOLAND

FROM SUBURBIA TO CITY

In the first residential area of Almere-Harbour, every family had a small house with a garden. Pedestrians, bikes and cars were stricktly separated. Every block had a small playground for children. In the 1990s, Almere started to develop it's urban core. It meant moving away from the purely 'suburban feel'. Instead, high-rise office towers, apartment blocks and iconic architecture were intended to give the inhabitants the impression of living in a big city. In the last years, there's a move away from top-down planning. New residential areas have been developed in a radical way: inhabitants can design their own houses according to their wishes.

从郊区到城市

在阿尔梅勒港口的第一个居住区，每个家庭有一栋带花园的小房子，人行横道、自行车道、机动车道都被严格分开，每个街区都有一个小的儿童娱乐场所。在 1990 年，阿尔梅勒开始发展城市核心，这意味着远离单纯的"城郊感觉"。摩天办公大楼、公寓区、地标建筑致力于给居民在大城市居住的印象。过去数年里，有一种远离组织管理的严密的计划，新的居民区已经发展为一种激进的方式：居民可以根据自己的喜好自行设计他们自己的房屋。

SHENZHEN

PHOTO: HE HUANGYOU　　摄影：何煌友

FROM FACTORY TOWN TO CITY

Most of the early inhabitants of Shenzhen moved into the enormous dorms at the premises of the factory where they worked.
As the city expanded, it swallowed more and more of the surrounding villages, leading to the emergence of "urban villages". The former farmers started to demolish their mud houses and construct apartment blocks of their own design.
After the year 2000, a lot of urban villages were demolished and replaced by modern compounds: gated communities of repetitive residential towers. Still, more than 50% of the Shenzhen inhabitants live in urban villages.

从工厂到城市

早期深圳居民居住在他们工作的工厂的大型宿舍中。
城市发展后，周围的村庄被侵吞，导致了"城中村"的出现，原住民推倒了他们的土房，建造了他们自己设计的公寓大楼。2000 年后，许多城中村被拆毁，被现代化的建筑取而代之：门控社区造型重复单调的住宅楼出现了。然而，仍然有 50% 的深圳居民住在城中村里。

西游项目：所有都市

LIVING IN SUBURBIA : IN THE OLDEST NEIGHBOURHOOD OF ALMERE , PEDESTRIANS , BIKES AND CARS WERE STRICKLY STRICKLY SEPARATED

住在郊区：在阿尔梅勒最原始的居民区，步行道、自行车道和机动车道被严格划分。

AFTER THE YEAR 2000, ALMERE STARTED TO DEVELOP MORE URBAN FORMS OF LIVING IN THE NEW URBAN CORE .

2000 年之后，阿尔梅勒在新的市中心逐渐开发出不同种类的城市化居住模式。

LIVING IN A FACTORY: MOST OF THE EARLY INHABITANTS OF SHENZHEN MOVED INTO THE ENORMOUS DORMS WITH BUNK BEDS AT THE PREMISES OF THE FACTORY WHERE THEY WORKED .

住在工厂：最早的一批深圳居民搬进了大型工厂中塞满上下床的宿舍中。

IN SHENZHEN , FARMERS STARTED TO CONSTRUCT THEIR OWN APARTMENT BLOCKS THAT THAT COULD REACH 10 OR 15 STORIES

在深圳，农民开始自建房屋，有的高达 10 到 15 层。

GLOBAL PERSPECTIVES / 全球视野　　GO WEST PROJECT: ALLMETROPOLIS

ALMERE, FLEVOLAND

AT THIS MOMENT, ALMERE IS EXPERIMENTING WITH LARGE RESIDENTIAL AREAS WHERE EVERYONE CAN DESIGN AND BUILD THEIR OWN HOUSES

这时，阿尔梅勒开始试验鼓励居民自己设计建造房屋的大规模居住片区计划

SHENZHEN

ALMOST ALL NEW RESIDENTIAL AREAS IN SHENZHEN ARE GATED GATED COMMUNITIES OF REPETITIVE HIGH-RISE

在深圳，大部分的居住区仍是围合小区加上造型重复单调的高层

CITY OF COMMUTERS

Every morning between 7am and 9am, there's a long traffic jam on the highway from Almere to Amsterdam. Every afternoon between 4pm and 6pm there's a traffic jam in the opposite direction.

The reason is simple: more than half of the working population of Almere works outside Flevoland, mainly in the Randstad. To them, Almere is merely a 'sleep city', a place that stands out in providing a good living environment.

城市通勤者：

每早七点到九点，从阿尔梅勒到阿姆斯特丹是长时间的交通堵塞，每天下午四点到六点，相反的方向同样的故事上演。

原因简单：有超过一半的阿尔梅勒上班族在弗莱福兰之外工作，主要在 Randstad。对他们而言，阿尔梅勒只是个"睡眠城市"，一个因为良好居住环境而突出的城市。

CITY OF MIGRANTS

Every year at the start of Chinese New Year and the October Holidays, there's a massive jam of workers queuing in front of Shenzhen train station to 'go home' to their families. And every year after Chinese New Year and the October Holidays, there's a massive influx of workers arriving to go back to work in Shenzhen.

The reason is simple: the majority of the population considers Shenzhen a 'working city', a place that merely offers them good jobs but never completely becomes home.

移民城市：

每年中国的春节要开始时和 10 月份黄金假期，大批在深圳工作的人挤在火车站排队等候"回家"，而在每年的春节和 10 月假期后，大批的打工者又开始返回深圳工作。

原因简单：大部分的人认为深圳是个"工作城市"，一个只为他们提供好工作但并不完全是家的城市。

GLOBAL PERSPECTIVES / 全球视野

GO WEST PROJECT: ALLMETROPOLIS

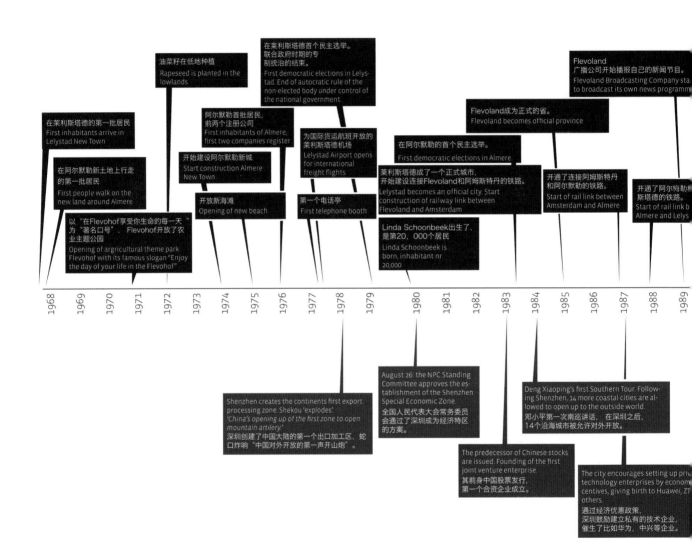

西游项目: 所有都市　　　　　　　　　　　　　　　　　　　　　　　　　　　357

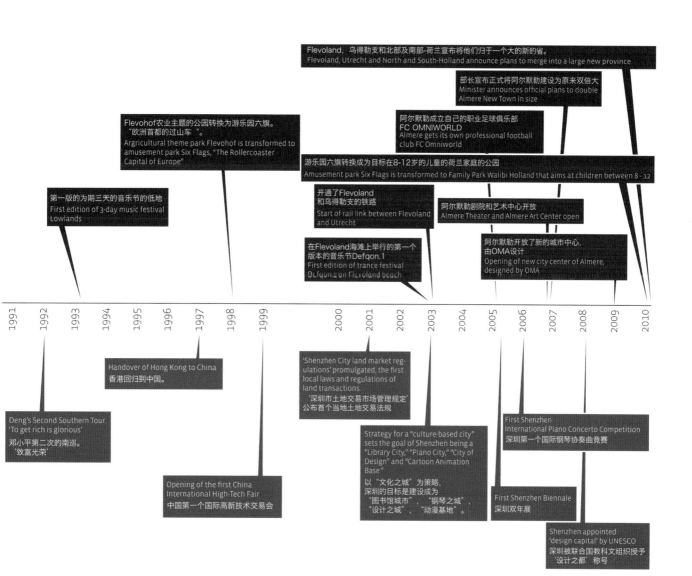

| GLOBAL PERSPECTIVES / 全球视野 | GO WEST PROJECT: ALLMETROPOLIS |

ALMERE, FLEVOLAND

SPACE
The biggest asset that Flevoland has is space. Right next to one of the most densely populated areas in the European Union, Flevoland offers vast amounts of space that can be used for space-intensive leisure, like the yearly Defqon.1 festival on Almere-Beach or play golf on one of the six golf clubs in Flevoland.

空间：
弗莱福兰最大的财产是空间。它临近欧洲联盟人口最稠密的地区之一，弗莱福兰有巨大的空间可用于如每年在阿尔梅勒海滩举行的 DEFQON .1 音乐节，或者在弗莱福兰的六个高尔夫俱乐部之一打打高球。

IN THE ORIGINAL PLANS FOR FLEVOLAND, THERE WAS A LARGE INDUSTRIAL AREA PLANNED THAT INTERCONNECTED THE NEW TOWNS OF ALMERE AND LELYSTAD.

在 Flevoland 初始计划中，有一个大工业区被计划为连接阿尔梅勒和莱利斯塔德新城。

SHENZHEN

PEOPLE AND MONEY
The biggest asset of Shenzhen is that it gives access to a vast and relatively cheap Chinese labour force, right next to one of Asia's most important financial centers. You can see the influence of Hong Kong everywhere.

人和钱：
深圳最大的财产是有大量的廉价劳动力，正临近亚洲最重要的一个金融中心。你随处可以看到香港的影响。

SHENZHEN WAS PLANNED TO BECOME A CITY OF WORKERS. AFTER MORE THAN 30 YEARS, THE FIRST GENERATION GENERATION OF FACTORY WORKERS IS AGEING RAPIDLY

深圳被计划为一个工人之城。三十多年后，第一代工人快速老龄化。

However, when production moved from Europe to China in the 1980s, the area remained empty.

然而，当制造业在 80 年代从欧洲转移到中国时，这里依然是空旷的。

Over the years, it has developed developed into a beautiful green area that that is now an official National Monument.

近些年来，它逐渐发展成绿色生态区域，如今它已经成为国家级自然公园

A lot of production has moved out to surrounding cities like Dongguan.

很多制造业已经搬到周边城市，例如东莞。

Shenzhen is becoming the hometown for a younger and higher educated generation of Chinese middle class.

深圳成为受过高等教育的人和年轻一代人的家乡，他们是中国的中产阶级。

GLOBAL PERSPECTIVES / 全球视野　　GO WEST PROJECT: ALLMETROPOLIS

ALMERE, FLEVOLAND

FUTURE
This year, the authorities of four Dutch provinces (Flevoland, Utrecht and North Holland) proposed to merge together into one new province of the Northern Randstad.
It is unsure if those plans will materialise, but if so, it would create the largest province of the Netherlands in both number of inhabitants, size and economy.

未来：
这些年，荷兰的四个省（弗莱福兰，乌得勒支和北部荷兰）被提议合并为北方 RANDSTAD 的一个新省。现在还不确定这些计划是否会实现，但真如此的话，无论在人口、面积还是经济方面，此地将会成为荷兰最大的省。

SHENZHEN

HONG ZHEN
There have been plans to merge the cities of Hong Kong and Shenzhen together into one big urban area, with the Second Boundary as its northernmost border. That would be a perfect way to make use of the existing border control system.

香圳：
已经有计划将香港和深圳并为最大的城区，用第二道边界线作为最北边的边境，这样，现有的边境控制系统仍将得到很好的利用。

THE STREET
街道

Curator: Terence Riley
策展人：泰伦斯·瑞莱

GLOBAL PERSPECTIVES / 全球视野

THE STREET / 街道

The modern architecture biennale started in 1980, when Paolo Portoghesi organized the first such an event in Venice. Conceived to alternate with the venerable Venice Art Biennale, the first International Architecture Biennale Venice also marked the public opening of the arsenale, the city's medieval armory and military shipyard. This Biennale, entitled The Presence of the Past, instantly made the recurring event into the place to understand current trends and achievements in the field of architecture.

For the event's core, Portoghesi devised a street of facades. Called the Strada Novissima, or Newest Street, it occupied the central nave of the arsenale's corderia, or rope making factory. Each of the twenty invited architects were asked to design a façade, which was constructed by scenic artists from Rome's Cinecitta studio, behind which they could display models, drawings, and photographs of their work. Behind the facade, the architects designed individual exhibits of their most current work.

The Street is organized in a similar way. Designed by the 12 architects selected by the curator in consultation with leading critics worldwide, the installations and facades create the "street" even as the "street" provides the structural basis of the architecture. Architecture Creates Cities. Cities Create Architecture. Unlike the 1980 Venice exhibition, however, the architects of the Street have been encouraged to design facades – or, as it has turned out, non-facades—that are spatial and material rather than two-dimensional.

Portoghesi successfully identified those architects—then in their 30's and 40's—who would become leaders in the coming decades: Rem Koolhaas, Frank Gehry, and Arata Isozaki amongst them. He did so at a time when the world of architectural innovation was spinning on an unusually unified intellectual axis—or, at least, so it seemed in the late 1970's. Despite a worldwide common emphasis on sustainability, a renewed interest in what might be called the "presence of the future", and a deeper bond between the practices of architecture, landscape design and urbanism, it is not clear that the work of the 12 architects featured here can be seen as unified—even loosely—under a single theoretical banner today. Nonetheless, it is certain that—like the 1980 Venice Biennale—a future perspective will see our contemporary architecture in ways that we cannot see ourselves today.

Participating Architects:
Alejandro Aravena, Arquitecto (Santiago, Chile)
Aranda Lasch (New York, US)
Atelier Deshaus (Shanghai, China)
Fake Industries Architectural Agonism (NYC, US and Barcelona, Spain)
Hashim Sarkis Studios (Beirut, Lebanon and Cambridge, US)
J. Mayer H. (Berlin, Germany)
Johnston Marklee (Los Angeles, CA)
MAD Architecture (Beijing, China)
Mass Studies (Seoul, Korea)
OPEN Architecture (Beijing, China)
SO-IL (New York, US)
Spbr (Sao Paolo, Brazil)

现代建筑双年展始于1980年，当时是在威尼斯，保罗·波多盖希（Paolo Portoghesi）第一次组织了这样的活动。在原本的设想中，第一届威尼斯国际建筑双年展将与具有悠久传统的威尼斯艺术双年展交替举行，它的开幕地点也选择了军械库（Arsenale）——自中世纪就存在的兵工厂和船厂。这届双年展以"过往的呈现"为主题，以这个循环事件来理解当下的潮流和建筑领域的成就。

波多盖希设计了一条外立面的街道，称为"新趋势"或"最新街道"，这条街道位于军械库的corderia，也就是制索厂。受邀的20位建筑师都要设计出一组外立面，然后由来自于罗马电影城工作室的美工师负责搭建。在立面背后，建筑师们可以展示他们作品的模型、草图和照片，建筑师们就这样设计出了他们的最新作品的展示。

在此次双年展中，"街道"也将以类似的方式进行组织。策展人向全世界的主要评论家征求意见后，选出了12位建筑师进行设计，"街道"提供的是建筑的结构基础，而这些装置和外立面创造了"街道"。建筑创造了城市，城市又创造了建筑。与1980年在威尼斯的展览不同，虽然"街道"的建筑师一直得到鼓励要设计出外立面，但结果是非外立面——在空间和材料上都不是二维的。

波多盖希的慧眼选择了那些建筑师——彼时他们都才三四十岁——都成为了下一个十年里的佼佼者，其中有雷姆·库哈斯、弗兰克·盖里和矶崎新。在波多盖希的时代，建筑界的革新仅仅围绕着一个很不寻常的统一的智识轴心旋转，或者说，至少在70年代后期它貌似如此。尽管对可持续性的强调是全球共识，尽管这是一种对"过往的呈现"的趣味复兴，尽管在建筑时间、地景设计及城市化之间有了更深的纽带，但这点尚待认定：这12位建筑师的作品能否被视为统一的——甚至不严格地说——今天是否还受单一的理论旗帜的指导。虽然如此，有一点是肯定的——正如1980年的威尼斯双年展——后人将以他们的观点看待我们当下的建筑，可我们在今天做不到这样自视。

参展建筑师：
Alejandro Aravena, Arquitecto（智利，圣地亚哥）
Aranda Lasch（美国，纽约）
大舍建筑（中国，上海）
Fake Industries Architectural Agonism（美国纽约及西班牙巴塞罗那）
Hashim Sarkis Studios（黎巴嫩贝鲁特及美国剑桥）
J. Mayer H.（德国，柏林）
Johnston Marklee（美国加州，洛杉矶）
MAD Architecture（中国，北京）
Mass Studies（韩国，首尔）
开放建筑（中国，北京）
SO-IL（美国，纽约）
Spbr（巴西，圣保罗）

| GLOBAL PERSPECTIVES / 全球视野 | THE STREET / 街道 |

ELEMENTAL / Alejandro Aravena

Santiago, Chile / 智利，圣地亚哥

POWER TO THE PEOPLE

Our idea is to present a project able to give a clue for how to respond to the urban challenge the world will face in the next 20 years. It's a fact that the migration of people towards cities will require to build a 1 million people city every week with an extremely limited budget.

For The Street exhibition, we would like to present what we have called, the AQUEDUCT HOUSE:
- A Structural Support (able to receive self-construction),
- carrying Basic Services (water, electricity, gas, sewage) and defining the Property Subdivision (formal ownership of land will be the key to overcome poverty)

This inhabited infrastructure will be infilled by people themselves: it is a radicalization of the approach Elemental has developed in the last 10 years. The scale and magnitude of the challenge will only be accomplished if people themselves are part of the solution.

人民的力量

我们的想法是提出一个项目，能够为如何应对城市在未来 20 年内将面临的挑战提供线索。事实上，在非常有限的预算内，人们迁往城市将需要每星期都造出能容纳 100 万人口的城市。

在"街道"项目中，我们想展示我们称之为"渡槽楼"的构造：支撑结构（能够接受的自身建设）、基本服务（水，电，气，污水），并界定产权细分（正式的土地所有权将成为战胜贫困的关键）。

这一居住的基础设施将由人民自己填满：这是 ELEMENTAL 在过去 10 年开发的一种激进的方法。这一挑战只有当人民作为解决方案的一部分时才能得以解决。

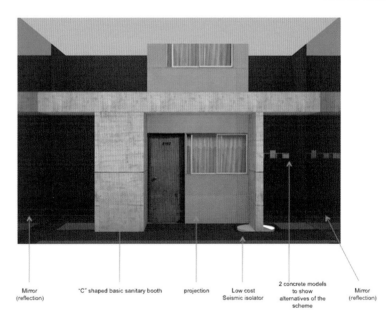

Design concept
设计概念

| GLOBAL PERSPECTIVES / 全球视野 | THE STREET / 街道 |

Aranda Lasch

New York, USA / 美国，纽约

Shadow Caster

Our installation for the 2011 Shenzhen Biennale draws on the Chinese tradition of Pi Ying Xi or Shadow Play. An ancient form of storytelling, Shadow Play is an inspiring way to tease animation, narrative and spatial depth from the simple layering of two-dimensional cutouts and the careful arrangement of lighting. On the outside of our exhibition space the movement of shadows from behind a screen will animate our facade on the Street. Inside, we borrow techniques from this tradition to unpack our projects as a series of layers; some of those layers will be made up of traditional photography and 3D models, others will only be light and shadow.

Each project is given a particular space in the room, which is wrapped by a large "screen" that is either hung from the ceiling. In the center of the room is the "Shadow Caster." The Shadow Caster is a single bright light source surrounded by three-dimensional models that casts a diffuse pattern over the projects that are associated with that geometry. Together the shadows represent a catalog of research and techniques that have evolved in our practice from project to project.

阴影投射

我们在2011年深圳双年展中制作的装置主要采用了中国传统的皮影戏，或者说"影剧"。这是一种古老的故事讲述模式，而且还是一种调笑动画片的有趣方式，通过简单的二维剪纸技巧以及精心的灯光安排，呈现出叙述和空间的深度。在我们的展厅外，屏幕后面的影子运动让街道的内部立刻有了生气，而在内部，我们利用了这项传统技法，将我们的项目作为一系列的层次展开来，一些层次构成了传统的摄影以及三维模型，其他层次则是光与影。

每个项目在空间中都被给予了一个特别的位置，被一面大的悬挂着的"屏幕"包裹着。房间的中央是"阴影投射"，即一个单独、明亮的光源被三维模型围绕着，在与其几何形态相关的项目图纸上投射出一个漫射的图案。这些阴影一起呈现着一本描述我们实践中各项目演变出来的研究和技术的画册。

| GLOBAL PERSPECTIVES / 全球视野 | THE STREET / 街道 |

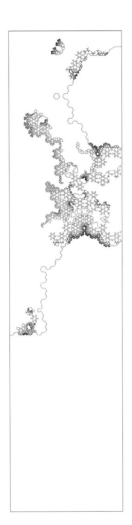

Aranda Lasch

| GLOBAL PERSPECTIVES / 全球视野 | THE STREET / 街道 |

Atelier Deshaus / 大舍建筑

Shanghai, China / 中国，上海

The 1980 Venice Biennale "Street" is a retrospective. From today's point of view, that was a banquet of the post-modern.

30 years later in Shenzhen, when we re-mention the "street", there must be some different meanings.

Because of the increase of population and the number of automobile, many streets have been lost or changed. Street might be still street.

However, we have to revise "façade".

The process of revision is related to where we stand.

Although it is not a new or a revolutionary standpoint that also has been concerned with "construction" for a long time, it is still enough to show our attitude.

Façade could be the presentation of interior space. Maybe "Literal Transparency" is not that boring as Colin Rowe said, or just the opposite, it shows our interests and relative abstractness and modernity practice.

In the construction of using "line" as main material, like a traditional space construction, the "solid" structure wall is constructed by using such material like "line". In the specific space: 5.80m(W)×3.55m(D)×4.00m(H), hanging "lines" constitute small spaces to exhibit our projects.

Between two small spaces or small space and the "street" there is a row of lines. Due to the variation of thickness, "wall" transfers into "curtain". The height of small spaces is defined by the "line-wall" and strengthened by the drooping "line-ceiling".

To see this façade from the street, it is not a surface, but a transparently overlying of the whole space. At this moment, a "wall" might not be a "wall". Its transparency is created by the alienation of "lines", and more importantly, "line-wall" will be in the adaptability of variation according to the mobility and changes of visitors.

1980 年威尼斯双年展的"街道"是一次回顾展，以今日的眼光来看，那是一场后现代的盛宴。

30 年后的深圳，重提"街道"，自是别样的意义。

尽管汽车与人口在增加，但很多街道正在消失或出现变化。也许街道仍是街道。

然而，我们不得不重新审视"立面"。

审视的过程和我们自身的立场相关。

尽管并不是新的或者革命性的立场，对"建造"的关注也是由来已久，仍足以表明我们的态度。

立面可以完全是内部空间的呈现，也许直白的透明性并不像柯林·罗说的那么无趣。或者恰恰相反，这正展现了我们的兴趣以及与之相关的抽象性和现代性的实践。

与传统意义的空间建造一样，在以"线"为主要材料的建造中，"实"体的围护结构墙由"线"这样的物质材料以悬挂的方式"砌筑"。在给定的 5.80m×3.55m×4.00m 的空间中，垂挂的"线"围合出几个用于展示我们作品的小空间。

小空间之间，还有小空间和"街道"之间由一排"线"隔离，由于厚度的变化，"墙"成为"帘"。小空间的高度既有赖于"线–墙"的围合，也因为下垂的"线–顶棚"而得以强化。

从街道上看到的空间"立面"并不是一个表面，而是整体空间的透明叠加。这时，"墙"或许并不是"墙"，它的透明性由"线"的疏离而产生，最重要的是，因为观展人群的流动性与多寡变化，"线墙"随时会处于变动的适应性之中。人群也将成为立面的一部分。这也是我们所希望建立的立面后空间与街道的关系。

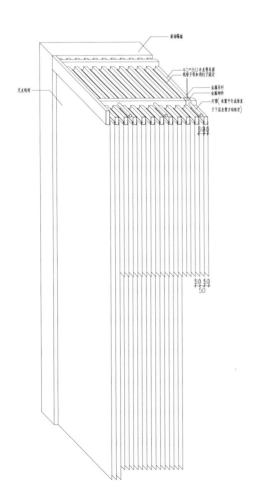

Node
节点

| GLOBAL PERSPECTIVES / 全球视野 | THE STREET / 街道 |

Fake Industries Architectural Agonism

NYC, US and Barcelona, Spain / 美国纽约及西班牙巴塞罗那

Architecture's innocence?

There is not original work on this installation. Not in the façade. Not on the table. The slide-show displays the evidence.

The four façade panels were stripped out from Ai Wei Wei's studio the night before it was demolished.The question is: What is the architectural value of this façade? Is it the fact that the fragments belong to a building of cultural relevance that was polemically demolish? Is it the fact that they display traditional vernacular Chinese construction in the context of an international biennale? Is it the fact that they have their place in the art market profitable business surrounding Wei Wei's public persona? Is it the fact that they refuse traditional notions of authorship assuming that to copy entails a radical reformulation of architectural imagination?

Hypothesis I: copies allow a radical renunciation to form—making—since form is defined a priory—in order to focus on architectural knowledge yet to be explored.

The models on the table appropriate despised typologies of the real estate bubble and re-visit their potential. The excess of architectural production generate knowledge that the speed of consumption condemns to oblivion. In this context, architectural knowledge can be public, yet undiscovered, if independently created fragments are logically related but never retrieved, brought together, and interpreted. We call them Domestic Dreams. They exist in an architectural unconscious, somewhere between the desire for forever-lost modes of inhabitation and the repressed pleasures of utopian typologies.

Hypothesis II: The revision of despised typologies enables the identification of issues that can be publically contested, fostering advancements in the field.

建筑的天真?

在这件装置作品中，所有物品都非原作。不在这些立面里。不在展台上。播放的幻灯片透露了线索。

在艾未未工作室被拆毁的前夜，这四个立面的嵌板就先被剥了下来。问题是：这一立面的价值何在？它是一座建筑的碎片，这座建筑与文化息息相关，并且被拆毁了，就是如此吗？这些碎片是在国际语境的双年展中展示传统的本土中国建筑，就是这样吗？这些碎片会因为艾未未本人的公共角色而在市场价值体系中占有一席之地，这是事实吗？这些碎片拒绝了作者身份的传统观念，而是假定了这一论点，复制需要对建筑想象进行彻底的重构，这就是现实吗？

假设 I：复制允许彻底放弃形式构成——因为形式已提前决定——于是，能够集中于尚未被探索的建筑知识上。

桌上展示的模型鄙视了房地产业泡沫的类型，并重新考察了其潜力。消费的速度将过量的建筑生产创造出的知识掩盖了。在这一语境里，建筑知识是公共的，且尚未被发现，而个体创造的碎片从逻辑上是相关的，但从未被找回，也从未放在一起。我们称其为"国内之梦"。建筑知识就存在于建筑的无意识中，处在对永远失去的居住模型的渴望以及各种乌托邦类型所压制的愉悦之间。

假设 II：重新修正一直被诟病的类型，这就将对问题的讨论提到了公共日程上来，促进了这一领域的发展。

GLOBAL PERSPECTIVES / 全球视野　　THE STREET / 街道

1. Everything started with the demolition of Ai Weiwei's studio in Jiading.

2. The psychological and symbolic effects of the resulting mountain of bricks were infinitely more powerful than its physical appearance.

3. When was the last time that architecture coalesced with art to produce such a powerful icon?

4. When was the last time architecture was so charg[ed with] contentious meaning that its challengers saw demol[ition as] the only appropriate punishment?

9. Where that fragments so powerful? We follow them through time and space to figure it out.

10. The pieces of the facades of La Strada Novissima were too bulky to be easily returned to their international origins. It was decided to leave them piled up as a gift to the city of Venice until a decision could be made.

11. So they stood -contemporary relics in a land of old ruins- while the political situation in Italy became tense. Bombs went off nearby. Kidnappings and detentions followed. Were the fragments forgotten as other problems became more pressing?

12. They were not. The Venetian section of the PCI h[ad] interest in historical fragments. They even had a pl[an to use] them in order to build shortcuts through the city. The results, however, were catastrophic.

17. The pieces finally arrived in San Francisco, camouflaged, some argue, as giant ducks that successfully crossed the Atlantic.

18. After long years of difficult journey, they stringed up the interest of architectural hunters.

19. Up to the point that the all-male family picture was taken and the Charlottesville Tapes recorded.

20. Even the world's most influential cultural instituti[ons] organized shows that would assure the immortali[ty of] the émigré fragments, just before they started to be[come] broken to be recognizable.

25. The increasingly diminished parts traveled in a borderless market with astonishing speed. They were mixed with millions of other bits that conformed the raw materials of an international real state bubble in rapid crescendo...

26. ...up to an ebullition point, with ever astonishing results proving its ideological flexibility...

27. ...And up to its ultimate 2008 definitive international crash.

28. By then, the travelling fragments, too liquid to e[xert] any resistance, had already reached China, engag[ing in a] traditional burn-off and reconstruction of cultural [...]

33. Nevertheless, the rumor seemed untruth. After some time, in an ancient souvenir shop, big pieces of facade were surprisingly found.

34. Negotiations were initiated, and three months later, within the context of the Shenzhen & Honk Kong Biennale of Architecture and Urbanism, the most complete fragments were displayed.

35. And here they stand, a Rosetta stone that attempts to decipher the complex ways in which architecture has collaborated with the current economic conditions, and the will of pointing out at possible paths of future action.

Fake Industries Architectural Agonism

GLOBAL PERSPECTIVES / 全球视野   THE STREET / 街道

There is not original work in this installation

Not in the facade.

Not on the table

And the slide-show on the right screen shows the evidence

The four pieces of facade were stripped out from Ai Wei Wei's studio the night before it was demolished

What is the architectural value of this facade?

Is it the fact that the fragments belong to a building of cultural relevance that was polemically demolish?

Is it the fact that they display traditional vernacular Chinese construction in the context of an international biennale?

Is it the fact that they have their place in the art market profitable business surrounding Wei Wie's public persona?

Is it the fact that they refuse traditional notions of authorship assuming that to copy entails a radical reformulation of architectural imagination?

(copies allow a radical renunciation to form-making– since form is defined a priory– in order to focus on architectural knowledge yet to be explored)

The models on the table appropriate despised typologies of the real estate bubble and re-visit their potential.

The excess of architectural production generate knowledge that the speed of consumption condemns to oblivion

In this context, architectural knowledge can be public, yet undiscovered, if independently created fragments are logically related but never retrieved, brought together, and interpreted

We call them **Domestic Dreams!!!**

They exist in an architectural unconscious, somewhere between the desire for forever-lost modes of inhabitation and the repressed pleasures of utopian typologies.

(The revision of despised typologies enables the identification of issues that can be publically contested, fostering advancements in the field)

Enjoy the Show.

Fake Industries Architectural Agonism

Fake Industries Architectural Agonism

| GLOBAL PERSPECTIVES / 全球视野 | THE STREET / 街道 |

Hashim Sarkis Studios

Beirut, Lebanon and Cambridge, US / 黎巴嫩贝鲁特及美国剑桥

"Quasi-Object"

The installation repositions the facade between architecture and city as the surface of the quasi-object. This new type of object engages the street, the city's prevalent compositional device, but also strategically disengages from it.

Conventionally, building a street facade usually implies: 1) that the street is the primary urban space to which architecture has to comply as context, and 2) that in defining the street the facade has to disengage from other formal determinants such as function and structure.

The rejection of postmodernism has come with a rejection of both the street and the facade as formal entities, and of contextualism as a strategy for linking architecture to city. Instead, and over the past two decades, we have seen the two being recast in a range of relationships from negation (e.g. bigness) to continuity (e.g. landscape).

The return to the street and to the proposition that "Architecture makes the City makes Architecture..." begs a reconsideration of this relationship beyond contextualism, negation, or continuity. It compels us to think of other possibilities that do not fetishize street or facade, and that do not impose a causal relation between architecture and city.

The installation consists of a square surface that floats in the plane of the street facade. It creates a silhouette with the ground and with the adjacent buildings. Simultaneously, this surface is part of a continuous form, set in structural and spatial tension with the 6x3.5 meter space. It deforms to let visitors in and out, to structurally lean on the existing walls, and to act as display space. In doing so, it acquires internal consistency and legibility. Importantly, it generates its autonomy out of the very contingencies of context, structure, and program.

"类似物"

这一装置将建筑与城市之间的立面进行了重新安置，成为了一种表面的"类似物"。这一新的类型物介入了街道——城市中最为普遍的构成物——但同时也策略性地从其中剥离开来。

通常情况下，建造街道的立面意味着：①街道首先是基本的城市空间，是建筑必须遵循的文目录；②于是这就决定了街道立面与其他例如功能和结构等形式上的决定性因素必须区分开来。

对后现代主义的摒弃逐渐拒绝了街道和立面同时作为形式上的实体，也拒绝把文脉主义当作连接建筑和城市的策略。相反，在过去的20年中，我们已经目睹了这两者在从否定（例如巨大）到延续性（例如风景）等关系领域中的重塑。

向街道的回归以及"城市创造"的主张都希望能够对文脉主义、否定或者延续性之外进行反思。它要求我们考虑其他的可能性，并非盲目地追随街道或立面，而且这些可能性也并非是在建筑和城市之间强加一种随意的关系。

这一装置包含了一个方形的平面，它漂浮在街道立面中。它为地面以及相连的建筑勾勒出了一层剪影。与此同时，这一表面是连续形式的一部分在6米×3.5米的空间中设定的结构及空间张力。它不断变形以使访者们进出，在结构上依靠在已有的墙体上，并作为展示空间。在这样做的同时，它要求了内部的联系性和可读性。重要的是，它在语境、结构和规划的偶然性之外产生了自主性。

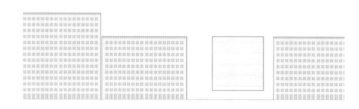

GLOBAL PERSPECTIVES / 全球视野

THE STREET / 街道

Conceptual Derspectives
概念透视图

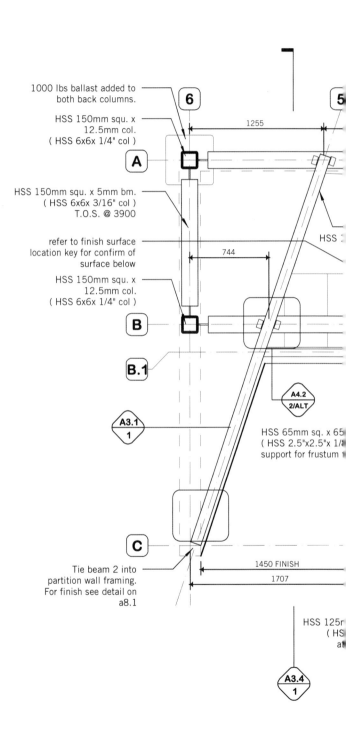

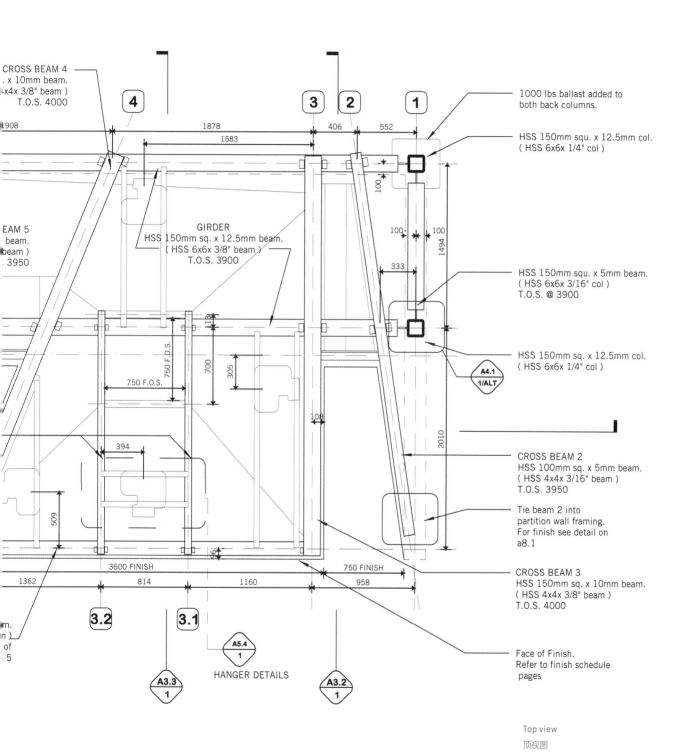

Top view
顶视图

| GLOBAL PERSPECTIVES / 全球视野 | THE STREET / 街道 |

J. Mayer H.

Berlin, Germany / 德国,柏林

Facade of contenance

Various lines of demarcation, or even better `facades of contenance´, have always separated the personal and the public. And in the case of information, the relationship between public and private becomes a complicated set of liabilities. It´s a contract of confidentiality. By the beginning of the Twentieth Century, information controll generated a visual pattern called Data Protection Pattern or DPP that helps to veil personal information in print media. Letter and numbers, ingredients of information construction, are used in excess to create a speechless and slurry form of covering text.

The pattern used to conceal private information have concealed their own technological development. Only a few traces remain to provoke my speculation about their origins: First, when printing a book, test prints are made using the same paper repeatedly during which text over text is formed. Second, because of mistakes during printing an offset grid can arise, resulting in a Moiré Effect. Third, the repeated use of carbon paper can create a pattern that can also be considered to be the predecessor of the Data Protection Pattern. Today, a new global network of unsecured data transfer remains to be resolved. While DPP continues to proliferate in print media, it provides the model for carriers of electronic information, which are physically erased by overwriting the entire data carrier, or at least the used sectors, with a confusion of pattern. An excess of information transforms the "private" into apparent nonexistence.

The FACADE OF CONTENANCE is making this relationship between public and privacy visible. The Facade is built in medium density fiberboard. The fireboard is painted in colour. The visitor is free to enter the exhibition space and walk-through the entrance in the middle of the structure. Inside all walls are painted with pattern. In front of the main wall a huge free standing board is placed showing J. MAYER H.'s recent projects. Theses printouts are laminated onto this board. A pedestal placed inside the "house" is showing the most important project of J. MAYER H.: The redevelopment of Plaza de la Encarnacion in Seville, Spain, actual one of the world largest timber glued constructions on the world.

态度的立面

"界限"由众多的线条来组成——或者更恰当地说则是"态度的立面"——区分了私人与公共。就信息层面来说,区分公共和私人之间的关系成为了一种复杂的义务,如同一项保密合约。20世纪初,信息控制衍生了名为"数据保护模式"(Data Protection Pattern, DPP)的视觉模板,它防止个人信息出现在印刷媒介上。作为信息构成的要素,字母和数字都被过度使用来创造一种无言的、如泥浆般掩蔽文本的方式。

这一用以隐蔽私人信息的模式也同样遮蔽了它们自身的技术发展。只有一些进程的痕迹唤醒了关于它们的衍生的想象:首先,印刷书籍时,打样用的是相同的纸张,从而使得文本覆盖文本。其次,因为打印时出现的错误,平板胶印的方格出现了,以至于产生了莫尔效应。其三,复写纸的反复使用可以创造出一种模式,这可以被认为是"数据保护模式"的前身。今天,一种新的不安全的全球网络信息传输的问题亟待解决。当DPP继续在印刷媒体中激增,它提供了电子信息传输的模型,即用模式的混乱从根本上通过重写来抹去全部的信息携带,或至少被用过的部分。一种信息的过量将"私人"转型为一种显然的不存在。

"态度的立面"让公共与私人之间的关系变得可见。这一立面是用上色的纤维板制成的。观众可以自由出入展示空间,并且可以从整个结构的中间通过。内部的墙面都以某种图案上色。在承重墙的前面,一块大型立式展板呈现的是J. MAYER H.的近期项目。这些印刷品都被印压在这块板上。一个安放在"房屋"中的基座展示的是J. MAYER H.最重要的项目:西班牙恩卡纳西翁广场改扩建方案,这是世界上最大的木材胶合结构之一。

| GLOBAL PERSPECTIVES / 全球视野 | THE STREET / 街道 |

J. Mayer H.

| GLOBAL PERSPECTIVES / 全球视野 | THE STREET / 街道 |

Johnston Marklee

Los Angeles, CA / 美国加州，洛杉矶

'1787'

For "The Street" exhibition of the 2011 Shenzhen Hong Kong Biennale of 城市 \ 建筑, Johnston Marklee propose an installation entitled '1787', named after two creations of the year 1787: the Potemkin Village and the Panorama. These two themes constitute the basic premise of the two-part project: a primitive facade facing the street with an immersive interior environment.

'1787'

Johnson Marklee 为 2011 深圳·香港城市 \ 建筑双年展中的"街道"展提出的装置方案名为"1787"，这一标题来源于 1787 年的两大发明：波将金村及全景图像。这两个主题成为了这两部分项目的基本前提：一个古老的街面，却能营造出一种身临其境体验的内部环境。

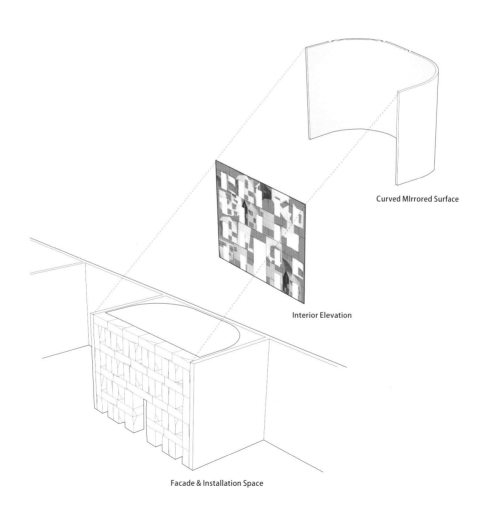

Curved MIrrored Surface

Interior Elevation

Facade & Installation Space

Design concept
设计概念

GLOBAL PERSPECTIVES / 全球视野

Potemkin Village

In 1787, Prince Grigory Potemkin supposedly erected expansive fake villages in Crimea to impress Russian Empress Catherine the Great during her visit to this recently conquered peninsula in the Black Sea. In the time since, the term 'Potemkin Village' has come to be used for any masquerade meant to fool naive outsiders about one's true intentions or to hide the dismal reality of a situation.

Paolo Portoghesi wrote that the idea for the 1980 'Strada Novissima' at the Venice Biennale of Architecture came from the thin, fake vernacular facades of a carnival creating a "closed and inviting space" at Alexander platz, the vast showpiece of postwar modern architecture in East Berlin. Today's new Biennale 'Street' could be seen as an equally Potemkin-like production of carefully constructed but temporary and hollow facades within a metropolis that is the case study for the rapidly changing architecture and urbanization of the present day.

While Portoghesi's Strada Novissima created a Potemkin street more than thirty years ago, the installation '1787' reaffirms the separation between exterior and interior environments while exacerbating their difference.

THE STREET / 街道

波将金村

1787 年，格里高利·波将金王子意欲在克里米亚广泛搭建起假村庄，以此来吸引到访的攻克了黑海地区的半岛的俄罗斯凯瑟琳女王。"波将金村"从此被用来描述愚弄天真的外来者的伪装，以此来展现真实的意愿或者隐藏某种悲惨的现实。

保罗·波多盖希为 1980 年的威尼斯建筑双年展"新趋势"写下了这一想法。用单薄和虚假的如嘉年华般的本地立面在亚历山大广场上创造出一个"封闭且引人注目的空间"，这是个放在东柏林的巨大的战后现代建筑展示作品。今天的新双年展中的"街道"被看成与波将金式相当，这一个细心地在都市中搭建的临时中空街面，也可以作为研究当前快速变化的建筑与都市化的案例。

在波多盖希创造了"波将金"街道 30 年之后的今天，这个装置"1787"重新确立了内部空间与外部空间的间隔，也加深了内外之间的差异。

1980 Strada Novissima
1980 新趋势

Johnston Marklee

The facade is constructed from a tectonic stacking of blocks that are inexact but self-similar, evoking a form that is at once primitive and futuristic. Something contemporary is created through the abstract approximation of something older.

所建造的立面由块状体堆叠而成,每一块都不完全一样,但却是自我复制般地雷同,由此唤起了一种古老又未来的模式。一些当代的事物正是通过对某些旧有之物的抽象化创造出来的。

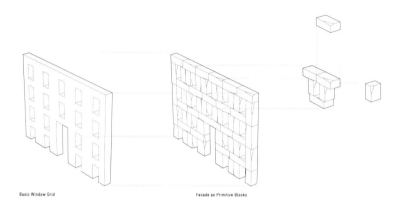

Basic Window Grid · Facade as Primitive Blocks

Potemkin's facades were a generic vision of what a prosperous village in Crimea might look like. Our facade itself exists in the ambiguous realm between generic and specific. The window grid for this installation evolved from studies of historical facades that have become generic tropes; these tropes were then deployed in a specific manner.

波将金的立面曾是对克里米亚地区村庄的繁荣景象的想象,而今天的立面则介于一般及特殊之间的模糊领域。这一装置中的窗户的位置是由已经成为一般性比喻的各种历史立面的研究演变而来的,这些比喻曾经通过独特的方式得到了运用。

STROZZI · RVCELLAI · FARNESE · ODESCALCHI

GLOBAL PERSPECTIVES / 全球视野	THE STREET / 街道
Panorama Immersion In 1787, Scottish painter Robert Barker patented a technique for creating 'panorama' paintings on the inside of large cylindrical rooms. At the tail end of the Enlightenment, before Romanticism and its exaggerated sense of emotion flourished, Barker ushered in a new era of immersive environments as urban spectacles, carefully controlled and distinct from the city around them. While the efficacy of the panorama depends on absolute separation from the outside, the interior graphic facade of the installation 1787 is interrupted by the apertures of the exterior facade—allowing glimpses of the street into the panorama.	全景沉浸 1787 年，苏格兰画家罗伯特·巴克尔获得了创造"全景"绘画的技术专利——他可以在大型圆柱形房间内绘制出 360 度的包围式全景图像。在启蒙运动结束和浪漫主义开始之前，这一发明放大了情绪的感知，巴克尔也迎来了都市奇观般的全景沉浸的新纪元，它被细心地控制并与其周围的城市区别开来。 全景的效力依赖于与外界的绝对隔离，装置"1787"所创造出的生动的内立面被外立面的缝隙所阻断——使得人们能从街道上一瞥全景。

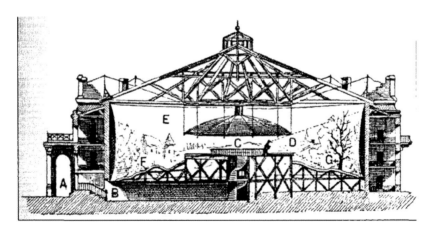

Typical panorama section
典型全景剖面

Johnston Marklee

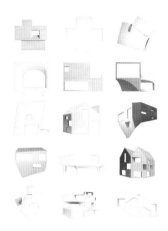

Johnston Marklee 'alphabet' of project primitives
Johnston Marklee 项目原语的"字母"

The street facade acts as a zone of stoppage, allowing the interior room to be a different realm entirely. The bricks from the front side of the facade become on the interior elevation specific yet still primitive representations in relief of the projects of Johnston Marklee, with the facade apertures punched through to allow brief moments of view and light to interface between street and interior.

街道的立面扮演了阻滞的角色，让内部空间成为了一个完全不同的领域。立面前的砖块同时也是室内的立面，构成了该项目特有的且依然延续的旧式呈现。立面上的缝隙是观看时刻与光线穿透街道与内面之间的接合界面。

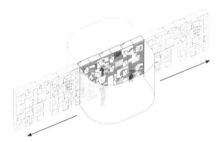

Horizontal multiplication of street facade
街道立面的水平复制

The other three walls of the space are lined with floor to ceiling mirrored surfaces, rounded at the corners, that reflect both the work of Johnston Marklee and the visitors themselves in a strange panoramic anamorphic effect.

The facade is specifically scaled to the given site, from a distance appearing normal scale but from up-close being completely indifferent to scale and confusing to the body's thoroughly ingrained notions of door and window. It seems instead to be ambiguous between the scale of furniture and the scale of a building.

Reflections of the interior elevation in the curved mirrored surface create a constantly-changing, immersive, and scaleless interior environment.

空间中的其他三面墙装有镜面，把顶棚与地面连接起来，在角落随之弯曲，这让 Johnston Marklee 的作品和观众们的图像都反射在了一个十分奇异的全景式失真效果当中。

这一立面根据场地定制而成，从某个特定距离看，这不过是一个普通的建造，但是当你走近，会发现其比例变得完全不同，打破了已经根深蒂固的门和窗的观念。它让家具的比例和建筑的比例都呈现出了模棱两可的状态。

弯曲的镜面反射着内部空间的立面，同样也创造出了一种不断变化的、沉浸式的以及无比例的内部空间环境。

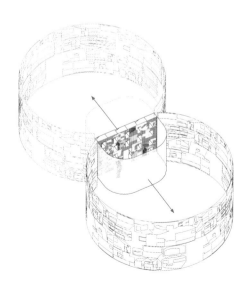

Reflection to create distored panorama effect
反射产生扭曲的全景效果

| GLOBAL PERSPECTIVES / 全球视野 | THE STREET / 街道 |

MAD Architecture

Beijing, China / 中国，北京

People often like to produce boundaries that do not exist, physically or psychologically. With this, a demarcation, ambiguity, conflict emerges, and then comes seduction and deceit. Regretfully, seen in the 1980's The Street exhibition; the role of the architect is often to 'aestheticise' this marginal zone.

This aestheticised zone instead allows for more and more flaws to be discovered, in response there is an increase echo of voices that are seeking to break free this boundary, allowing the world to restart in a equalize flat plane. But is it so attractive to eliminate the characteristics, ambition, and the desire of an open world? May be this isn't better than the extreme opposite of the full of boundary zone.

'Boundless' is installation exploring the evolution of the single marginal boundary into a space. It allows the pedestrian to discover existence of this boundary at a different level. An ovum shaped object floats between the 'the street' and 'the stall'. From the illuminating pinhole, the audience discovers a boundless 'universe', the nine MAD works floats within this 'non-existent universe'.

人们总是喜欢制造出原本并不存在的边界，物理的，或心理的。随之而来的就是隔阂，矛盾，竞争，然后便是谄媚和欺骗。悲哀的是，1980年版本的"街道"展中，建筑师的工作通常就是关于如何"美化"边界的。

美丽的边界反而导致越来越多的丑恶被发现，有更多的声音要求打破边界，让世界重新变成平的。但是一个消除了特色、野心和欲望的开放世界真地那么吸引人么？也许它并不会比一个充满隔阂的另一个极端好到哪去。

"无边"是一件将边界引申为空间的作品，它让人们去发现存在于边界中的另一个层次。一个卵形的物体悬浮于"街道"和"房间"之间，从透光的孔中，人们发现的是一个无边的"宇宙"，9个MAD的作品则漂浮在这个"并不存在"的"宇宙"中。

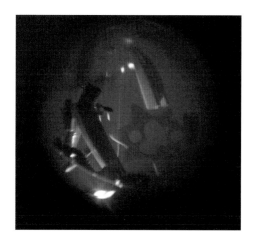

| GLOBAL PERSPECTIVES / 全球视野 | THE STREET / 街道 |

Artist illustration
方案表现图

MAD Architecture

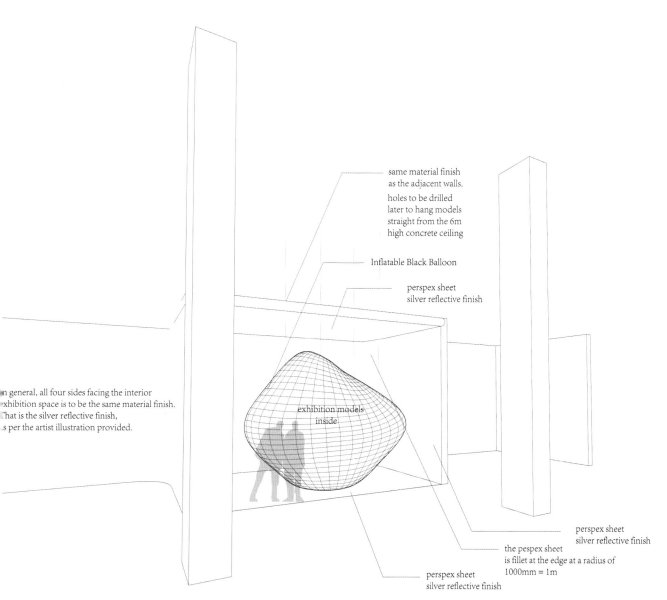

same material finish as the adjacent walls.
holes to be drilled later to hang models straight from the 6m high concrete ceiling

Inflatable Black Balloon

perspex sheet
silver reflective finish

In general, all four sides facing the interior exhibition space is to be the same material finish. That is the silver reflective finish, as per the artist illustration provided.

exhibition models inside

perspex sheet
silver reflective finish
the pespex sheet is fillet at the edge at a radius of 1000mm = 1m

perspex sheet
silver reflective finish

design proposal
设计方案

| GLOBAL PERSPECTIVES / 全球视野 | THE STREET / 街道 |

Mass Studies

Seoul, Korea / 韩国,首尔

(N)ON Façade

Our proposal is to create a "(n)on facade": a façade that is differentiated from the traditional architectural facades by allowing people to occupy the façade itself. Unlike the more conventional façade, which is intended primarily as something to be looked at, the "(n)on facade" offers an opportunity to occupy and experience the façade as a social space.

To achieve this, we have protruded the given 5.8 x 4 meter facade plane into the 3.55 meter-deep exhibition space. Through a pixelization-like process, the 23.2 square-meter facade becomes a multi-faceted three-dimensional surface totaling 79.2 square meters in surface area.

The proposal consists of two basic types of components —"Light faces" and "Dark faces", which are supported by a steel frame. Light faces contain exhibition content—including drawings, photos, scale models, and LCDs—presenting fifteen projects in total, both built and un-built. The "faces" displaying drawings and photos, will be back-lit panels of printed translucent acrylic. One of the two scale models will be placed on a back-lit white acrylic face, while the other will be the face itself (a 3-D print), also lit from behind. There is also an LCD playing looped content of diagrams and animations. "Dark faces" will accommodate social content, engaging the users to climb or rest on these surfaces. Cladded in various light-absorbing black materials, the receding faces will create a very tactile experience while allowing the "light faces" to float.

(无)立面

我们的参展方案是创造一个"(无)立面":这一立面不同于传统的建筑立面,它可以让人去占据它。它也不同于更加传统的立面,即出发点就是给人们看。"(无)立面"提供了让人们将该立面作为社会空间来占领以及体验的机会。

为了达到这一目标,我们突出了这个给定的 5.8 米 ×4 米的立面,将其拉伸为深度达 3.55 米的展览空间。通过像素化的过程,23.2 平方米的立面成为了多相面的三维表面,即 79.2 平方米的表面区域。

Mass Studies 的方案包含了两个基本要素——"光面"与"暗面"——这两种面由钢构进行支撑。光面上展示的内容主要包括图纸、照片、结构模型和屏幕,总共展示了该团队的 15 个建成及进行中的项目。这些"面"所展示出的图纸及照片装裱在有背光的有机玻璃上,其他展出内容则通过"面"本身(3D 打印)展示在灯箱上。展览中还有一块循环播放图表与动画的液晶屏幕。"暗面"主要是社会性内容,邀请用户在这些表面上攀爬与休息。暗面是用多种黑色的吸光材料包裹起来的,在亮面漂浮起来的同时,往后退却变化的表面则创造出了一种触知体验。

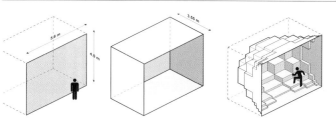

Concept diagram
概念示意图

23.2 sq m

79.2 sq m

79.2 sq m

Surface area comparison diagram
表面面积比较示意

表面之上

MASS S

(N)ON FACADE

邁思斯特地斯

GLOBAL PERSPECTIVES / 全球视野 | THE STREET / 街道

OPEN Architecture / 开放建筑

Beijing, China / 中国，北京

OPEN NATURE

We are living in an era of a looming environmental crisis. After 3 decades of unprecedented urban growth in China, we are seeing remarkable prosperity on the surface while facing the worst environmental problems ever seen in history.

OPEN began its practice at this very moment. We see it as our mission and responsibility to find the ultimate balance between nature and people. We take the special opportunity at the STREET exhibition in 2011 Shenzhen & Hong Kong Bi-City Biennale of Urbanism\Architecture to showcase our hidden yet consistent agenda in a series of recent architectural and urban projects – "building nature or anti-building in the process of building".

In Chinese, the word "park" is made up of two characters, the first means "public" and the second means "garden". This series of our work all evolved around these two common threads. More specifically, the relationship among people, and the relationship between people and nature. We are interested in public space that is filled with nature, and nature that is enjoyed by everyone!

In OPEN's space on the STREET, nature dominates and façade disappears. In an era of architecture dominated by fashion, shapes and images, our answer to the given proposition – the making of a street façade – is actually "Façadeless". Nature is the space and the façade. The 3 surrounding walls are skinned with polished stainless steel sheets that dematerialize and dissolve the space by reflecting the nature indefinitely. On the surface of these walls 7 selected projects are each showcased by a multimedia presentation and a conceptual model floating on the walls.

It is an OPEN Garden on the very special STREET!

公·园系列

我们生活在一个环境岌岌可危的时代。在经历了将近 30 年史无前例的高速城市化发展之后，中国在一片欣欣向荣的同时也正面临着有史以来最严重的环境问题。

开放建筑 /OPEN Architecture 的实践便始于这一特殊时刻。我们寻求人与自然的平衡共生，把这一诉求看作自己的使命和责任。以 2011 深圳·香港城市\建筑双城双年展的"街道"展作为契机，我们得以通过这个装置和近期的一系列作品来展示其中潜在的和共同的目标——"创造自然，或者说在建造的过程中反建造"。

我们展出的这些作品被称为"公·园"系列，因为它们的核心思想都是围绕着"公共空间"和"营造自然"这两条主线，更具体地说，都是在探索人与人之间的关系和人与自然之间的关系。我们感兴趣的是充满自然的公共空间，和每个人都可以享有的自然！

在本届双年展"街道"展上，我们的空间里，立面完全消失。在建筑以潮流、形式和图像为主导的当下时代里，面对"在'街道'上营造一个立面"的命题，我们选择了"无立面"。自然是我们的空间，也是我们的立面。三面半包围的墙面以镜面不锈钢为完成面，空间里的树木植物在镜面里无限反射，空间的边界被完全消融掉了。在镜面的墙上，展示的是凸出于墙面的 7 个抽象的不锈钢模型以及 7 个嵌入墙面的多媒体录像展示。

特殊的"街道"上的开放 (OPEN) 的公园欢迎人们的参与！

| GLOBAL PERSPECTIVES / 全球视野 | THE STREET / 街道 |

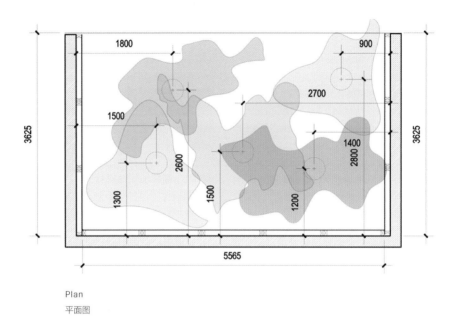

Plan
平面图

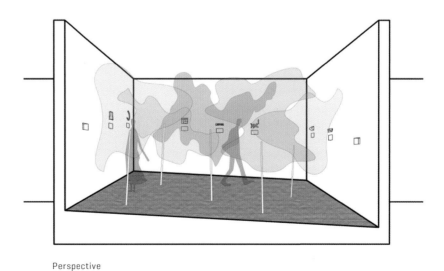

Perspective
效果图

OPEN Architecture / 开放建筑

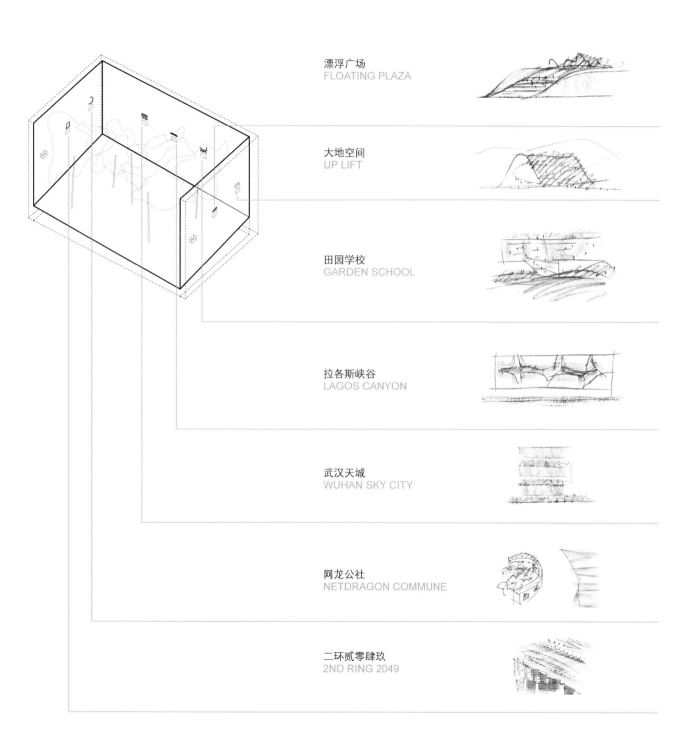

漂浮广场
FLOATING PLAZA

大地空间
UP LIFT

田园学校
GARDEN SCHOOL

拉各斯峡谷
LAGOS CANYON

武汉天城
WUHAN SKY CITY

网龙公社
NETDRAGON COMMUNE

二环贰零肆玖
2ND RING 2049

| GLOBAL PERSPECTIVES / 全球视野 | THE STREET / 街道 |

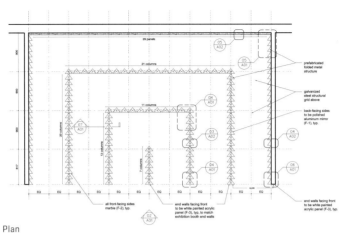

Plan
平面图

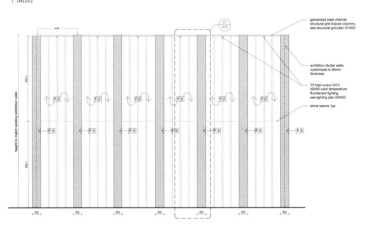

Elevation
立面图

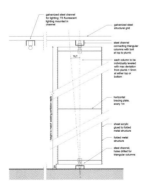 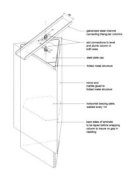

Column section　　　　Column axonometric
柱子剖图　　　　　　　柱子轴测

SO-IL

New York, US / 美国, 纽约

Architecture through facade
Architecture of facade

As the historic 1980 exhibition has come to represent a moment of architecture's fixation on the symbolic facade, 2011 marks an anxiety that architecture itself has collapsed into nothing more than skin.

Our proposal, through a kind of reinvented colonnade, seeks to spatialize and instrumentalize what has become implicitly flat. The 'facade' unfolds, becomes itself the space of occupation and exhibition. Its prismatic materiality—'marble' on each column's front face, and mirror on its two back faces—creates an instable space always changing in relation to its viewers.

When approaching the facade, the marble surface seemingly expands into infinite depth. Crossing the threshold, the image of the marble is refracted and fills the space as visitors wander through. Multiplied and reflected, flatness is scrutinized and made spatial, and becomes charged with public engagement.

穿越立面的建筑
立面的建筑

1980年的那次历史性的展览，是建筑对象征性立面的一次固执的表态，而2011年的展览展示了一种焦虑，建筑本身沦落为表皮一张。

我们的提案是通过柱廊重新构建，以此将平面事物进行空间化和工具化。"立面"被打开，成为了其所占空间和展览空间本身。它以三棱镜的形态呈现，每一个柱廊的正面都装上了大理石，另外两面则安装上镜子，由此创造了与其观者的关系时刻都在改变的不稳定空间。

逐渐走近这立面时，大理石的表面看上去似乎延伸至一个无限的深度。跨进入口大理石的图像被镜面折射，充满整个空间，观众们便可以漫游其间。平面经过重叠、反射之后，形成了被公众的参与影响的空间。

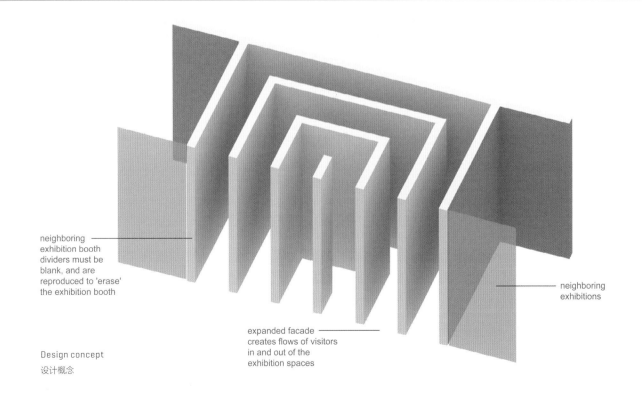

neighboring exhibition booth dividers must be blank, and are reproduced to 'erase' the exhibition booth

expanded facade creates flows of visitors in and out of the exhibition spaces

neighboring exhibitions

Design concept
设计概念

| GLOBAL PERSPECTIVES / 全球视野 | THE STREET / 街道 |

SPBR

Sao Paolo, Brazil / 巴西，圣保罗

One single line makes the facade.

This line is the edge of a floating prism that is hung from the ceiling.

A slightly rotated prism related to the axes of la strada, thus its vanishing faces suggest to the visitors a path to inside the exhibition space.

The prism defines three different fields for exhibition:
- Surrounding it for the drawing
- Inside it for pictures
- Underneath it for the models

The prism is 2m wide, 4m long and 3.2m high; it is placed 0.8m above the ground. It is made with MDF panels structured on wood frame from inside and finishing in matte white paint. With an estimate weight of 600 kg, it is hung from the ceiling slab by a few steel cables.

Surrounding the prism the three walls are covered with 3.2 high MDF panels slightly detached from the walls and 0.8m detached from the ground. These panels display drawings showing sections and plans of some selected projects. All sections are connected by the same horizon line and scale, plans are showed below it, and some detail or highlight above it.

Inside the prism there are six positions for showing pictures of some projects. The darkness inside the prism is quite convenient for projections, thus in each position of those positions there is a projector showing several images in looping. The visitor is able to watch it through a horizontal 2.5cm gap continuously provided in the whole extension of the prism.

Underneath the prism we show a few models. The field for the models exhibition is delimitated by a 2x4m dark grey rectangle painted on the floor. These models could be highlighted by using spotlights attached on the bottom of the prism.

一条线造出一个立面。

这条线是从顶棚上垂下来的悬空的棱镜的边缘。

在"街道"的轴心，悬挂着微微旋转的棱镜，它渐隐的面为观众指出了通向展览空间内部的道路。

这一棱镜确定了展览的三个面向：
——外围的绘画；
——内部的图片；
——底部的模型。

这个棱镜 2 米宽，4 米长，高度达到 3.2 米。它离地 0.8 米。它由密度板构成，从里面架在木板上，表面是光滑的白色喷漆。总重量约 600 千克，它将从顶棚悬挂下来，以几根钢索固定。

在棱镜旁边的三面墙上都覆盖了 3.2 米高的密度板，微微地与墙面分离，离地 0.8 米。这些嵌板上将展示的是图纸，呈现的是部分项目的剖面和平面。所有的剖面将由相同的水平线和区域来联系，平面会展示在下方，而细节或亮点则是安排在上部。

在棱镜内部，有六个用来展示项目图片的位置，棱镜中的暗部对于展示而言可以说是非常恰当的，因此每一个位置都有一台投影仪，循环播放相关图片。访问者可以通过水平的、2.5 厘米宽的连续绕棱镜一周的缝隙观看。

在棱镜下方，我们展示了一些模型。模型的展示将被固定在地面上面 2 米宽、4 米长的暗灰色方块上。这些模型可以利用安放在棱镜下方的聚光灯来提亮。

| GLOBAL PERSPECTIVES / 全球视野 | THE STREET / 街道 |

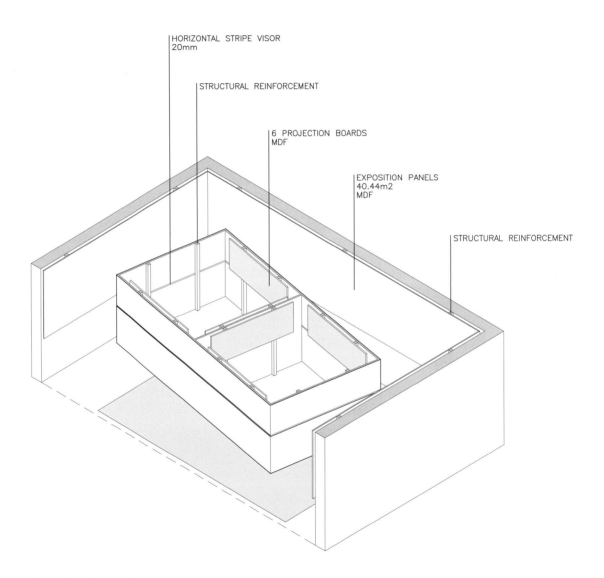

Perspective - horizontal cut
横切透视图

GLOBAL PERSPECTIVES / 全球视野　　　　THE STREET / 街道

"STRADA NOVISSIMA" AND "THE STREET"　　"新趋势"和"街道"

Liu Lei　　　　　　　　　　　　　　　　刘磊

In 1980, under the direction of Italian Paolo Portoghesi, the first Venice Biennale of Architecture took place. An architect, theorist, history scholar, and professor at University La Sapienza in Rome, Portoghesi then became President of the Venice Biennale of Architecture and lead its planning during the 1980, 1982, and 1992 editions of the Biennale. During the 1980 inaugural Biennale, the exhibition based on the topic of "Strada Novissima" was one of the important components.

Portoghesi, who specializes in classical architecture, keenly grasped the late-1970's post-modern architecture movement, and made "The Presence of the Past" the main theme of the first edition of the Biennale. In his opinion, the popularity of modernism in the post-war period slammed the door on tradition, nationality, regionalism, and the individuality of international forms. Dialogue between ordinary people was lost, and thus the continuity of architectural history was cut short. Architecture should not just be a recreation of the elite; it should be designed for the masses. These views conform to those of the time during the post-modern architecture movement, and they show a diverse aesthetic taste and respect for and emergence of historical context within design. Similarly, in designing urban spaces, "Strada Novissima" constitutes an element of the basic construction of urban life and is re-emphasized. Against this backdrop, the purpose of the exhibition "Strada Novissima" is to be able to make visitors' witnessing of "return to Strada Novissima" act as a key component of urban formation and one of the ultimate goals of post-modern research.

After a period of hard work and successfully persuading the military, Portoghesi had the first edition of the Biennale housed within the Corderie dell'Arsenale, which was a cable factory that served as the first pavilion. The enormous dimensions of the space and its excellent location given its proximity to the center of Venice were Portoghesi's reasons for selecting the site. He firmly believed that only with this site would it be possible to get ordinary viewers to more easily visit the exhibition space. For this famous "Strada Novissima" exhibition, Portoghesi invited 20 architects from around the world to participate, among them were figures well-known within the architecture world such as Michael Graves, Frank O. Gehry, Christian de Portzamparc, Rem Koolhaas, and Arata Isozaki. The exhibition space for "Strada Novissima" was 70 meters long, and based on the placement of the pillars, the hall was divided into a central communal space "Strada Novissima" (4.5 meters wide) with ten compartments along its two sides. Each architect was allocated one compartment and was required to address the topic of "the presence of the past" in designing a full-scale architectural façade that ran along the central "Strada Novissima", the façade of which was approximately 7 meters wide and 7.2 to 9.5 meters high. The architects could express their own unique sense of form through design. The whole "Strada Novissima" connected to the hall's entrance, which was designed by famous Italian architect Aldo Rossi.

Portoghesi hoped that this Strada Novissima exhibition could be seen as residences built side by side. These "residences" could be the architects' own houses; they could also be ordinary, everyday, private residences, or even places for recreation or entertainment. The designs needed to allow for viewers to be able to freely pass through the façade and enter the compartments' interiors to see the architects' individual work within these places where they could show or "show off" their thoughts. From a design point of view, many of the architects used classical elements in designing the façade and directly addressed the "presence of the past" topic.

"Strada Novissima" played a decisive role in determining the significance of this first Venice Biennale of Architecture. As opposed to the blueprints, photographs, and models usually displayed in traditional exhibitions, "Strada Novissima" for the first time brought full-scale architecture spatial form into the interior of the exhibition space. For just this reason, Portoghesi thought ordinary viewers could now better engage with the works and participate in discussions about architectural and urban design.

GLOBAL PERSPECTIVES / 全球视野

"STRADA NOVISSIMA" AND "THE STREET"

The first International Venice Biennale of Architecture had a profound and lasting influence on the outside world, including then-architecture student Terence Riley. Thirty years later, when Riley served as the Chief Curator of the Shenzhen Hong Kong Urban/Architecture Biennale, he naturally connected the two Biennales. For the Shenzhen Biennale, this engaging of a non-Chinese to serve as Chief Curator for the first time truly embodied the steps the city has been making to gradually move towards internationalization. Its purpose and that of the 1980 Venice Biennale of Architecture in building an even larger platform are one and the same. The "Strada Novissima" exhibition was brought back again by Riley within the Shenzhen Biennale.

After soliciting the opinions of many architectural curators, critics, and scholars, Riley invited twelve young and middle-age architects rising to international prominence to exhibit in the Biennale. Each architect was required to develop a vision within a space approximately 6 meters wide, 3.5 meters deep, and 4 meters high, and to display their own works there. Similar to the "Strada Novissima" exhibition thirty years before, each architect also had to consider how to design a "façade". While configuring the inside of the exhibition hall, six facades were made into a group, with a total of two groups placed on the two sides facing the "Strada Novissima" in between them (on the pillars within the exhibition site "The Street" was an interview between Aaron Betsky and a large number of the original participants who were still active during the time between the "Strada Novissima" exhibition and the present Bienniale). At the same time, in their designs the architects were required to address the topic of "cities create architecture, architecture creates cities" and to display to their best ability works that were related to both urbanism and sustainable development.

Finally, the twelve methods of displaying the works varied. In most of the works, the representational form of the traditional façade was softened, and on the contrary, in considering how to answer the question of urban development, what the architects conveyed were new trends in the expression of present-day architectural language. Among these new trends were the application of new technology and parameterized designs in architecture (Aranda Lasch, J. Mayer H., and MAD Architecture), the examination of architecture form and structure and its relationship to city "street" space (Hashim Sarkis and SPBR), consideration of public space within the city (open architecture), the relationship of architecture and Politics (Fake Industries Architectural Agonism), the structural design of low-income housing (ELEMENTAL / Alejandro Aravena), and many different spatial experiences in which construction was contingent on materials, form, and structure all together. (Johnston Marklee, Mass Studies, SO-ILand Atelier Deshans)

If you say that the "Strada Novissima" exhibition thirty years before was at a turning point striding towards the post-modern architecture movement, then ought the 2011 Shenzhen Biennale's "The Street" also be expected to have some sort of importance? Or perhaps having excessive expectations for the works in "Street" is a sort of misconception about the exhibition itself, because when however many years from now we look back on this exhibition, we ought not only to evaluate these works themselves, but we ought also to look at the influence that these architects and their works had on the world of architecture, cities, and the public, and this influence is one that cannot be estimated or measured. Additionally, there are at least two points within China's present environment that are especially worth investigating. First, the majority of the participating architects are young or of middle-age. The Chief Curator's original intention was to encourage emerging architects. When Riley was chief curator of architecture at the MoMA in New York City, he always actively promoted PS1's emerging architect program (every year, one architect was selected to build a temporary installation within the museum's courtyard), thus giving many emerging American architects an opportunity to reveal their design abilities on an international platform. In discussions between Riley and the curatorial team, he often spoke of giving emerging architects opportunities to create, as making them learn construction skills through practice is very important. Yet China's present-day situation is just the opposite, as most emerging architects engage in "paper architecture", and practical opportunities are very few.

Second, that "The Street" serves as the most basic urban public space has been reiterated through the exhibition. Currently, the concept that the street acts as a space for interaction has already become very hazy, our impression of the street is mostly as a roadway on which cars travel. Yet during today's popularization of sustainable development, ought the street be designed in order to increase humanization and traditional activity? In addition to the exhibition, this edition of the Biennale also advanced the investigation of the potential for sustainable development of "the street" through activities held during the exhibition period. On January 7th, 2012, during the Bienniale's "Street Theater" activity, dancers and the viewing public interacted between the twelve façade works, making us aware of the possibility for the rebirth of the street's life.

As Chief Curator Riley says, these twelve works seem to not be able to be defined by simply one theoretical banner, but we are never able to view ourselves in the present, and only in the future will we be able to judge and define through various methods today's architecture. Because of this, the recollection of "The Street" exhibition and each of the works within it perhaps will be both the summary and expectation of present and future urban development. Although one exhibition cannot predict future trends, it is likely that as long as the life of "the street" becomes richer, then the city can finally have vitality.

References:
Borsano, Gabriella. eds. 1980. *Architecture 1980: The Presence of the Past*. Rizzoli: New York.
Levy, Aaron & William Menking. 2010. *Architecture on Display: On the History of the Venice Biennale of Architecture*. London: Architectural Association.

GLOBAL PERSPECTIVES / 全球视野

"新趋势"和"街道"

1980 年，在意大利人保罗·波多盖希（Paolo Portoghesi）的推动下，首届威尼斯国际建筑双年展诞生了。作为一名建筑师、理论家、历史学家和罗马大学的教授，波多盖希随后成为了威尼斯建筑双年展的主席，并主导策划了 1980 年，1982 年和 1992 年三届威尼斯建筑双年展。在 1980 年的首届双年展中，以"街道"（Strada Novissima）为题的展览成为了重要的组成部分之一。

崇尚古典建筑的波多盖希敏锐地捕捉到了 20 世纪 70 年代末期的后现代建筑运动，并以"过往的呈现"（The Presence of the Past）作为首届双年展的主题。在他看来，"二战"后普及的现代主义建筑摒弃了传统、民族性、地域性和个性的国际式风格，失去了和普通市民之间的对话，从而打断了连续的建筑历史。建筑不应该仅仅是精英们的游戏，而应是为大众而设计的。这些想法顺应了当时的后现代建筑运动，即呈现的是多元的审美情趣和在设计中对历史文脉的尊重与呈现。同样，在城市空间设计中，街道这一基本的城市生活组成元素，也被再次强调。正是在这样的背景下，"街道"这个展览的目的是让参观者直接地证实"回归街道"是城市的形成要素和后现代研究的根本目标之一。

经过一番努力并成功地说服了军方之后，波多盖希将首届双年展安家在第一次作为展馆的兵工厂（Corderie dell'Arsenale）内的缆绳车间里。巨大的空间尺度和毗邻威尼斯市中心的极佳位置是波多盖希主要的选址理由，他坚信只有这样才能让普通观众更加容易地到达展览场地。在知名的"街道"展览中，波多盖希邀请了来自全世界的 20 位建筑师参展，其中包括之后被建筑界熟知的格雷夫斯（Michael Graves）、盖里（Frank O. Gehry）、鲍赞巴克（Christian de Portzamparc）、库哈斯（Rem Koolhaas）和矶崎新（Arata Isozaki）等。当时的"街道"共 70 米长，根据柱子的位置，大厅被分为中间的公共空间——"街道"（4.5 米宽）及其两边各十个隔间。每位建筑师都被分配到一个隔间，并被要求针对"过往的呈现"这一主题，沿着中间"街道"做个全比例的建筑立面，立面宽约为 7 米，高在 7.2 米到 9.5 米之间，建筑师可以通过设计来表达自己独特的形式感。整个街道也连接着由意大利著名建筑师阿尔多·罗西（Aldo Rossi）设计的展厅入口。

波多盖希希望这条由罗马电影工作室搭建完成的街道能被看作并排相连的"住宅"。这些"住宅"可以是建筑师自己的家，也可以是一个普通的日常的、私人的住所，还可以是一个供聚会和娱乐的地方。设计要求观众可以自由地穿过立面，进入到隔间内参观建筑师设计的一个他自己了作、展示或"卖弄"想法的地方。从设计上来看,很多建筑师在立面设计上都使用了古典的元素,直接回应了"过往的呈现"的主题。

"街道"对于首届威尼斯建筑双年展的意义可谓举足轻重。相对于传统展览中经常出现的图纸、照片和模型等元素,"街道"首次将全比例的建筑空间形式带入到室内的展厅中。正因为这样,波多盖希认为普通观众"新趋势"和"街道"能更好地与作品互动,并参与到建筑和城市设计的讨论当中。

首届威尼斯国际建筑双年展对外界的影响尤其深远,包括当时还是建筑系学生的泰伦斯·瑞莱(Terence Riley)。30年后,当瑞莱担任2011深圳·香港城市\建筑双城双年展的总策展人的时候,他自然而然地将两者联系在了一起。对深圳双年展来说,第一次聘请非华人的总策展人正体现了其逐步走向国际化的步伐,这与1980年威尼斯建筑双年展的要搭建更宽广的平台的目的是一致的。"街道"展览也被瑞莱重现在深圳双年展中。

在征求了很多建筑策展人、评论家和学者的意见之后,12位崭露头角的中青年建筑师被邀请参展。每位建筑师都被要求在大约6米宽、3.5米深、4米高的空间里发挥空间想象力,并展示他们自己的作品。和30年前的"街道"展一样,每位建筑师也要考虑设计一个"立面"(facade)。在展厅内排布时,每六个"立面"一组,共两组,分在两边,并对着中间的街道(展览现场"街道"中间的柱子上是Aaron Betsky对当年"街道"项目中的至今仍活跃的绝大多数参与者的访谈)。同时,设计要求建筑师能够回应"城市创造建筑,建筑创造城市"的主题,尽量展示与城市和可持续发展相关的作品。

最终,12个作品呈现的方式是多样的。在大部分的作品中,传统立面的表现形式被弱化了,相反,在如何去应对城市发展这个问题上,建筑师表达出来的是当前建筑语言所呈现的新趋势。其中有新技术和参数化设计在建筑中的应用(Aranda Lasch, J. Mayer H. 以及MAD Architecture),建筑形体与城市街道空间关系的反思(Hashim Sarkis 和 SPBR),城市中公共空间的思考(开放建筑),建筑与政治的关系(Fake Industries Architectural Agonism),低收入住宅的结构设计(ELEMENTAL / Alejandro Aravena)以及各种材料和形体构筑而成的新的空间体验(Johnston Marklee, Mass Studies, SO-IL 和大舍建筑)。

如果说30年前的"街道"(Strada Novissima)处在一个向后现代建筑运动迈入的转折点上,那么2011年深圳双年展上的"街道"(The Street)是否也应该被期望有着某种重要性呢?又或许对"街道"作品的过度期待是对展览本身的一种误解,因为若干年后我们回顾这次展览时,不应该仅仅评价这些作品本身,还应该看到这些建筑师和作品对建筑界、城市和大众的影响,而这些影响是不可预估和测量的。此外,至少有两点在中国当前的环境下尤其值得探讨。首先,参展的建筑师多为中青年,这也是总策展人希望鼓励年轻建筑师的初衷。瑞莱先生在纽约当代艺术博物馆(MoMA)当建筑策展人时,便一直积极推动PS1年轻建筑师项目(每年都会挑选一名建筑师在博物馆的庭院里搭建一个临时性装置),从而让很多年轻的美国建筑师有机会在国际舞台上展露自己的设计才能。在瑞莱与策展团队的交谈中,他多次谈及给青年建筑师创造机会,让他们在实践中学会建造的技能是非常重要的。而当前中国的状况正好相反,大多年轻建筑师从事于"纸上建筑",实践机会非常少。

其二,"街道"作为城市中最基本的公共空间,在展览中被重申。当前,街道作为交往空间的概念已经变得非常模糊,对它的印象多为行车的马路。在推广可持续性发展的今天,街道是否应该被设计得更加人性化又充满活力呢?本届双年展除通过展览之外,还结合展期内的活动进一步探讨了"街道"的可持续发展潜力。在2012年1月7日举办的双年展活动"街道剧场"中,舞者与观众在12个立面作品之间的互动,让我们感受到了街道生活重生的可能性。

如总策展人瑞莱先生所说,这12件作品似乎不能以单一的理论旗帜来定义,但我们总是无法自视现在,只有未来能以各种方式来评断和定义当前的建筑。因此,对"街道"这个展览及每件作品的回味或许是对当前和未来城市发展的总结和展望。虽然一次展览是无法预测未来趋势的,但有一点是肯定的,只有拥有更丰富的街道生活,城市才会有活力。

参考文献:

Borsano, Gabriella. eds. Architecture 1980: The Presence of the Past. Rizzoli: New York. 1980.

Levy, Aaron & William Menking. Architecture on Display: on the History of the Venice,Biennale of Architecture. London: Architectural Association. 2010.

STREET THEATRE
街道剧场

CURATORS: ZHANG YUXING, ZHOU HONGMEI, WU XUEJUN
策划：张宇星 周红玫 吴学俊

Artistic Directors: Teng Fei, A Fei
Execution: Shenzhen Old Heaven Cultural Broadcasting Company
Participating Artists: Long Yunna (Guangzhou), Samuel (Chile), Zhu Fangqiong(Guangzhou), Tian Zhilin (Guangzhou), Deng Boyu (Shenzhen), Ma Mu'er (Beijing), Sheng Jie (Beijing), Song Yuzhe (Beijing), Anthony (USA)
导演：滕斐 阿飞
执行：深圳市旧天堂文化传播有限公司
街道剧场参演艺术家：龙云娜（广州），Samuel（智利），朱芳琼（广州），田志林（广州），邓博宇（深圳），马木尔（北京），盛洁（北京），宋雨喆（北京），Anthony（美国）等

In January 2012, the "Street Theater" activity held following the Biennale's opening ceremony once again raised great public interest. Through the exhibited works, the actors and the reconstruction of the third-party relationship with the audience, the "Street Theater" reinvented and reinterpreted the Biennale's exhibition space, overturned the exhibition hall, only for professionals and transformed passive works that were previously only able to be viewed into works that could be participated in, creating an entirely new, interactive experience.

"Street Theater" was one of a series of activities within this edition of the Biennale, drawing out the main theme of "Architecture creates cities, cities create Architecture" from narrative form and the depth of the space. Actors used abstract thought within the exhibition space of "The Street" in order to create an exchange between the human body and sound and architectural space, drawing on lighting and sound effects to evoke the narrative power of mythology. Actors used both improvisational and non-improvisational methods of acting, reminding us to be conscious of the present interactions that cannot be ignored that take place within the space. Make the city return to the life of the streets!

2012 年 1 月中旬举办的"街道剧场"活动继双年展开幕后再次引爆了公众热潮。街道剧场通过展品、演员和观众三者关系的重构,重新解读了双年展的展览空间,颠覆了仅供专业人士参观的陈列厅,将被动地被观看的展品转化为可以参与的、全新的互动体验。

"街道剧场"作为本届双年展的系列活动之一,从叙事形式和空间深度上延伸了"城市创造"的主题,演员将用抽象的理念在"街道"展品空间中用肢体和声音与建筑空间交流,借助灯光、音效唤起神话的叙述力量,演员即兴和非即兴的演出,提醒我们意识空间里正在进行中的现在和不可忽视的过往。让城市重回街道生活!

The host Tian Na and the Organizing Committee Secrefary Xue Feng,
Huang Weiwen (Director of Shenzhen Public Art Center),
Liu Hongjie (General Manager of OCT-LOFT)
Curators: Zhang Yuxing, Zhou Hongmei, Wu Xuejun
Artistic Directors: Teng Fei, A Fei
Photograph: Niu Niu

主持人天娜和主办方代表:组委会秘书长薛峰
协办方:深圳市公共艺术中心主任黄伟文
深圳华侨城创意园文化发展有限公司总经理刘洪杰
策划张宇星、周红玫、吴学俊及导演滕斐、阿飞
现场摄影:牛牛

THE PRESENCE OF THE PAST REVISITED
再造访"过往的呈现"

Curator: Aaron Betsky
List interviewees: Frank Gehry, Michael Graves, Allan Greenberg, Leon Krier, Thomas Gordon Smith, Robert Stern, Stanley Tigerman, RemKoolhaas, Venturi Scott Brown

策展人：Aaron Betsky
访问对象：Frank Gehry, Michael Graves, Allan Greenberg, Leon Krier, Thomas Gordon Smith, Robert Stern, Stanley Tigerman, RemKoolhaas, Venturi Scott Brown

| GLOBAL PERSPECTIVES / 全球视野 | THE PRESENCE OF THE PAST REVISITED |

Allan Greenberg

Rem koolhaas

Michael Graves

再造访"过往的呈现"

Leon Krier

Robert Stern

Stanley Tigerman

Frank Gehry

Venturi Scott Brown

Thomas Gordon Smith

Staging Postmodernity:
The Strada Novissima and the First Architecture Biennale

Aaron Betsky

Was it a shopping mall, a church, a trade fair, or a stage set? Those are some of the models the participants in the so-called Strada Novissima, the parade (to use yet another image) of facades stretching down the middle of the Corderie, or rope-making space, of Venice's Arsenale, during the first official International Architecture Biennale in that city in 1980, used to describe the results. Assembled by the Italian architect, critic, and historian Paolo Portoghesi and a cast of advisors, the displays fronted small rooms in which the architects were asked to show their wares. Whatever the right model, the result was a full exposition of the architecture of Postmodernism, but what that meant or means is as open for debate as to whether the work was there to be sold, exposed, adored, or consumed as spectacle.

The very ambiguity over both nature and function within our image culture marks the nature of Postmodernism and is its hallmark. It delighted in the conflation of different images and forms from different times and places. It mixed metaphors and materials. Its references were multivalent and its meaning ambivalent. Collage was its most dominant discipline. The modes in which it appeared also mixed cultural, consumerist, and image industry aspects. The Strada Novissima thus exposed Postmodernism for what it was – which was many different things at the same time.

Looking back on the Strada and on Postmodernism, however, we should be able to come to some definition. On one level, Postmodernism is simply everything that comes after modernism. The problem with that statement is that modernism aims always to be the representation of modernity, so you have to first be able to say that modernity had transformed into postmodernity to allow Postmodernism to appear. It is difficult to define the difference between modernity and its successor, though some sense that progress, the always new or modern, was not a given, and the subsequent loss of innocent or optimism, would seem to be central to such a division. By that token, however, modernity may have ended with the Depression and the rise of totalitarian economic, social, and political systems. In fact, you could argue that the dominant modes of both socio-economic structures and their representations since 1929 have been an attempts to bring back past structures, re-establish static order, live with limits, and create myths of modernity that could be a operational, i.e., could form a functional modernism that contributed to nation- and company-building. Planning replaced progress, resource allocation invention, and bureaucracy entrepreneurial action.

Postmodernity has thus been a long process and, as always, it has taken all forms of art quite some time to catch up with this reality, with architecture lagging the furthest behind. In many ways it took the sheer and guerilla-oriented violence of the 1960s, with its social unrest, assassinations, and Vietnam War, the subsequent turmoil in the American markets and social structures, the Club of Rome report of 1972 and the Oil Shock of the following year to make architects realize that they could not go on proposing ever bigger, more abstract, more radically expressive structures.

Postmodernism as a word has a more limited history, and one perhaps more germane to the discipline narrowly considered. Though there were obvious precedents in literary criticism, the first public use of the phrase came with the publication in 1977 of Charles Jenck's The Language of Postmodern Architecture. Jencks defined Postmodernism above all else as a linguistic movement: reality had become reduced to signs, both by science and by the efficiencies of capitalism that, as Karl Marx and Friedrich Engels had predicted, systematically removed all weight, reality, and friction from the economic system. By the 1970s we indeed lived in what Roland Barthes called "The Empire of the Sign." Jencks' approach was to embrace this reality, and to call for an architecture composed purely of signs. Coherence would be grammatical or syntactical, elaboration would be through tropes, metaphors, and analogies, and the forms of expression would be as myriad as the languages of mankind.

It was not that Charles Jencks invented Postmodernism, but he did give a name to a divergent set of developments within the discipline. The first and most obvious of these was the return of history, and it is significant that Portoghesi titled his Biennale "The Presence of the Past." In Europe, the past came back in part because a new generation had not lived through its horrors, and, as a practical matter, Europe's whole-scale reconstruction had reached it limits, leaving it to architects to think about re-use and reevaluation, rather than just replacement. Moreover, the public, both in Europe and in the United States, not only had lost faith in the new and progress, but had also made clear their rejection of modernism, finding it to be the imposition of alien forms by faceless bureaucracies. Architects responded as much to changed political and economic conditions that foregrounded consumerist pressures and financial and resource limitations as they acted from a sudden re-appreciation of past forms and images.

A second wellspring of Postmodernism was popular culture-again, as much in response to the change from a production-oriented to a consumer-oriented economic and political system as anything else. "Fine" artists had, of course, embraced everything from packaging to television images, from poster "art" to cartoons, a full generation before, and were already moving onto the next cycle of artistic appropriation (that of their own forms). It was not until Robert Venturi's Complexity and Contradiction in Architecture, however, that architects began to take the environmental aspects of popular culture, namely the road, the strip, and all its signs, seriously. Even then, virtually none of them went so far as to turn to the reality of the built environment in terms of mass-produced housing or commercial spaces for inspiration –though Venturi and his wife and partner, Denise Scott Brown, did call for such an appropriation in their seminal 1976 exhibition Signs of Life.

Finally, a third component of Postmodernism as a movement within architecture was uncertainty itself. The breakdown of systems, whether of the state, of the family, of planning, or of science, and the realization of the paramount importance of an uncertainty that could only be controlled through scenarios or embraced as productive chaos, led many architects to try to come up with an architecture that would contradict itself, destroy itself, or otherwise negate its very construction. This self-negation was not always somber or serious, but could take the form of a delight in play or what Venturi, in Complexity and Contradiction, had called "the nearness of chaos…that vitiates."

It was in this scene that Portoghesi staged his Biennale. The occasion was the "off-year" of the Venice Art Biennale, a venerable institution that had been presenting art from around the world in odd-numbered years since 1895. While architects in Italy had been lobbying for a chance to present their work on equal footing, and had held several small exhibitions in conjunction with the art event, the establishment of the Architecture Biennale was as much the result of the fair organization's realization that they should use their brand and their spaces more effectively by reaching beyond their core field, in-between the widely-spaced art events, as it answered a call by designers. The recovery of a large section of Venice's former armory, which had for centuries been used by the Italian navy, was itself part of a world-wide movement, then only just beginning, to reuse industrial and military sites. The Architecture Biennale, in other words, was first and foremost itself a mark of the victory of consumerism, branding, and retailing.

Portoghesi's format was simple, and was the result of discussion he had with his advisors, Eugenia Fiorin, Paolo Cimarosti, Charles Jencks himself, Christian Norberg-Schulz, and Vincent Scully. Each architect was to be given one of the bays in the section of the armory once devoted to rope making; massive brick columns created a central nave along which the ropes were once strung, with side aisles that were just about as broad as the distance between those columns, creating roughly square modules. There

| GLOBAL PERSPECTIVES / 全球视野 | THE PRESENCE OF THE PAST REVISITED |

would be thus be simple and expository line along which visitors could travel, leaving the sides to be marked off as cubicles in which the architects could show their work. Whether the architects Portoghesi and his cohorts selected chose to make these spaces into shrines to what they felt were sacred principles or objects, used them to explain what they felt were operative principles, or simply took the occasion to present their designs in the most attractive and consumable manner possible, was entirely up to them. None of the architects today recall being given clear instructions on these matters: they were merely assigned a space and told to fill it.

Thus the Strada Novissima was already Postmodernism in its very conception. It was part of a consumerist project, in which buildings were to be reduced to signs. There was not one message, beyond the availability of all of the past to be reused, and the results were highly mixed in form and content. The visitor was offered a choice, as were the architects: there was no dominant order, metaphor, or system, but only the many fragments available to the consumer.

The facades were the one place where the architects could express themselves or allude to larger ideas. For these billboards of design, they had to produce sketches; scene builders from Cinecitta, the Rome movie studio, created the actual facades. These craftsmen by all accounts did a remarkable job at using minimal means to translate the architects' visions into something that was not so much real, as it gave the illusion of being so.

This pseudo-reality was itself a revelation to many of the participants. It turned out that one of Postmodernism's greatest achievements was the liberation of architecture from built form. If almost all theories of architecture until then had depended on justifying whatever one experienced of a building as being the absolutely necessary result of form, function, structure, or some other essential aspect of the construction of a specific object on a site, it now became clear that architecture, as a language, as a consumable, as a form of self-reflection, and as a stage set, was completely free to do what it wanted. It only had to be able to signify or sell, not to make.

It was perhaps no accident that the most pervasive mode in which Postmodernism appeared on the Strada Novissima was as various forms of neo-classicism. Though this was also the result of Portoghesi's predilections, and was one of the reasons why the critic and historian Kenneth Frampton resigned from the advisory group, classicism was the mode of architecture had left its relation to structure or any kind of site-specific and temporal reality behind for the longest period. It was and is, indeed, a language, though one with the particular property of being univalent. It is, for better or worse, a form of Latin, not spoken by everybody, but underlying and founding many forms of signification and coherence, from language to legal systems to scientific classifications, in the Western world. Classicism came back into its own, in 1980, as a form of meta-architecture as Latin is a meta-language. If architecture is that which is about building, then classicism is what is about architecture.

In a concrete form, classicism was also the most recognizable connection to a pre-modern past, certainly to what was, at the first biennale, still an overwhelmingly Italian audience. It was also the easiest to reproduce and to explain. Unfortunately for its most committed participants, however, it was and is, like Latin, also the most restricted in its expressive possibilities and the most difficult to do right. The most committed neo-classicists who participated in the Strada Novissima either went on to have careers in which they constructed relatively little, such as Thomas Gordon Smith or Allen Greenberg, or moved on, like Michael Graves and Robert A.M. Stern, to more varied practices.

What the Strada Novissima did beyond presenting Postmodernism in its fullest form to that date was also, quite simply, to make a street. Many of the participants recall the linear, enclosing aspects strongly to this day. Making streets and traditional urban patterns became the hallmark of Postmodernism as it moved from architecture into urban planning, and you could argue that its influence there was even larger than on the peculiarities of facades. The Strada demonstrated the strength of a defined space along which you could move, having discrete experiences, and encounter repetitive spaces in a rhythm that would ground you, while it opening up to radically different realms within that sequence. Instead of extensive, multi-layered, but abstract and sometimes labyrinth-like spaces that were supposed to reflect and materialize the open-ended, democratic, and limitless possibilities of modernity, you now walked down a familiar line and found yourself in a confined and consumable space.

The Strada Novissima's strength came as much, however, from the exceptions to the general turn towards neo-classicism and order as it did from the designs that set its tone and rhythm. From G.R.A.U.'s tomb or columbarium, to Hans Hollein's destruction of classical columns through mimicry, covering them with vines, turning them into buildings, and cutting them off at the midriff, classicism came into question as soon as it was stated. Other architects, such as Frank Gehry or the Office for Metropolitan Architecture, went beyond classicism to the kinds of concerns that would mark their work in the next generation, and which mainly involved coming up with a critical alternative to what they felt was the descent of architecture into Postmodern consumerism.

The Strada Novissima did not appear in a void. The year 1980 was also when Time Magazine displayed Philip Johnson on its cover, holding the AT&T headquarters like the Pop Art toy it became. Postmodernism was triumphant. It was also under construction, in the work of architects such as Michael Graves (the Portland Government Building), Frank Gehry (Loyola Law School), or Aldo Rossi (the Theater of the World, which appeared at this Biennale). Several architects related their experiences in Venice to those they were having at the same time in Berlin as participants in the Internationale Bau Ausstellung (IBA) there. The Strada Novissima itself moved on, first to Paris, and then to San Francisco, increasing the impact of its forms while losing the relation to its site, both immediate and proximate, that had created the legitimacy for its structure.

None of the architects I have interviewed remember receiving any commissions because of their participation. They report no revelations or innovations. Instead, the Strada Novissima affirmed their interests and predilections, and helped them understand that, at least for a moment, a group of then mainly young architects were all groping to come to terms with Postmodernity and to construct scenarios for that reality. In Venice, they had a Postmodern moment. After thirty-one years, they still believe in what they were trying to do, and they still remember the 1980 Biennale as a stage set in which they could act out their roles as architects as they saw fit.

Note on Sources

The preceding is based on conversations the author held with most of the surviving participants in the 1980 Biennale between September and November of 2011. Additional information comes from the exhibition catalog's English version, published by Rizzoli International Publications in 1980, and from Aaron Levy, William Menking, Architecture on Display: On the History of the Venice Architecture Biennale (London: Architectural Association, 2010).

GLOBAL PERSPECTIVES / 全球视野　　THE PRESENCE OF THE PAST REVISITED

上演后现代主义：
"新趋势"和第一届威尼斯建筑双年展

Aaron Betsky/ 著

我们看到的是购物中心？教堂？商品交易会？还是舞台布景呢？这些都是关于 1980 年第一届威尼斯建筑双年展的主展"新趋势"（Strada Novissima）中的参与者们所创造的各种模型的描述。这些作品以外立面的方式长长地排列在威尼斯"军械库"(Arsenale) 制绳空间中央的两侧。当时的意大利建筑师、批评家和历史学者保罗·波多盖希（Paolo Portoghesi）的策划和召集下，建筑师们在各自的小展厅的前方组装了"立面"式设计来展示他们的"货物"。无论这种模式是否正确，其结果是展览成为了后现代主义的盛会，但对其意味与意义的争论就像当时这些作品是否被当做某种景观被售卖、曝光、崇拜和消费的讨论一样开放。

后现代主义的本质特点正是我们的图像文化本身模棱两可的特性与功能。它乐于把不同时间与空间的图像及形式进行异文合并；它混合各种隐喻及材料；它本身的参照来源丰富多样，其含义也很矛盾多元。拼贴是它最重要的原则。在"新趋势"展览中的这些模型和作品本身也表现出了多元混合文化、消费主义及图像工业等方面的特点，就这样，"新趋势"展览本身揭示了后现代主义的特点：将众多不同的元素同时呈现。

当我们回顾"新趋势"展览及后现代主义时，我们应当涉及一些定义。一个层面上，可以简单地说后现代主义就是在现代主义之后的那些事物。这种陈述的问题在于现代主义被认为是现代性的代表，这意味着只有当现代性得以转变为后现代性的时候，后现代主义才得以现身。要区分现代性与其后继者的区别是困难的，对某种进步或者新的事物的感知并不是区分两者的假定条件，但对于天真及乐观主义随后的丧失，仍被视为此区分的核心。然而，照此来说的话，现代性可以说是终结了大萧条与之后集权主义经济、社会和政治体系的出现。事实上，你也可以说自 1929 年之后具主导地位的社会经济结构及其代理人们企图将过去的结构带回给世界，重新建立某种静态秩序和受限制的生存，并创造那些可以被操作的以及能够对国家或公司建构有贡献的功能现代主义的现代性神话。规划取代了进步、资源分配发明和官僚企业行为。

因此，后现代主义的形成经历了长期的过程，就像过去一样，各种艺术形式需要更长的时间来追赶这种变化中的现实，而建筑更是远远地滞后。从 20 世纪 60 年代那种游击队式的暴力开始，持续的社会动荡、暗杀、越南战争，到美国市场和社会结构的震动、1972 年罗马俱乐部发布的《增长的极限》（The Limits to Growth）报告以及接下来多年的石油危机，都使得建筑师认识到他们无法继续像从前一样去提出总是更宏大、更抽象，同时也更激进的建筑的构想。

"后现代主义"是一个有着有限历史的名词，这种说法或许对这个狭义重量的学科更恰当。尽管在文学批评的领域有更多的案例，但第一次公开使用这个名词是在查理斯·詹克斯的著作《后现代建筑的语言》中。詹克斯将后现代主义定义为一场语言学的运动：现实本身被科学和资本主义效率简化成为符号，正如马克思和恩格斯所寓言的一样，科学和资本主义效率系统地将重量、现实和摩擦力从经济体系中剔除。在 20 世纪 70 年代，我们确实生活在罗兰·巴特所声称的"符号帝国"中。詹克斯的著作正是对这一现实的拥抱，他在召唤一种全然由符号构成的建筑。建筑的协调性将是文法和句法性的，设计细节将通过修辞、隐喻和类比的方式来实现，建筑的表达形式将会如同人类的语言一样无限。

查理斯·詹克斯并没有发明后现代主义本身，但是他的确给此范畴内分散的各种发展提供了一个共通的名字。这些发展中最早和最显著的就是回归历史，这也是波多盖希把他策划的建筑双年展的主题定为"过往的呈现"（The Presence of the Past.）的深意所在。在欧洲，过去的历史部分回归的原因在于：新一代欧洲人并没有经历欧洲历史的可怕年代。同时，从现实的方面来说，欧洲的整体战后重建已经达到了它的极限，因此对于建筑师而言，与其完全用新的建筑替代旧建筑，不如思考如何重新评估、使用和再造这些建筑。此外，当时欧洲及美国的公众，不仅仅对所谓的"革新"与"进步"失去了信心，而且很显然地，他们排斥"现代主义"本身，将其视为通过不公平的官僚机制强加于人们的异域的形式。在此情形下，建筑师尽可能地去回应在消费主义压力下以及金融和资源有限的改变的政治经济条件，就像他们突然地表现出对于过去的图像及形式的钟爱那样。

后现代主义的第二个发展动力是流行文化，就像其他所有的事物都对从生产导向转为消费导向的政治经济体系有所回应一样。艺术家理所当然地接受了这些变化，从包装到电视画面，从广告海报到卡通，整整一代的艺术家已经进入到新一轮的艺术形式的挪用中。直到罗伯特·文丘里发表《建筑的复杂性和矛盾性》（1966 年）后，建筑师才开始接受和

再造访"过往的呈现"

认真运用流行文化的环境方面的因素：公路、商业街和其他所有的符号。即便如此，最终他们也并没有过多地求助于从以批量住房和商业空间为依据的环境建造中获得灵感，尽管文丘里和他的妻子兼拍档丹尼斯·斯科特·布朗在他们 1976 年的展览"生命的符号"中号召这样的挪用。

最后，后现代主义的第三个组成部分是建筑领域的运动本身的不确定性。国家的、家庭的、计划的、科学等系统的崩溃以及那只能为情境控制的或将其视为"建设性混乱"的不确定性的最重要部分的实现，引领许多建筑师去尝试提供某种自我矛盾、自我毁灭或自我否定的建筑。这种自我否定不总是阴郁的或严肃的，也可以采取某种轻松的方式，如文丘里在《建筑的复杂性和矛盾性》中所说的："靠近混乱，使之无效"。(the nearness of chaos…that vitiates)

在上述情形之下，波多盖希筹划了他的建筑双年展，恰逢不是威尼斯艺术双年展这个自 1895 年始的百余年中享誉艺术世界的艺术盛会举办的奇数年。意大利的建筑师们其实一直要求在同等基础上展出自己作品的机会，并与艺术双年展相结合举办了一些小型展览。当时，建筑双年展形式的确立更多地是因为双年展组委会意识到要通过那些远离自身核心领域、更为中立和领域更为开放的艺术事件来更有效地使用自己的品牌和空间。威尼斯古老的"军械库"得到修复，这里曾是古意大利海军使用了几个世纪的地方，而刚刚开始重新使用工业和军事场地也是全球类似运动的一部分。换句话说，建筑双年展本身最先给自己打上了消费主义、品牌化及零售业获得胜利的记号。

波多盖希的展览形式很简单，这也是他和他的顾问 Eugenia Fiorin, Paolo Cimarosti, Charles Jencks, Christian Norberg-Schulz 及 Vincent Scully 商议的结果。每一个建筑师可以获得一个从前是军械库制绳作坊的小空间，军械库壮丽的石砖立柱与这些位于两侧的制绳空间是平行的，在军械库中形成了一个中庭，旁边的通道的宽度巧好是立柱之间的间隔距离，大致形成了一群方形空间。于是，以这样的方式，展场给了观众一条清晰、简单且具有说明性的路线，留下两边的小隔间给建筑师展出他们的作品。这些空间的使用方式完全取决于由波多盖希及其顾问们选出的设计师本人，你可以把这里制作为承载你觉得神圣的原则或物件的神龛，也可以用来解释和呈现你认为更有效的原则，或者，你想很简单地利用这个机会来展示你最吸引人及最具消费潜力的设计也无妨。没有一个当年参与其中的建筑师记得有人提供过什么指示。基本上，他们就是被给予了一个空间，并被告知要把它填满而已。

可以说，"新趋势"展览在它的概念阶段就非常的后现代主义了。它本身是一个消费主义项目的一部分，建筑本身被简化为了符号。这里没有任何明确的信息，除了可以把"过去"本身重新使用，其结果和形式也高度地混合。观众们和建筑师一样被赋予各种选择：这里没有任何占据主导地位的秩序、隐喻或系统，而是提供给消费者的诸多片段。

立面成为建筑师可以表达自我和暗示他们作品背后更宏大概念的地方。这些广告牌似的立面设计，先由建筑师绘制草图，再由电影院的造景师和罗马电影工作室的工作人员来完成。这些手工业者的工作令人惊艳，他们用最有限的方法将建筑师的愿景转译成了某种具有幻觉特质的不那么真实的场景。

这种"伪造现实"本身对于很多参与者而言是一种启发。它成为了后现代主义最伟大的成就之一：将建筑（architecture）从建造形式 (built form) 中解放出来。如果说直到那时各种建筑理论有赖于将任何人对于建筑的体验证明为基于某种形式、功能、结构，或其他关于某地某物之建造基本方面的必要性结论，那么现在，建筑本身成为了一种语言，一种可以消费之物，一种自我反映的形式，就像舞台布景一样完全自由的。惟一重要的是能够销售，而非是否能够建造。

新古典主义是在"新趋势"中出现的最为普遍的后现代主义模式，这或许并不让人意外。这是策划人波多盖希自身偏好的结果，也是批评家、历史学家 Kenneth Frampton 会退出顾问团队的原因，古典主义只是一种将自身与结构的关系、各种特定场地要求及世俗社会律令早就抛诸脑后的建筑模式。古典主义曾经是，现在仍是一种建筑语言，不过，是一种有着单一特定属性的语言。无论好坏，它就像某种拉丁语，不是每一个都会说，但是在西方世界，拉丁语是所有科学分类、法律体系语言以及所有意义和相关性的潜在基础。像拉丁语作为元语言一样，古典主义在 1980 年作为一种元建筑回归了到其自身。如果说建筑是关于建造的，那么古典主义就是关于建筑的。

在具体的形式上，古典主义也是最可被识别的与前现代性的过去的联结，在第一届建筑双年展上，仍有很多意大利的观众。它也是最容易被复制和解释的。可惜的是，对于大部分的参与者而言，古典主义就像拉丁语一样，也是最受到其自身表现力局限的，同时，也是最难做得好的。大部分参与了"新趋势"的新古典主义建筑师们，或者像 Thomas Gordon Smith 和 Allen Greenberg 一样，用他们相对少建造的方式继

GLOBAL PERSPECTIVES / 全球视野   THE PRESENCE OF THE PAST REVISITED

续自己的职业生涯或者像 Michael Graves 和 Robert A.M. Stern 一样，进入更多样的实践。

"新趋势"除了在当年以最充分的形式展示了后现代主义以外，更重要的是，它创造了街道。许多参展人至今仍强烈地记得那线性、围合的部分。建造街道和传统城市模式成为后现代主义的一个重要特点，并从建筑本身转向了城市规划，你甚至可以说它的影响要比那些独特的立面带来的影响大得多。"新趋势"呈现了某种定义的空间的优点，在这种空间中，你可以在移动中获得不同的经验，在围绕着你的节奏中与反复出现的各种空间相遇，而它在那样的秩序中打开了一个全然不同的领域。它不是广阔、多层面的，而是抽象且有时候像迷宫一样的空间，以物化形式来反映现代性之开放、民主和无限可能。你行走在自己所熟悉的那些路线中，并发现自己处于一种相对狭窄的可消费的空间。

尽管"新趋势"的优点突出，然而那些当年整体的新古典主义转向中的例外，他们的设计创造的规则设定了它本身的格调和节奏。有时候，古典主义的再现使其本身立即成为一个问题，比如 G.R.A.U 设计的墓地和公墓，汉斯·霍莱因以模拟的方式摧毁了古典主义的立柱，在其表面布满葡萄藤图式，将它们移到建筑的内部，甚至是拦腰截断。其他的建筑师，比如弗兰克·盖里、大都会建筑事务所，都已经超越了古典主义，将他们的建筑作品更多地面向未来的世界，他们主要的关切是如何找到一种针对他们过去所认知到的后现代消费主义中的建筑本源的批判性替代形式。

"新趋势"没有出现并不是空洞的。1980 年，菲利普·约翰逊捧着 AT&T 公司总部的建筑模型出现在《时代周刊》的封面上。当时的后现代主义是胜利者。很多建筑师的项目正处于进行时，比如迈克尔·格雷夫斯设计的波特兰政府大厦、盖里设计的罗耀拉法学院、阿尔多·罗西设计的世界大剧院等。许多建筑师将他们在威尼斯的经验与他们参与的同期的德国柏林国际性城市规划与建筑设计展（IBA）联系起来。"新趋势"本身也移到了巴黎，然后是旧金山，在与它最初相关的地点渐渐无关之后，新趋势形式的影响持续增强，直接或间接使它的结构获得了某种合理性。

没有一个参与了"新趋势"的建筑师记得他们因为参加这个展览而获得任何项目委托。他们认为自己在当时并没有做出什么革新和启示。相反，"新趋势"让他们确认了自己的兴趣和偏好，同时，帮助他们更好地理解了这一点。至少有那么一刻，这些当时都很年轻的建筑师曾一起摸索什么是后现代主义，一同为当时的现实建构某种场景。在当时的威尼斯，他们的确经历了后现代的时刻。在 31 年之后，他们依然相信他们曾经想要做的事情，他们依然记得在 1980 年的第一届威尼斯建筑双年展上，他们就像站在舞台上那样，以建筑师的角色做任何他们相信并认为应该做的事情。

资料来源：

上述资料主要来自 2011 年 9 月至 11 月期间，作者与仍健在的 1980 年威尼斯建筑双年展的亲历者的对话。其他信息来自当时展览的英文版画册（由 Rizzoli International Publications 出版，1980）以及《展示中的建筑：威尼斯建筑双年展历史》（作者：Aaron Levy, William Menking，由 London: Architectural Association 出版，2010）

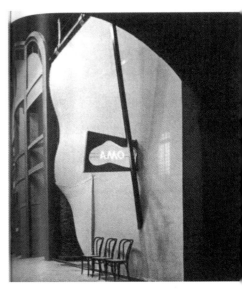

R. Koolhaas / OMA, OMA pavillion, Strada Novissima, Architectural Biennal, Venice, 1980
R. Koolhaas / OMA, "新趋势"威尼斯建筑双年展 1980

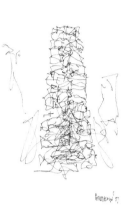
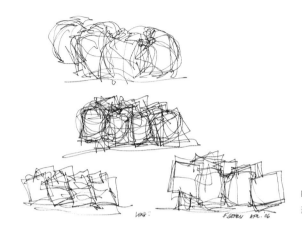

Frank Gehry's sketches
盖里的手稿

Rem Koolhaas / OMA
库哈斯 / OMA

FAVELA PAINTING
彩绘都市

Curator: Jeroen Koolhaas, Dre Urhahn
策展人: Jeroen Koolhaas, Dre Urhahn

GLOBAL PERSPECTIVES / 全球视野

FAVELA PAINTING / 彩绘都市

As people migrate to urban areas in masses, many cities are growing at an unprecedented rate. A majority of these new city dwellers live in informal additions to the urban landscape, often unwanted, neglected or simply forgotten. Our world faces the challenge to find creative solutions to include these areas and their inhabitants into society.

Art is a unique messenger, crossing borders and building bridges. If implemented in an intelligent way it can be powerful weapon to catalyze social change. This is the main objective of the Favela Painting Project, founded by Dutch artists Dre Urhahn and Jeroen Koolhaas. Turning public urban spaces in deprived places into inspiring and monumental artworks.

Offering local youth education and job opportunities, while making their community a nicer place to live in. A strong social acupuncture that could unlock local potential, boost the pride and self esteem and send a strong positive message to the outside world.

The Favela Painting Project started in 2005 in Rio de Janeiro. A series of projects was set up in Vila Cruzeiro and quickly spread to other places like Santa Marta. Together with local youth they created amazing results that became monuments in Rio's urban landscape, changed public opinion and attracted massive coverage by international media.

Favela Painting is not about cosmetically 'camouflaging' an area but aims at creating lasting effects. Using creativity and imagination to rethink, redesign and thereby rebrand a community as a whole. At the moment Urhahn and Koolhaas are working on a large-scale community project in North Philadelphia while working on several proposals for Rio and other cities around the world.

The Favela Painting Foundation supports the Favela Painting Project and works on funding, education programs and the maintenance of the murals. It also creates extra opportunities that rise from the project like the possibility of setting up small-scale facilities for paint production in the favelas they work in.

As the possibilities are endless the foundation is always on the lookout for partners to help make this dream possible around the world.

当人们大量涌入城市，城市便开始以前所未有的速度不断发展。大部分来到城市的新居民都生活在城市的非正规场域，这些地方往往被人嫌弃、忽略或者遗忘。我们面临这样一个挑战：如何找到创意的方式把这些场所以及住在其中的居民融入到社会中？

艺术是一位独特的信使，跨越边界，架起桥梁。如果以巧妙的方式加以利用，艺术就能够成为激励社会进行改变的有力武器。这正是"彩绘都市"项目的主要宗旨，这个项目由荷兰艺术家 Dre Urhahn 和 Jeroen Koolhaas 创立，把贫困地区的城市公共空间变成了一件巨幅且鼓舞人心的艺术作品。

这个项目为当地的年轻人提供了教育及工作机会，把社区建设成了一个更适合居住的地方。这是给社会实施的一剂针灸疗法，释放当地潜力，增加自豪感和自尊心，并向外面的世界发出强有力的积极信号。

"彩绘都市"项目始于 2005 年的巴西里约热内卢，这一项目先是在里约热内卢北部的维拉克鲁塞罗贫民区开展，随后迅速扩展到其他地方，比如哥伦比亚的圣玛尔塔。策展人与当地的年轻人一起创造了很多不可思议的成果，这些成果成为了里约热内卢城市景观的纪念碑，改变了大众的观念，并吸引了来自国际媒体的大量报道。

"彩绘都市"项目不是用化妆技术对一个地区进行"伪装"，而是旨在创造持续的影响力。通过使用创造力和想象力，从总体上对一个社区进行重新思考、重新设计以及重新标识。目前，两位策展人在里约热内卢及世界上其他城市实施新方案的同时，也在美国北费城开展了一个大型社区项目。

"彩绘都市"基金对贫民窟绘画项目提供支持，也在筹款、教育项目以及壁画维护等方面展开工作。该项目同时也创造了另外的可能和机会，比如在艺术家们工作的贫民窟建立一些小型的工坊进行涂料生产。

由贫民窟绘画项目衍生出的可能性不胜枚举，"彩绘都市"基金会也一直在全球范围内寻求合作者以实现更多的梦想。

| GLOBAL PERSPECTIVES / 全球视野 | FAVELA PAINTING / 彩绘都市 |

Toy guns
玩具枪

Boy with kite
放风筝的男孩

Santa marta work
圣玛尔塔工作中

Santa marta
圣玛尔塔

Rio cruzeiro work
里约克鲁塞罗项目

Rio cruzeiro detail
里约克鲁塞罗项目

GHANA THINKTANK: DEVELOPING THE FIRST WORLD
加纳智库：发展第一世界

Curator: John E. Wing
策展人：John E. Wing

GLOBAL PERSPECTIVES / 全球视野

GHANA THINKTANK: DEVELOPING THE FIRST WORL

The Ghana ThinkTank is a worldwide network of think tanks creating strategies to resolve local problems in the developed world. The network began with think tanks from Ghana, Cuba and El Salvador, and has since expanded to include Serbia, Mexico, Iran, Gaza, Ethiopia, Taiwan and a group of incarcerated young women in the U.S. Prison system.

These think tanks analyze the problems and propose solutions, which we put into action back in the community where the problems originated – whether those solutions seem impractical or brilliant. Members of the host city are invited to get involved by helping to implement the solutions. Documentation of the actions is then sent back to the think tanks allowing them to evaluate the success or failure of their proposals and another round of dialogue ensues.

The Ghana ThinkTank process applies typical methods of community development against the grain, inverting traditional hierarchies, creating intercultural dialogue, and connecting people from disparate communities. Our approach works against typical power flows by sending problems from "developed" countries to think tanks in "developing" nations for resolution.

The first phase in any iteration involves collecting problems from residents of the host city. This is generally done on one-to-one basis with local helpers going out into the community and simply asking for complaints. These problems can range from the deeply personal such as, "I can´t bring myself to believe in the power and beauty of me. Yesterday I pretended to be bisexual to be accepted," to the societal, " the elderly are treated like a burden," to the nonsensical, " the city of Karlsruhe is too flat and boring." As the problems accumulate, a portrait of the city begins to emerge, as the values, needs and desires of a community become apparent through their discontents. In a sense, the host community holds up a mirror up to itself.

Problems are then delivered to an international network of think tanks, who discuss the problems and propose solutions that are, of course, necessarily based in their own experiences and cultural perspective. Their discussions are recorded and exhibited in the host city. Stereotypes and (mis)understandings are tossed about with thoughtful commentary and intercultural critique. The conversations are often peppered with playful and sometimes flippant remarks as one culture contemplates the woes of another. For example, Mexico has said of the US: " People in the US emphasize education in order to maintain a system of worldwide domination." Ghana has asserted that "Westerners stress too much individualism." This exchange again offers a mirror as the residents of a host city are able to see themselves in the reactions and commentary of the international groups analyzing their problems.

At the conclusion of discussion each think tank proposes solutions to the problems they have reviewed. Unlike with the conventional processes of international development, no effort is made to be unbiased. The result is an uncomfortable and often hypocritical social contract between disparate communities. The solutions reflect the attitudes of the think tanks towards the host city, so they range from effective to useless to positively embarrassing. Some of these actions have produced workable solutions, but others have created intensely awkward situations, as we play out different cultures' assumptions about each other.

In the next phase, we partner with members of the host community to devise actions based on the given solutions and implement them in the places where the problems originated. These actions are human scale, direct interventions that can be accomplished with minimal resources and with but a hand full of people. Despite scant resources past actions have proven to be remarkably effective, even poetic, and show that brilliant solutions to seemingly insurmountable problems can come from entirely unexpected places.

In 2007, residents of the wealthy (and predominantly Caucasian) town of Westport, Connecticut, complained to the Ghana ThinkTank that their neighborhood lacked diversity. The El Salvador Think Tank responded with the observation that Westport was probably very diverse; the problem was that the residents weren't counting certain people. El Salvador guessed (correctly) that the people hired to tend their lawns and repair their houses were not also wealthy, white people, and suggested that the immigrant day laborers who normally work for the wealthier and whiter residents, be hired to attend those same people's social functions.

The solution was adopted and immigrant day laborers were paid cash in a parking lot to drink wine, eat sushi, and add diversity to Westport social functions. This solution was awkward and incisive. It revealed a blind spot in American society, and presented an opportunity to see, and even socialize with some of the invisible people of wealthy U.S. society. This sort of interpersonal solution to large-scale problems has become a staple of the Ghana ThinkTank process.

For instance, in 2009, a Welsh person complained that the people of Cardiff are unfriendly, "A large number of people have no concept of social interaction with other passing strangers. They look through you, or worse at you like you're something they've trodden in, with disgust and bitter resentment." The Iranian think tank suggested that people should wear images of their own smiling faces to counteract the negative visage they presented to passersby. Accordingly, a sidewalk print station was set up with a camera, a borrowed printer, iron and ironing board. Scowling pedestrians were asked to have their picture taken. In allegedly typical Welsh form, people were standoffish at first but with a good amount of coaxing many allowed us to take a picture. An iron on t-shirt was then made on the spot. In under 10 minutes participants were handed a brand new piece of clothing with their own smiling face on it. Anyone who allowed themselves to be photographed was also given an invitation to a close of exhibition party at the Ffoto Gallery where they could mingle with all the other people who helped make the exhibition possible. This very effective action brought a number of initially grimacing and suspicious pedestrians together for a friendly mixer with others they may have not otherwise met.

A problem that has recurred many times is that the elderly are not respected in society. ""There seems to be a perception that older people are a burden to society." Somewhat cheekily, Iran suggested that we collect funny dirty stories from the elderly. This is a solution that has been repeated many times in countries including the US, Wales, Israel and Germany and each time it has been especially effective. The stories are collected by individuals on a one to one basis in bars in the afternoon, parks on the weekends, and senior centers during tea time. In the beginning it may take some coaxing to convince the older generation to talk about their naughty past but the stories always come and they come in a whirl of grins, far away looks and nostalgic sighs. The fascinating thing about this action is that it has the capacity to evaporate time. As seniors recount their youthful gallivanting they are transported to the long gone past, to perhaps better days. When we share the stories and young people listen to the elderly reminisce about their juvenile licentiousness they are reminded that everyone was young once. For a brief period everyone dwells in a common place of youthful transgression.

The Ghana ThinkTank process is adaptable and transportable. In the Summer of 2011 the Ghana ThinkTank worked in Corona, Queens. This intensely diverse city shared many politically charged problems centering on the issue of immigration. The socio-economic gulf between residents of this community created an opportunity to work domestically allowing estranged residents to discuss and solve each other's problems.

One resident submitted that "the police harass Latin American immigrant workers on the street, especially all over Roosevelt Avenue. They make use of racial profiling to have an excuse and give them tickets." The think tank of Incarcerated Boys in the New York Penal System suggested to "highlight the differences between the cop's world and the real world." Members of the 110th Precinct's police force were asked to comment. One officer responded with, "Look, you have to understand, we see the worst side of people every day. No-one calls the police to say 'It's a beautiful day, come have a picnic.' We are lied to so often, it changes your perspective." Another officer emphasized the conundrum of his role when he stated "We are hired to protect, so if we see someone waiting around in the same place for a long time, and we come back a few times and he is still there, we have to question him. If we did nothing, and he does something, then people will ask 'Where are the police!.' And if we talk to him, then it's 'Why are the police questioning this man!?!" Despite its simplicity the solution worked to give police a chance to share their perspective and residents a glimpse into the police's experience in an honest and constructive way.

A common complaint in Corona, Queens centered on the issue of immigration. On one side, the more established residents complained about newcomers, making statements like "I am becoming a minority in my own neighborhood." On the other side, new Immigrants to Corona were beset by issues of racism and profiling. "I feel like old-timers are racist to newcomers." One of the solutions we received was to organize a promotional campaign on behalf of Immigrants in Corona. A focus group from the Corona Veterans of Foreign Wars, a predominantly white organization, was asked to consider the experience of immigrants and the value of their contributions to society. They came up with the slogan, "I came here to be and American." The slogan was paired with a photograph of a smiling Latino man and the image was made into a large poster for affixing to the side of a NY MTA bus. This inspirational reminder of the hope and optimism of immigrants to the United States was so seamlessly incorporated that it was confused for an official campaign. It was effective not only because of the feelings it inspired but because it grew out of the "old-timer's" identification with the "newcomer's" experience.

Recently, the Ghana ThinkTank process has expanded its scope. This summer it was used as part of a reconciliation process between Serbs and Albanians in the disputed town of Mitrovica. Mitrovica is a town divided by a river in the disputed territory of Kosovo. To the North of the river live Serbs, and to the South live Albanians. We used the Ghana ThinkTank process to collect problems on one side of the river and bring them to the other to be solved. It was a tense process—these two sides went to war in 1999, and have had violent outbreaks even after the war ended. But we learned that being able to look at the problems of someone you have stereotypes about can help break down those stereotypes. In the end, people who had not crossed that river since the war ended in 1999, came across to work with the other side for this project.

Whether successful or not these actions and reactions are carefully documented and shared again with the think tanks who are then able to evaluate the success or in some cases failure of their solutions and another round of dialogue ensues. Some solutions produce useful results while others create awkwardly tense situations. In the former case we find that even seemingly insurmountable problems can have a human scale solution and that genius can come from unexpected places. In the latter case we see how difficult it is to grasp the socio-cultural context of another culture and how preposterous it is to impose ready-made solutions on a community from the outside.

In this way, the Ghana ThinkTank not only draws attention to the optimism and efficacy of small-scale direct interventions but a way to work directly with the misplaced assumptions that often plague international development and cross-cultural innovation.

The project has been commissioned in cities worldwide including NY, NY; Cardiff, Wales; Liverpool, England; Bat Yam, Israel and Karlsruhe, Germany.

The Ghana ThinkTank is John Ewing, Carmen Montoya and Christopher Robbins. It was founded in 2006 with Matey Odonkor. Carmen Montoya joined the project in 2009.

Ghana ThinkTank is made possible with the generous help of many people: Custom technology for the project by Kevin Patton.

GLOBAL PERSPECTIVES / 全球视野

GHANA THINKTANK: DEVELOPING THE FIRST WORL

加纳智库是为解决发达地区的本地问题提出应对策略的世界智库网络。此网络由在加纳、古巴和萨尔瓦多的智囊团发起，目前扩展至包括塞尔维亚、墨西哥、伊朗、加沙地带、埃及利亚及中国台湾地区的参与者以及一群被美国监狱系统所囚禁的女孩子们。

各地的智库分析问题并提出各种解决方案，无论这些方案看上去是不切实际的，还是优异杰出的，都将在这些问题的社群中进行实施，主办城市的成员都将受邀协助实施各种方案，参与进智库项目中。关于这些行动的记录和资料将反馈到智库，使得方案的结果评估和随之而来的进一步对话成为可能。

加纳智库的过程运用典型的社区建设方式，倒置了传统的等级制度，创造了各文化间的对话，并联系不同社区的居民。我们的方式颠覆了传统的权力流动方向，而将"发达"国家的问题发送到"发展中"国家寻找解决方法。

第一阶段总是反复地从主办城市的居民那里采集各种问题。这部分的工作通常由本地参与者进入社区，通过一对一的方式征询各种投诉来完成。这些问题范围宽泛，既可以是非常私人的问题，比如"我无法让自己相信自身的力量与美丽"，"我昨天试图接受自己是双性恋者"，也可以是社会问题，比如"老年人竟然被视为社会负担来对待！"甚至可以是些毫无意义的怨言，比如"卡尔斯鲁厄（德国）这个城市太乏味太无聊了！"当问题累积的时候，一个城市的自画像便开始浮现，通过这些不满的描述，一个社群的价值、需要和诉求将变得更显而易见。在某种意义上，主办城市和社区给自己竖立起了一面镜子。

这些问题将会被交付给一个国际网络智库。他们讨论相关问题并提出必然带有其个人经验和各自文化视角的解决方案。讨论将在主办城市被记录和展出。各自刻板的陈词滥调或误读将在具有深度的评论和跨文化的批评中被颠来覆去。由于这是一种文化对另一种文化的问题的关注，对话本身常常充满了幽默甚至是轻率的评论。比如墨西哥说美国："美国人强调教育是为了维护它自己的世界主导体系。"加纳则声称："西方人过分地强调个人主义。"这种互换再次提供了一面镜子，让主办城市的居民从自身对国际性小组关于他们问题的分析的反应和评论中看到自己。

讨论的最后，每一个智库就他们评估的问题给出解决方案的建议。与传统国际发展的过程不同的是，我们并不试图实现某种公正和毫无偏见。

结果常常是让人不舒服的不同社群间的伪善的社会契约。解决方案折射出了各地智库对于主办城市的态度，有的有效，有的无用，甚至注定是尴尬的。这其中的一些行动产生出了切实可行的方案，但另一些则制造了非常难堪的局面，因为我们将各种文化对彼此的假想公开呈现了出来。

接下来，我们与主办社区的人们合作，根据那些解决方案来制定行动计划，并在问题的发源地实施。这些行动是以人为尺度的，以最小化的资源及个人可实现的行为实施干预。不论资源如何有限，过去的行动被证明是非常有效的，甚至是诗意的，并且让我们看到，那些看似坚如磐石的问题的优秀解决方案可以来自于你完全想象不到的地方。

在 2007 年，康涅狄格州韦斯特波特市一个富裕城镇的居民（以白人为主）向加纳智库抱怨他们的邻居缺少多样性。来自萨尔瓦多的智库以他们的观察作出了回应：他们发现韦斯特波特可能非常多元化，问题在于这里的居民没有将特定人群计算在内。萨尔瓦多的智库猜想那些受雇来打理草坪和修缮房屋的人们本身并不是像雇主那样的富裕白人。于是，他们建议的方案是，聘请那些通常受雇于富裕白人的移民临时工来参与当地富裕白人的正式社交集会。

方案被采纳之后，便花钱聘用移民临时工来一个停车场饮酒、吃寿司，以此增加了韦斯特波特市正式社交集会的多样性。这个解决方案犀利、棘手，它揭示了美国社会的一个盲点，并提供了一个看见美国富人社会所无视的人们并与之交往的契机。这种以人际互动方式来为宏大社会议题提供解决方案的行动成为了加纳智库工作方法的典型案例。

例如，在 2009 年，一个威尔士人抱怨卡迪夫（英国港市）的居民很不友好，"那里的大部分人没有与过路陌生人社交的观念，他们就那么看着你，好像你是什么让他们反感厌恶并唾弃的玩意儿。"来自伊朗的智库建议应该让卡迪夫的居民将他们自己的笑脸的图像戴在身上，以中和他们呈现给过路人的消极形象。于是，一个由照相机、借来的打印机、熨斗和熨衣板构成的街边打印站设置在了人行道旁，眉头紧锁的当地行人被要求拍照。按照典型的威尔士人的方式，他们会一开始很不友好，经过一番软磨硬泡，很多人让我们拍了照，并在现场制作印有他们本人照片的T恤，不到 10 分钟，参与者就可以得到一件印有他们自己笑脸的新衣服。参与并接受拍照的人们也同时收到了参加 Ffoto 画廊展览闭幕式聚会的邀请，在那里，他们可以与其他参与这个项目的人混在一起。这个很有效的行动将那些一开始面露难色、多疑的路人与他们可能永远不会相识的其他人友好地聚到了一起。

加纳智库: 发展第一世界

老年人在社会中不被尊重是一个重复又重复的问题。"似乎有某种将老年人视作社会负担的认知。"多少有点厚颜无耻的是,伊朗智库建议我们从老一辈那里收集他们的风流韵事。这个方案在威尔士、以色列和美国反复实施,每一次的效果都很好。我们以个人面对面的方式去收集这些故事,比如某个下午的酒吧、周末的公园和老年人社区的下午茶时段。一开始,可能总是要连哄带骗说服老一辈分享那些年轻时候的荒唐事,但故事还是会随着他们似笑非笑的表情、远眺的眼神和怀旧的叹息娓娓道来。这项行动的迷人之处在于它有能力让时间消失,当这些老人在重构他们的青葱岁月时,他们仿佛被带回到了很久以前,或是那个他们认为美好的时光中。当我们和年轻人分享这些故事的时候,在聆听老一辈对于似水年华的追忆的同时,他们也意识到每一个人都曾经年轻过。在某些时刻,每一个人都在共同的年少轻狂中存在过,停留过。

加纳智库的工作方式有很强的适应能力。2011年的夏天,我们在皇后区的科罗娜展开工作。这个极度多元化的城市有很多集中在移民问题上的政治性难题。社区居民之间的社会经济分歧创造了一个在内部实施行动的机会,也使得彼此隔阂的居民之间可以去讨论和解决彼此的关切。

一个当地居民提到:"警察总是在街上骚扰拉美移民工人,特别是在整个罗斯福大街。他们以种族成见的方式获得了这样做的借口和许可。"由止在纽约监狱服刑的男孩们组成的智库建议:"应该突显警察世界与现实世界的不同点。"第110区的警员们被要求给出回应。其中一位官员的回答是:"你们得明白,我们每天都要看到人们最坏的那一面。没人打电话给警察是来说'今天天气不错,一起去野餐吧'。我们被谎言欺骗的太多了,这会改变一个人的视野。"另外一位官员则在开始就强调他自身角色的两难:"我们是受雇来保护人们的。如果我们发现一个人在同一个地方长时间徘徊,我们就必须上前盘问。如果我们什么都不做,而之后这个人犯了罪,人们就会质问:'警察死去哪啦!'如果我们盘问了他,人们又会抱怨:'为什么警察只查他呢?!'"无论这个方案是不是过于简单,但警察得到了一个机会去分享他们的视角,同时,居民们也可以以某种真诚的、建设性的方式来稍微了解警察的经验。

在皇后区科罗娜的怨言都集中在移民问题上,那些已经在本地立足的老居民常会抱怨新移民:"我在自己的地盘都成了少数民族了。"另一方面,新移民则会被种族主义和成见的问题困扰:"我觉得老居民对待新居民都像种族主义者。"我们收到的其中一个方案是建议代表新移民们在科罗娜地区做广告。一个来自科罗娜地区的以白人为主的退伍老兵的代表性团体被要求回忆新移民的经历以及他们对于社会的贡献。他们提出了一个口号:"我来到这里,我成为了美国人。"与这个口号相应的是一张面带笑容的拉丁裔男子照片,并且被制作成大海报张贴在纽约的公交系统上。这个对于新移民来说是积极乐观的、令人鼓舞的提示,紧密的和正式的运动结合了起来。整个方案的有效不仅仅是因为令人备受鼓舞的感觉,还因为它以"新移民"的经验带出了"老居民"的身份认同。

近期,加纳智库也进一步扩展了自己的工作范畴争论,在这个夏天参与到了塞尔维亚与阿尔巴尼亚关于城市米特罗维察的和解过程之中。米特罗维察被一条河流所分割,是科索沃地区一个有争议的城镇,北岸是塞尔维亚人的居住地,南岸则属于阿尔巴尼亚人。我们采用加纳智库的工作方式,收集一方的问题并交给另一方来解决。这是一个紧张的过程,因为双方曾在1999年爆发战争,战后至今仍不时有暴力事件发生,但是,过去的经验告诉我们,当你可以了解那些你持有成见的人的问题时,成见本身也有可能被击破。最后,可使自1999年战争结束后就不再往来的河岸的双方,跨越分割彼此的河流来一同工作。

无论成功与否,这些行为和反馈都被仔细地记录下来,使方案的结果再评估和随之而来的进一步对话成为可能。有的方案产生了有用的效果,同时,也有方案造成了很棘手的情况。在过去的一些案例中,我们发现那些看似坚如磐石的问题的解决方案和提供者可以来自于你完全想象不到的地方。近期的一些项目中,我们认识到要把握另一个文化的社会情境是多么的困难,因此,强行将一个从外部设计好的所谓的解决方案在社区内实施是多么的荒谬啊!

通过这种方式,加纳智库不仅可以引起人们对于小规模的干预性行动的乐观与效能的关注,同时,也可直面那些经常困扰国际发展及跨文化革新运动中的假想和误判。

该项目曾接受过来自世界不同城市的委托,如美国的纽约,英国的卡迪夫、利物浦,以色列的巴特亚姆以及德国的卡尔斯鲁厄。

加纳智库的人员包括:John Ewing, Carmen Montoya 及 Christopher Robbins。于2006年与Matey Odonkor一同创建。Carmen Montoya于2009年加入此项目。

加纳智库的实践得到了很多人的慷慨支持。本项目的技术支持为Kevin Patton。

GLOBAL PERSPECTIVES / 全球视野　　　　GHANA THINKTANK: DEVELOPING THE FIRST WORL

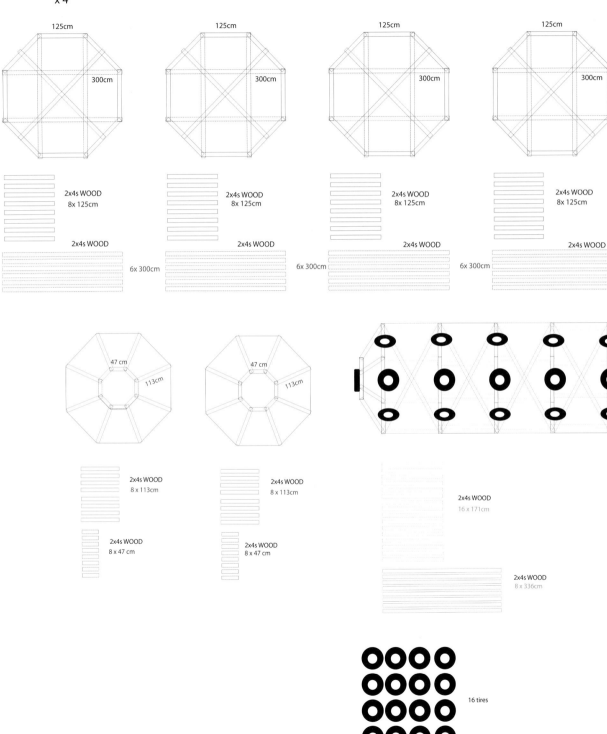

加纳智库: 发展第一世界 　　　　　　　　　　　　　　　　　　　　　　　　　　　453

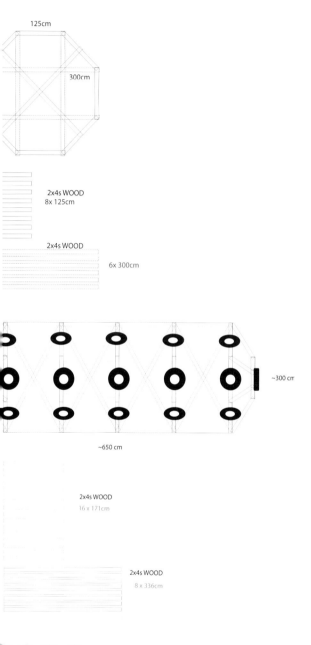
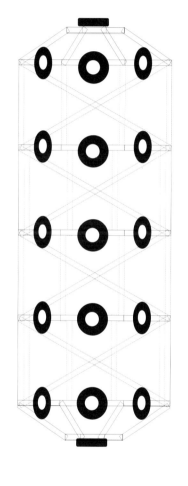

Intsallation Structure
装置结构

5 韦斯特波特，康涅狄格，美国
5A
问题：狗叫 "我的邻居认为我的狗总是在吠，但事实并不如此。"
解决办法：给你的狗命名为爱，给你的狗一个平静的名字，例如 "爱"。
（Ghana）
5B
问题：多样性 我们所居住的社区缺乏多样性
解决办法：在同源化的社区雇用来自各地的日工来参与社会功能
(El Salvador)
5C
问题：农药在我们美丽的草坪和高档的高尔夫球场伤害着广大的动植物及野生自然还有我们人类。在我们这个区域内有很高比例的癌症率。
解决办法：杂草也是花朵。寻找 "收养家庭" 的蒲公英，都种植大片土地蒲公英并保护他们的人们。有一个蒲公英比赛，并评选最佳艺术和手工艺蒲公英制品奖 (Mexico)
蒲公英日：拥有一个蒲公英庆祝仪式来促进杂草也能像草坪植物一样不需要通过化学药剂来破坏。有一个蒲公英比赛，并评选最佳艺术和手工艺蒲公英制品奖，教大家如何用蒲公英烹饪 (Mexico)

GLOBAL PERSPECTIVES / 全球视野　　　　GHANA THINKTANK: DEVELOPING THE FIRST WORL

What are the problems in your city?
你们城市里的问题是什么？

停车费已经很贵了还要上涨
Increasingly expensive parking

那边郊区晚上治安问题很不好
Safety issues at night in the suburban area

出租车的租金费用很高
High maintenance fees for taxi drivers

孩子上学越来越难
More difficulties of entry to school

感觉像一些艺术类的演出比较少
Very few art performances

深圳交通比较拥挤
Traffic congestions in Shenzhen

我觉得深圳对比广州缺少文化底蕴
Lack of culture compared to Guangzhou

问题就是房价高
High property price

关内的绿化和街道规划都非常整齐，非常完整，关外就没那么好
Poor landscape design and street planning in Guan Wai

一些规划问题
Probably some planning problems

加纳智库：发展第一世界

幼儿园收费太贵
High childcare fees

物价与自己打工的收入不相符
A huge gap between high expenses and low incomes

人与人之间的感觉不那么亲密
Lack of support between people

超市也会卖些伪劣商品
Fake products even in the supermarkets

有些社区没有人有热心与热情来组织活动
Lack of community events

有些公共场馆没有充分利用起来
Underused public facilities

我觉得应该关注一下盲道的问题
Poor design of blind sidewalk

就业问题
Unemployment

我觉得现在深圳房价很贵啊
Extremely high property price

交通是最大的问题吧
Inefficient transportation

年轻人压力挺大的吧
Young people under big pressure

每个人在深圳的压力都比较大
Everyone in Shenzhen is under a lot of pressure

INTERNA
PARTICI
国际会与

NATIONAL
OCCUPATION

HOUSING IN VIENNA
维也纳住房

Curator: Diemar Steiner
策展人: Diemar Steiner

| INTERNATIONAL PARTICIPATION / 国际参与 | HOUSING IN VIENNA / 维也纳住房 |

Housing as an essential cultural expression of life.

At the turn of the last century, the housing situation in European cities was deplorable. This was a result of the Industrial Revolution, which drove millions of workers into the cities. Vienna, the capital of the Austro-Hungarian monarchy, had just completed the Gründerzeit, an era of rapid industrial expansion that had started around 1870. The razing of the medieval bastions opened the city to the surrounding villages and the ornate Baroque structure of the old city was supplanted by rational tenement buildings. This structure continues to exist to this day.

Social housing established itself in Europe and especially in Vienna in order to counteract the miserable housing conditions. Social housing means the government assumes public responsibility and control over the level of rents and the quality of housing architecture. Housing is thus extricated from the free market, while lower rents allow for lower wages and hence higher economic productivity. Housing as a system is understood as an important instrument that within a city allows for social equilibrium and prevents gentrification and slumification. In addition, the public control of housing ensures a certain degree of architectonic quality.

Red Vienna has attracted international attention with its "super blocks" designed by architects who for the most part trained under Otto Wagner. While these architects still followed a traditional language of design, the new residential complexes represent a Modernist paradigm.

After the Nazi era and the period of reconstruction after World War II, the system of housing construction in Vienna was further refined in the 1980s to suit present-day needs. These complexes responded to the various urban development situations with a high degree of architectonic quality. Exemplary solutions were realized. Public support, political responsibility, committed developers, and outstanding architects continue to guarantee livable apartments for all population groups with an architectonic attitude that proclaims housing as an essential cultural expression of life.

住房是一种生存之必需的文化表达。

欧洲城市的住房状况在 20 世纪的转向是不幸的。这是工业革命所驱使的数以百万计的工业人口进入城市造成的结果。维也纳——奥匈帝国的首都，刚刚完成自身的经济繁荣期（Gründerzeit，德语，意指 19 世纪中后期开始的德国、奥匈帝国的经济成长时期，于 1873 年股灾后的世界大萧条时期终结——译者注），即始于 1870 年的高速工业化扩张的年代。随着中世纪的古堡被夷为平地，将城市向周遭的乡村开放，老城绚丽的巴洛克构造被理性的经济型出租公寓替代，这一结构一直持续到今天。

社会福利性住房（Social housing）在欧洲，特别是在维也纳的出现是为了抵消当时悲惨的住房条件。社会福利性住房意味着政府应承担公共责任，调控租金水平和房屋质量。由此，住房供给得以游离于自由市场，使较低的租金和较低的工资成为可能，并因此提高了经济生产力。住房供给作为一个系统被视为实现社会平衡的一个重要工具，同时避免了一个城市的中产阶级化和贫民窟化。此外，住房的公共调控也在一定程度上保证了建筑质量。

红色维也纳（Red Vienna，意指 1918~1934 年社会民主党执政期间，实行了一系列社会主义式的城市和社会改革——译者注）因那些受教于奥托·瓦格纳的建筑师们所设计的"超级街区"而为世界所瞩目。尽管这些建筑师仍然遵循了传统的设计语言，但仍是现代主义典范的代表。

20 世纪 80 年代，维也纳住房建设体系进一步得到改善，以适应现时的需求。这些住宅区以高水准的建筑质量回应了多样的城市发展状况。可仿效的解决方案得以实现。公共支持、政治责任、有承担的开发商和杰出的建筑师以宣告住房作为一种生存之必需的文化表达的建筑学态度，确保所有不同的人群获得宜居的公寓和住宅。

RECLAIM, THE KINGDOM OF BAHRAIN
再生－巴林王国

Curators: Noura Al Sayeh, Fuad Al Ansari
策展人: Noura Al Sayeh, Fuad Al Ansari

| INTERNATIONAL PARTICIPATION / 国际参与 | RECLAIM, THE KINGDOM OF BAHRAIN |

More sea. More land. More public.

The much-publicized urban transformations of the Gulf region have been radical in their reshaping of the urban form. Nowhere is it more apparent than along the Bahraini coastline, where 80 years of accumulative land reclamation have significantly transformed the relation to the sea. An island nation once completely dependent on the sea, through its fishing and pearling activities has today nearly turned its back on it. Nearly, albeit for some high-rises competing for a postcard view of the sea and a few disseminated fishermen's huts searching for a slice of coast along the temporary coastline.

Reclaim is an investigation into the socio-political changes that have lead to the current state of affairs in view of stimulating a debate on future planning policies. The geographical retracing of national boundaries has been accompanied by a more profound social transformation- a decline of sea culture in favor of a more generic urban lifestyle. Beyond the ecological implications of land reclamation, it is an investigation into these resulting social implications through the value given to the coast as a public space

Three fishermen's huts disrupted from their original sites in Bahrain form the focal point of the exhibition. The awkwardness of their situation, disconnected from their coastal scenery, speaks of the discomfort of our current relation with the sea and draws parallels between the fates of the two former fishing villages of Bahrain and Shenzhen. This architecture without architects, through the immediacy of its architectural form, speaks of the quest for a more direct relation to the sea. It offers the visitors the chance to experience rather than observe architecture and, through a series of interviews allows them to engage with the anonymous architects and fishermen of these huts as they speak about their relation to the sea.

In the 1920s, similar informal coastal structures, el door, were the gathering places of pearl divers hosting the first organized syndicates in Bahrain. Today, scattered here and there, at the edge of the reclaimed and soon to be claimed sea, the huts host five o'clock tea sessions and backgammon games; a small attempt to reclaim a zest of leisurely coastal space.

再生 - 巴林王国

更多海洋。更多土地。更多公众。

在海湾地区,已经被大量宣传的城市转型对其都市形态的重塑十分彻底。没有任何一个地方能比巴林海岸线的沿线更加显著,对 80 年里所积累的土地进行再利用已经很大程度上改变了这些土地与海洋的关系。一个曾经通过渔业以及珍珠采集业依附于海洋的岛国,在今天,这些产业已然不能再继续。除了为获得明信片式海景的住宅,还包括四散的渔民棚屋,都极力想要在当前的海岸线上分一杯羹。

以激发一次关于未来规划政策的讨论为目的,"再生"是一项探讨导向当前各项事务状态的社会政治变革的调查。这一对国家边界的地缘性回溯伴随着一个更加深刻的社会转型——海洋文化的衰落,转向的是更加普遍化的都市生活形态。除了在关于土地开垦的生态学上的可能结果之外,"再生"还是一次对海岸公共空间的价值和产生的社会影响的探讨。

这一项目将巴林的三间渔屋原地拆解,并将它们搬到展览现场,成为了展览的焦点。这些处在窘迫状态的渔屋,与它们原本的海岸景观相脱离,述说着当前我们与海洋之间的不适的关系,并以此对应前身都是渔村的两个城市——巴林和深圳的命运。通过其直接的建筑形式,这些无建筑师的建筑显示出了需要与海洋建立更加直接的关系的诉求。它为访问者提供了具体体验而非观察建筑物的机会,并通过一系列的采访,允许介入与这些渔屋的无名建筑师和渔民一道,讨论其本身与海洋的关系。

在 20 世纪 20 年代,类似非正规的海岸结构,例如埃尔门(el door),是巴林的采珠者们组织的第一个联合型辛迪加的聚集地。而在今天,曾经的联合体却四散各地,分布在被开垦填海的地区边界,这些渔屋组织了下午五点茶叙和西洋双陆棋游戏聚会,这都是对再生休闲海岸空间的乐趣的小尝试。

INTERNATIONAL PARTICIPATION / 国际参与 | RECLAIM, THE KINGDOM OF BAHRAIN

再生 – 巴林王国

GIMME SHELTER! CHILE IN
给我避难所！智利馆

Curators: Hugo Mondragón, Sebastián Irarrázaval
策展人：Hugo Mondragón, Sebastián Irarrázaval

INTERNATIONAL PARTICIPATION / 国际参与

Various elements included in the display
展览装置中的各种元素

Shelters are emergency places that people turn to in times of natural disaster. They are places that offer the most essential relief, places that people resort to when in search of protection. Inside the shelter, the emergency landscape unfolds, with piles of mattresses and blankets, security cones and barriers, flashlights and bottles of water.

The different forms of expressing this essential relief is the central theme that governed the selection of the projects included in this exhibition. It was our decision to focus a central part of this show on expressions of Chilean cultural patrimony that refer to the essential.

The poetic expression of these emergency landscapes has also oriented the construction of the Chilean pavilion. To achieve this, we chose to overturn the conventional relationships of the elements that comprise it: mattresses positioned vertically become screens for projecting images; security cones and water bottles, cut up and then reassembled, become lamps; emergency tape and water bottles become tensors and counterweights. Once this mechanism was set in motion, we provocatively introduced certain conventionally used forms: a massive bed with mattresses placed in the center of the pavilion, and a window display with large water drums and dispensers at the far end of the pavilion, promising visitors a bit of rest and relief.

We are interested in the shelter's literal and symbolic nudity, the way it facilitates a return to essential forms of individual and collective habitation, the social and material ingenuity it promotes, its poverty and material precariousness, but most of all we are interested in the shelter as a place where people dream of new beginnings.

For the exhibition, we selected architectural works, visual pieces and technological innovations that experimented with the concept of the essential and the ingenious in precarious contexts. On the other hand, and in keeping with the project mechanism put into action through the formalization of the pavilion, we also decided to select projects that exhibited a certain degree of disruption to some element of the cultural or material patrimony of Chile.

To this end the show is comprised of projects, visual pieces and technological innovations that don't espouse any rhetoric, that evade canonical languages and procedures, and that in fact explore with experimental languages, culturally rooted in Chile's cities and landscapes.

GIMME SHELTER! CHILEIN / 给我避难所! 智利馆

避难所是在遇到自然灾害等紧急状况时人们可以寻求保护的场所，是可以提供最"基本"救济的地方。避难所内部呈现的是一派处在紧急状态下的景观：成堆的睡垫和毯子，路障锥筒以及手电和瓶装水。

使用怎样不同的形式来表现"基本救济"这个概念是主导本展览选择展品的中心主题。我们决定把该展览的核心部分集中于体现智利的文化遗产上。

智利馆的建设也旨在体现如何对这些紧急状态下的景观进行诗意的表达。为了达到这个目的，我们推翻了构成该景观的元素的内部传统关系：睡垫垂直摆放成为播放图片的屏幕；安全锥和塑料瓶被切割并重新组装成为灯具；紧急胶带和塑料瓶成为装置的张量和平衡力。一旦这个装置开始启动，我们将呈现一些传统的使用形式，比如放置于场地中央的由睡垫组成的巨床以及放置在场地远端的大水桶和取水机构成的展示，还能给参观者带来一丝休息和放松的感觉。

我们对避难所的兴趣体现在很多地方，比如其直白又富有象征意义的直观，其回归个人和集体生活基本形式的方式，它在社会以及物质层面所推动的独创性，它的一穷二白以及物质上的不稳定性等。但是我们感兴趣的地方在于避难所是人们拥抱新梦想的地方。

在这次展览上，我们使用了建筑作品、视觉物件和技术创新，所有这些元素都是对不安全语境中的"基本"以及"创新"概念的一种实验。另外，为了与场馆付诸实施的形式保持一致，我们也选择了一些在一定程度上与智利的文化和物质遗产元素不太一致的项目。

整个展览由不同的项目、视觉物件以及技术创新组成，所有这些元素并不迷信华丽的辞藻，也与教规性的语言和程序相去甚远，实际上它是用实验性的语言来进行探索，在文化上根植于智利的城市和景观。

Photographer: Cristóbal Palma
摄影：Cristóbal Palma

NEWLY DRAWN – EMERGING FINNISH ARCHITECTS
新绘图——芬兰新生代建筑师

Curator: Martta Louekari
策展人: Martta Louekari

INTERNATIONAL PARTICIPATION / 国际参与

NEWLY DRAWN - EMERGING FINNISH ARCHITECTS

What is Finnish architecture? This question arose in a round-table discussion with young Finnish and Spanish architects in Madrid in 2009. Finding an answer was suprisingly hard and trying to define it felt a little dangerous. For an hour or so of discussion ranged from sustainability and environmental issues to practical design solutions, all of which have much to do with Finnish architecture. Eventually someone, probably a Spanish student, proposed that Finnish architecture has always had a humble, human approach. By looking back on our architectural history and at more recent buildings, I can say it's more and less true. From big corporate buildings to schools and private houses, we seem to have an understanding of what is required to make person feel good and how to orient things, as well as how to create spaces that are both functional and pleasant. It's hard to explain, but Finnish architecture doesn't feel aggressive or loud. In most of the cases, buildings fit into their cities, their environments and into human life in a very natural way.

"NEWLY DRAWN—Emerging Finnish Architects" introduces the most interesting young, upcoming Finnish architects, their latest projects, visions and ways of working. It is a joint project by the Hollmén Reuter Sandman, Verstas, NOW, Anttinen Oiva Architects, Lassila Hirvilammi, Avanto, ALA, AFKS, K2S Architects resulting in publications, exhibitions, workshops and other events in both in Finland and abroad. Social interaction, pleasant user experience and transparency have emerged as key elements in architecture of Newly Drawn offices.

In 2012 Helsinki is World Design Capital. The title is an initiative by the International Council of Societies of Industrial Design (Icsid) that celebrates the merits of design. Held biennially, it seeks to highlight the accomplishments of cities that are truly leveraging design as a tool to improve the social, cultural and economic life of cities. The vision of World Design Capital Helsinki 2012 extends the concept of design from goods to services and systems. That means finding solutions to peoples needs through innovative design and a user-driven perspective. Usability, sustainability and desirability are the themes of the year.

With the help of different design solutions in architecture and services, Finland is reaching to wellbeing and sustainable future. Newly Drawn – Emerging Finnish Architects exhibition offers a fresh look to upcoming projects and latest developments of the Finnish architecture.

And what's our connection with China?

As a part of Finland's cultural programme for Shanghai World EXPO in 2010 we organised, together with Chinese architecture offices and China-based think tank Movingcities, the Snowball Architecture seminars in Shanghai and Helsinki. That same year, we met more Chinese architects at Tianjin University where we organised a roundtable discussion in co-operation with the university. In the autumn of 2010 we published a special issue together with Beijing-based Art & Design Magazine, featuring interesting new agendas in Finland and China, as well as collaborations between Chinese and Finnish architects, journalists, artists, photographers.

In recent years, we've witnessed the rise of a new Chinese architecture. During our recent visits to Beijing, Shanghai and Shenzhen, we realised that there are similarities between the new generation of Chinese architects and Finnish architects. These emerging offices are developing their own ways of working, as well as their proper/official agendas and approaches, all the while actually creating and erecting new, real architecture. In Finland, the unique competition culture provides new talent with opportunities to land significant commissions, while in China the sheer volume and speed of transition, urbanisation, investment and building creates a rich and diverse terrain for different players.

The objective of Finland's programme for the Shenzhen Hong Kong biennale is to offer insight the latest achievements of Finnish architecture through exhibitions and events. The aim of bringing professionals from both cultures together is to establish and deepen Finnish and Chinese networks. I addition to exhibitions and events, Finnish delegation of 20 architects visited local offices and institutions.

Finland will carry on this dialogue in the coming years. Numerous events will take place in China, mainly in Beijing and Shanghai, in collaboration with universities and medias. The programme is set up to establish potential business and creative matches between Chinese and Finnish architects, as well as to increase the exposure and business opportunities for Finnish architecture in China.

Young architects belong to a global generation, internationally conscious, experienced and informed. While many ambitious architects have worked and studied abroad, and there is ample evidence of global trends and concepts applied in local contexts, it seems they seek to maintain a critical and enthusiastic relationship with both past and present traditions. At the same time they are focusing their efforts on enhancing and crafting their own approaches in practice.

"With broadened horizons and heightened expectations, we would like to invite all of you to participate in an ambitious architectural future one that balances and connects bold goals with a humble quality, sensitive responsibility with fresh concepts, local traditions with new technology, common sense with elaborate skill, complex diversity with enlightened clarity, big ideas with intelligent details, good questions with relevant answers…"

INTERNATIONAL PARTICIPATION / 国际参与

NEWLY DRAWN - EMERGING FINNISH ARCHITECTS

什么是芬兰建筑？这个问题在 2009 年马德里的芬兰和西班牙建筑师圆桌会议上被提出。找到答案出奇地难，而试图去定义它又有点危险。在大约一小时的讨论里，大家从可持续性谈到环境议题，又谈到实际的设计方案，所有这些都跟芬兰建筑有着密切的关系。最后有一位貌似西班牙学生的人提出，芬兰建筑总是有着谦虚和人性的特征。回顾我们的建筑史，再看看最近的建筑，我可以说，他说的基本正确。从大型的公共建筑到学校和私人住宅，我们似乎在什么让人感到舒适、如何引导事物以及如何创造实用和舒适兼备的空间方面有着共识。解释起来很难，但是芬兰建筑看上去并不大胆高调，大部分建筑与城市、环境和人类生活自然地融合在一起。

"新绘图——新兴芬兰建筑师"介绍了最有趣的一批年轻的芬兰建筑师以及他们最新的项目、愿景和工作方式。该项目的参加者包括 Hollmén Reuter Sandman 建筑事务所，Verstas 建筑设计公司，NOW 建筑设计公司，Anttinen Oiva 建筑事务所，Lassila Hirvilammi 建筑事务所，Avanto 建筑师事务所，ALA, AFKS 建筑设计公司，K2S 建筑设计公司，成果来自于它们在芬兰及国外的出版、展览、工作坊及其他事件。社会互动、愉快的使用体验和透明度成为了新绘图的各个事务所在建筑上的要旨。

赫尔辛基是 2012 年的"世界设计之都"。"世界设计之都"是由国际工业设计协会理事会（Icsid）发起的设计成果庆祝活动。该活动每两年举办一次，旨在突显那些真正将设计作为工具来利用以改善社会、经济、文化生活的城市。2012 年赫尔辛基"世界设计之都"将设计的概念从物品延伸到服务和制度，这意味着要通过创新的设计，在用户主导的前景下找到解决人民需求的办法。可用性、持续性以及合意性是这一年的主题。

在建筑和服务上的各种不同设计方案的帮助下，芬兰正迎来幸福的可持续性未来。"新绘图——新兴芬兰建筑师"展览提供了芬兰建筑的未来计划和最新发展的新鲜图景。

我们与中国有什么样的联系？

作为 2010 年上海世博会芬兰雕塑项目的一部分，我们与中国的建筑事务所、设在中国的智囊团"移动城市"一起，在上海和赫尔辛基组织了"雪球"建筑论坛。同年，我们在天津大学与更多的中国建筑师会面，并与

Espoo hospital / K2S
埃斯波医院 / K2S 建筑事务所

Courtyard space and typical patient wing
庭院空间和典型病房单体

天大合作举办了圆桌论坛。2010年秋天，我们与北京的《艺术与设计》杂志联合推出了专刊，介绍芬兰和中国有趣的新项目以及中国和芬兰两国建筑师、记者、艺术家和摄影师之间的合作。

近几年，我们在参观北京、上海和深圳时看到了中国新建筑的涌现。我们发现，中国和芬兰的新一代建筑师身上有着相似之处。一些新兴的事务所以自己的方式、进度和途径工作，他们都在创造并建造新的、真正的建筑。在芬兰，独特的竞争文化为重要的建筑工程提供了天才和机遇，而中国的幅员辽阔和步伐迅速的转变、城市化、投资和建设为不同的人提供了不同而丰富的平台。

芬兰在本届深港双年展上的项目，试图通过展览和相关活动让大家看到芬兰建筑的最新成果。通过融合两国文化上的专业人士建立和加深中国和芬兰之间的合作。除了展览和各种活动之外，由20位建筑师组成的芬兰代表团还参观了当地的事务所和机构。

芬兰在未来几年将把这场对话持续下去。我们将与大学和媒体合作，主要在北京和上海举行各种活动，这个项目意图开发潜在的商业合作，建立中国与芬兰建筑师之间的创造性结合，同时增加芬兰建筑在中国的展示和商业机会。

年轻建筑师是全球化的一代，有着国际意识、经验和资讯。许多有抱负的建筑师都曾在海外学习和工作，他们将全球动态与观念与当地语境相结合，似乎试图在过去和现在的传统之间保持一种具有批判性而又热情的关系。同时，他们专注于在实践中提高和优化各自的方法。

"我们带着广阔的视野和高度的期待，邀请你们参与满怀雄心的建筑未来，它平衡并连接着大胆的目标与谦虚的品质、敏感的责任心与新鲜的观念、当地传统与新技术、常识与专业技能、复杂的多样性与高度的清晰性、伟大的思想与聪明的细节，好的问题及其答案……"

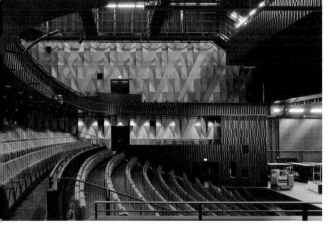

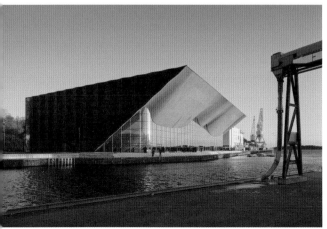

Kristiansand performing arts centre / ALA
Photographer: Iwan Baan
克里斯蒂安桑表演艺术中心 / ALA 建筑事务所
摄影师：Iwan Baan

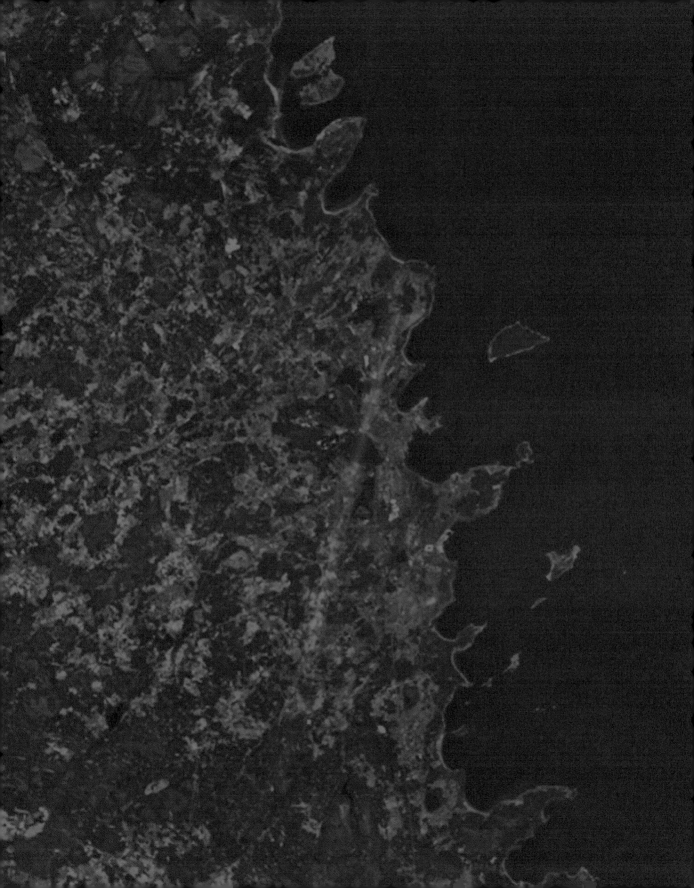

SOLUTION FINLAND
THE WELFARE GAME
芬兰方案: 福利博弈

Curator: Martta Louekari
策展人: Martta Louekari

| INTERNATIONAL PARTICIPATION / 国际参与 | SOLUTION FINLAND THE WELFARE GAME |

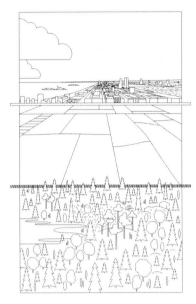

The Wall and the Fence create a division of the country into three distinct territories: City, Field and the Wild North. Illustration by Tuomas Toivonen.
"墙与篱笆"将这个国家分成了三个单独的区域：城市，土地和野性的北方。由 Tuomas Toivonen 绘制。

In the Solution series, edited by Ingo Niermann and published by Sternberg Press, select authors are asked to develop an abundance of concise and original ideas for countries and regions, contradicting the widely held assumption that, after the end of socialism, human advancement is only possible technologically or requires a yet-to-be-established world order.

Published in December 2011 the title Solution Finland: The Welfare Game by architect Martti Kalliala with writer and curator Jenna Sutela and architect Tuomas Toivonen, addresses the Nordic country's numerous predicaments. The three authors propose eight and a half solutions to their native country's quandaries, ranging from the practical (rescuing ailing public space through climatization and the introduction of the Winter Garden City), to the absurd (dividing the country into two interlocking sub-nations: City and Wilderness) and the earnest, if far-reaching (the repurposing of the country to host the world's nuclear waste). Solution Finland elucidates the northern country's modern history as a nation under construction, proposing that its identity remain a malleable myth, in which designing a more tenable future is the conduit for crucial adaptation.

At the SZHK Biennale the authors present in collaboration with Z.A.K. Studio a visual installation depicting three diagrams, or constellations of words, related to the the solutions proposed in Solution Finland: The Welfare Game. The book itself is present as a pre-recorded audio narration.

The Happy People of boreal forest live a self-made life. Illustration by Susumu Mukai.
北方森林里幸福的人民，过着自给自足的生活。由 Susumu Mukai 绘制。

芬兰方案: 福利博弈

在由 Ingo Niermann 编辑、斯滕伯格出版社出版的《解决方案》（Solution）系列丛书中，被挑选出来的作者们被要求为国家和地区开发大量简洁、新颖的点子，以对抗这样一个广泛被认定的假设：在社会主义终结后，人类的进步仅在技术上可能或者需要一个有待建立的世界秩序。

于 2011 年 12 月出版的《芬兰方案：福利博弈》一书，由建筑师 Martti Kalliala 联合作家兼策展人 Jenna Sutela 和建筑师 Tuomas Toivonen 一道著述，道出了北欧国家的诸多困境。三位作者为他们本国的困境提出了八个半解决方案，有实用性的（通过气候调解和引进"冬季花园城市"来拯救境况不佳的公共空间），有荒诞的（将国家切分为两个连锁的次级国家：城市和荒地），还有虽非意义深远却最为热心的（将国家用于收集全世界的核废料）。"芬兰方案"将北方国家的现代史阐述为一个建造中的民族国家，同时提出，其身份保留着一个可塑的神话，在这个神话里，设计一个更加待定的未来是重大调整的出路。

在深港双年展上，作者们与 Z.A.K. 工作室合作呈现了一个视觉装置，描述了跟《芬兰方案：福利博弈》一书中所提出的解决方案相关的三个图表或者词语群。这本书也被事先录制成音频进行呈现。

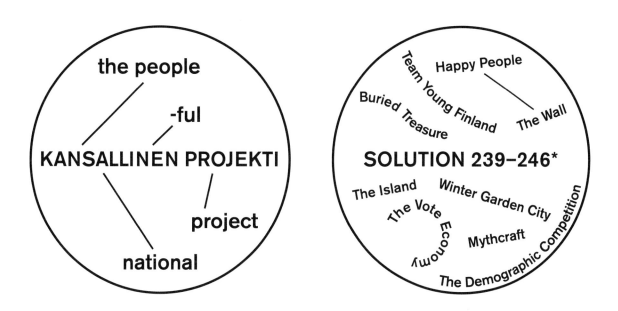

*S239–246 (WEALTH ÷ PEOPLE) × WELL-BEING = WELFARE GAME

HOUSING WITH A MISSION, DUTCH AND CHINESE ARCHITECTS' DESIGNS FOR THE ANTS TRIBE, THE NETHERLANDS
住宅的使命，荷兰与中国建筑师为蚁族而设计，荷兰

Curator: Ole Bouman   **Joint Curator: Jorn Konijn**
策展人: Ole Bouman   联合策展: Jorn Konijn

INTERNATIONAL PARTICIPATION / 国际参与

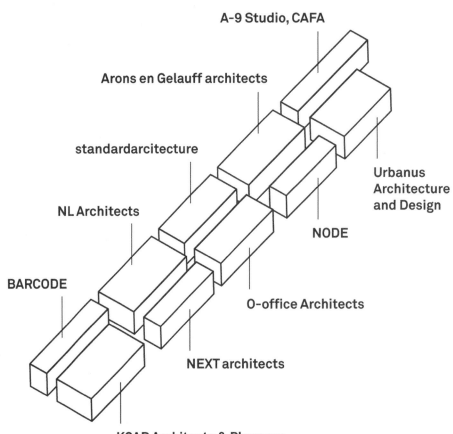

HOUSING WITH A MISSION, DUTCH AND CHINESE ARCHITECTS' DESIGNS FOR THE ANTS TRIBE, THE NETHERLANDS / 住宅的使命, 荷兰与中国建筑师为蚁族而设计, 荷兰

Dutch architects have been active in the Chinese market ever since the opening up of the Chinese economy. What has been the effect of their presence in China for their practice and how did it affect their architectural language in this fascinating country? Have we seen a process of integration, or did their work by and large remain isolated? And if they worked together, did their common efforts lead to creating solutions for specific problems in Chinese society today?

In the exhibition Housing with a Mission, The Netherlands Architecture Institute (NAi) tries to find an answer to these questions. The exhibition is showcasing a unique project in which the NAi brought five Dutch architecture offices together with five Chinese, to work on an actual project to be realised in Beijing by VANKE real estate developing. The NAi selected from the Netherlands NL Architects, Arons & Gelauff architects, NEXT architects, Barcode architects and KCAP. Counterparts from China were URBANUS, Standard architects, NODE, O-Office and CAFA University. These offices vary in scale, background, experience and working methods, but one thing they all share: a strong desire to contribute to the social challenges of their time. They all want to make an architecture of consequence. The ten architects have worked in complete equality to develop new concepts of housing for this group. All architects derive from different historical backgrounds. The Netherlands have a long and high quality tradition in social housing. In China social housing is understood in different terms and has also greatly developed over recent years. Housing with a Mission shows these historical traditions, its past, the present and the future.

This future is constructed by the ten selected firms. They are developing new housing for a group that is often overlooked in Chinese society: low income graduates, also known as "ants tribe". They are large in size, smart, talented and vocal: the future of China lies in their hands. Up till now this group often lives in poor circumstances. Good quality housing for this group is in many ways a priority for the Chinese government but also for project developers and for architects. Clearly, it's an investment in the future of a peaceful society.

The exhibition Housing with a Mission focuses on the process of working together to find a common language, but also showcases the low living conditions in which low-income graduates are living today and what their improved housing of tomorrow might become.

自中国经济开放以来，荷兰的建筑师就活跃于中国市场。他们在中国的出现和实践产生了哪些影响？在这个迷人的国度他们又是怎样表达他们的建筑语言的？我们是否看到了一个融合的过程，还是他们的工作总的来说仍是孤立的？如果共同工作，他们的共同努力是否为今日中国社会的特殊问题创造了解决办法？

在"住宅的使命"这个展览上，荷兰建筑协会（NAi）试图为这些问题找到答案。这个展览只展出一个项目，NAi带领五个荷兰的和五个中国的建筑事务所一起，就将由万科房地产开发公司在北京实施的一个真实项目展开工作。NAi选择了来自荷兰的NL建筑事务所，Arons & Gelauff建筑事务所，NEXT建筑事务所，Barcode建筑事务所以及KCAP事务所。相应的中国团队有都市实践、标准营造、南沙原创、源计划以及中央美术学院。这些事务所的规模、背景、经验和工作方式各不相同，但是有一点是共同的：强烈渴望为他们这个时代的社会挑战做出贡献。他们都希望创造出有意义的建筑。十位建筑师在完全平等的基础上为这个团体开发新的居住概念。每个建筑师都有不同的历史背景。荷兰建筑师在社会住房方面有着长期优质的传统。近年来，在中国，以不同方式理解的社会住房也有很大的发展。"住宅的使命"展示的正是这些历史传统以及它们的过去、现在和未来。

未来便由这十个被选出的公司建造。他们正在为经常被中国社会所忽视的一个群体开发新的住房：被叫做"蚁族"的低收入毕业生。他们已长大成人，聪明、有才能并且能表达自己的意见：中国的未来在他们手上。这个群体通常生活在穷困的环境中。为这个群体而设计的优质住房从很多方面来讲都是中国政府的优先考虑，也是开发商和建筑师们的优先考虑。显然，这是对未来和谐社会的一项投资。

"住宅的使命"展览的重点在于通过一起工作找出共同语言的过程，同时展示低收入毕业生们现今的低水平居住条件以及他们改良后的明日住房的样子。

INTERNATIONAL PARTICIPATION / 国际参与

KCAP Architects & Planners

Exploiting the building's unique position in the master plan, as well as the high intellectual capacity of its future residents, were the two guiding ambitions in the building designed by KCAP Architects & Planners.

Simple, robust and voluminous building envelope secures a strong presence at a such a prominent location in the master plan, the building being a cornerstone of the envisioned ensemble and the final destination at the end of the main public space axis through the proposed urban quarter. At the same time a long staircase through the middle of the building extends the central street into the building, seamlessly connecting public, retail and residential realms into a complex and truly urban condition.

Given the resourceful character of the future residents and the small size of the living units, a strong emphasis was put to the creation of high quality communal space within the building that would add to the quality of living and working conditions. The result is in an environment where the inhabitants can work, rest, read, meet and socialize. As such the proposed communal space not only makes up for the small size of the living units, but also inspires intensive exchange and active participation of the residents in the communal life. It articulates their contribution to its continuous improvement, securing the long-term vitality of the whole community.

KCAP Architects & Planners 在其建筑的设计中，以两点为指导原则：充分利用该建筑在总体规划中的独特位置以及未来入住该建筑的居民的较高文化程度。

简单、坚固和分段式的建筑围护结构确保在总体规划中地处重要位置的该建筑给人留下深刻的印象。该建筑是总体规划构想中的奠基石，也是穿越这片城市区的主要公共空间轴线指向的最终目的地。与此同时，长长的楼梯穿过该建筑将中央街道延伸到了建筑中，将公共、商用和住宅区天衣无缝地与复杂、真实的城市环境连接起来。

考虑到未来居住者头脑灵活的特点以及居住单元的较小尺寸，设计者们花大力气强调在建筑内营造优质公共空间，借此提高生活和工作环境的品质。结果他们设计出了一个能让居住者在其中工作、休息、阅读、聚会和交流的场所。他们推出的这片公共空间不仅弥补了居住单元较狭小的缺点，还能鼓励居住者更积极主动地参与公共生活，进行人与人之间的交流。该设计强调的是居住者不断投入以持续改善整个社区，确保其拥有长久的活力。

HOUSING WITH A MISSION, DUTCH AND CHINESE ARCHITECTS' DESIGNS FOR THE ANTS TRIBE, THE NETHERLANDS / 住宅的使命，荷兰与中国建筑师为蚁族而设计，荷兰

Arons en Gelauff architects

Arons and Gelauff architects aim to find a high quality design in low-cost housing. The architects use their Dutch experience and translate this to Chinese circumstances with the idea of 'standardization' as a point of departure.

The Hui Long Guan project is assembled from a prototypical building block of high quality units for one or two person households. These standardized units can be arrayed and / or stacked in a wide variety of combinations, depending on the site, brief or demand. The possibilities are unlimited - slabs, towers, large city blocks or low-rise suburban apartment buildings all are conceivable with this prototype. The type adapts to any urban setting and to every climate zone throughout China.

The apartment units have deep floor plans with a zoning of different functions. From inside to the outside, the unit is gradually getting more natural light and the functional lay-out is organized accordingly. To generate a sensation of space in the hyperefficient dwelling type, the units have a flexible lay-out. Sophisticated fixtures that are inspired by sliding and folding systems that are found in temporary accommodations are use to change the space in accordance with the activities of the daily dwelling routines. The residential floor plans are mirrored at every level. This generates double height and unobstructed light on all balconies. The dynamic form of the glass façade helps reflect natural light into the narrow commercial streets. The architecture achieves a crystalline quality.

Arons & Gelauff Architects 的设计目的在于为低成本住宅寻找高质量的设计方案。建筑师们借鉴了荷兰的经验并将其用于中国式的环境中，同时以"标准化"概念作为出发点。

回龙观村项目是由仅以一至两人为居住单位的高质量样板建筑楼群组成的。根据建筑地点情况、主旨或要求的不同，这些标准化单元可按照多种不同的组合方式进行排列和／或堆叠。组合的可能性是无限的——扁平状、塔状、大型城市街区或低层郊区公寓式建筑都可借由这种样板单位的组合实现。这种建筑适用于中国各地任意的城市布局和气候带。

在套房房间的设计上，建筑师们对每一楼层进行了功能分区。每个单元获得的自然光照量从内到外逐渐增大，并以此为基础进行了功能型布局设计。为了在高度紧凑的居住区营造空间感，各个单元使用了灵活多变的布局。建筑师借鉴临时住宅中的滑动和折叠式结构，使用了复杂的固定装置，从而按照居住者的日常活动来改变空间。每一楼层的住宅都被镜向复制。这一设计方式产生了双倍的空间，并使得所有阳台可获得毫无阻碍的光照条件。大量使用玻璃幕墙有助于将自然光反射到狭窄的商业区街道上。整体建筑实现了水晶般的效果。

INTERNATIONAL PARTICIPATION / 国际参与

Standard architecture / 标准营造

The design strategy for Standard Architecture is to combine a number of basic small housing units together with a grand semi-outdoor public theatre-like space. Their project is called 'social theatre' housing. The theatre space would serve as a great public living room for the building as well as for the entire community. The concept for the design for the residential units is based on the idea of the shared wall. The division between each unit consists of a wide wall that works as an infrastructure where all the appliances and services are hidden until used. This maximizes the amount of available free space in the room. There are two units one of 14 m² and 22 m². These units only differ in their depth. The height and width of each unit follows a standard module layout of 2.8 m x 3m which is projected onto the east and west façades. The area of each unit is enlarged by small balconies that enrich the quality of the interior space.

The building program is organized through six levels. The first three levels accommodate commercial functions, with the possibility of having double height floors to enhance spatial quality and flexibility. The three top floors are residential. The rigidity of the internal unit module is offset through the varied placement of balcony openings on the east and west facades. This cutting operation of solid and void gives a dynamic sense of rhythm to the façades and simultaneously operates to mask the placement of facilities required on the exterior of each room.

标准营造的设计被称为"社会剧场"式住宅。其设计宗旨是将一些基本的小型住宅单元与大型的半露天式公共剧场型空间相结合。该剧场空间将是整体建筑中的社区全体居民的公共起居室，而住宅单元则是基于公共墙的概念进行设计的。每个住宅单元之间的区隔由一堵厚厚的墙构成，它既是房屋的基础结构，又能让您将所有的家用电器和服务设施收纳其中，待需要时再取出使用。这一设计能最有效地利用房间里现有的空间。目前，建筑中有两种尺寸的单元：14米和22米。这些单元仅在深度上有差异，这是由于走廊的形状导致楼层平面不对称的结果。每个单元的高度和宽度都设计成2.8米×3米，并以标准化模块的形式分别向东、西两个方向的外墙铺展开。每个单元还附加了小阳台，为其内部空间增添更多层次。

整个建筑规划分为6层。最下面三层容纳商业功能区，并保留了余地，可将楼面高度扩建为原来的两倍，从而增强空间感和灵活性。最上面的三层为住宅区。通过在东、西两边的外墙上变化阳台的开口位置，楼内单元模块千篇一律的缺点得以弥补。这种虚实结合的分隔手法赋予了建筑外墙充满动感的韵律，同时遮盖住了每个房间外所安装的各种必要的生活设施。

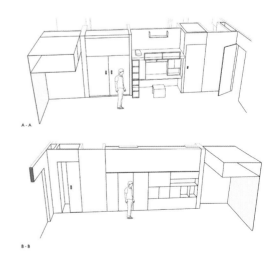

HOUSING WITH A MISSION, DUTCH AND CHINESE ARCHITECTS' DESIGNS FOR THE ANTS TRIBE, THE NETHERLANDS / 住宅的使命，荷兰与中国建筑师为蚁族而设计，荷兰

BARCODE ARCHITECTS

BARCODE Architects has created a design that will provide small, efficient and affordable prototypes for living and emerging businesses within a dense masterplan of small buildings. Their design focuses on three main strategies: continuity, mixed program and light. By creating openings through the slab, the design introduces a high level of permeability. Residents, workers and visitors can freely move through the public spaces, meet up with friends and colleagues or visit the shops and restaurants at grade.

The design also proposes to raise the height of the retail space on the ground and first floor, to just less than 6 meters, providing the opportunity to add a mezzanine in the future and to replace retail spaces on the second level with flexible office space for entrepreneurships and creative studios. Lastly, BARCODE Architects paid special attention to the light. Since the neighbouring building is only 6 meters away, creating voids in the slab, light access to the pedestrian street and the adjacent building is enhanced. Each unit has two exterior walls and therefore receives light from two directions.

BARCODE Architects 的设计是要在小型建筑的密集式总体规划中为居住者和新兴的商业单位提供规模小、效率高且价格低的样板房间。其设计侧重于三大策略：连续性、混合功能及采光。该设计方案通过在各建筑层之间设立开口而获得了高度的通透性。居住者、工作人员和访客可在各个公共空间内自由往来，与朋友和同事见面或前往同一层的店铺和餐厅。

该方案还建议将第一层和第二层的商用空间加高至略低于 6 米，从而预留出将来增加夹楼的空间，并将第二层的商用空间以灵活的办公空间取代，留给企业和创意工作室使用。最后，Barcode Architects 还特别用心地进行了采光方面的设计。由于周围的建筑群仅相隔 6 米，因此他们在设计上于建筑楼层中留出天井并在行人通道和相邻的建筑上留出采光口以增强采光效果。每个单元都有两面外墙，因此双方向均可采光。

INTERNATIONAL PARTICIPATION / 国际参与

A-9 Studio, CAFA / 中央美术学院

For the architectural design concept for the plot that A-9 studio did, the first thing they did was to soften the adjacent edges. The ground floor and second floor commercial spaces are located within the setback line. There are outdoor balconies on the second floor. The 3-5 stories are set back to make roof terraces as the major vertical public spaces, serving the apartment tenants. There is a half partitioned wall surface to define public spaces, and to solve the visual problem caused by narrow streets. The void produced by mass-reducing at 3-5 floors is kept as a space of outdoor public living room serving all tenants vertically, and provides semi-private public space comparing to the street. For units design, use doublestorey suite to increase ratio of living area, and provide more open feeling for interaction among apartments and outdoor public spaces.

整体建筑设计概念基于 1 号建筑在总图上的位置和与相邻 2 号单体建筑的位置关系，在相邻的界面作软化处理。首层和二层商业尽量退让建筑控制线，二层设置外廊，三层至五层退出以屋顶平台为主导空间的垂直公共空间，服务于公寓的住户，并设置一面半穿透性的墙面，限定公共空间，改善因狭窄的街道空间所导致的视觉干扰问题。这个起缓冲作用的公共空间对两侧的公寓都有益处。3~5 层消减的体量被留作为室外公共空间——大起居室。垂直向度的公共空间为 3~5 层的住户提供了一个相对于街道而言半私密性的公共起居室。采用跃层户型，面积为 21~23 平方米，尽最大的可能提高了居住面积使用率。跃层户型与室外公共空间的互动提供了区别于集合宿舍单调空间的开放感觉。

HOUSING WITH A MISSION, DUTCH AND CHINESE ARCHITECTS' DESIGNS FOR THE ANTS TRIBE, THE NETHERLANDS / 住宅的使命, 荷兰与中国建筑师为蚁族而设计, 荷兰

NODE / 南沙原创

NODE choose as a starting point a sketch and a poem on the word 'home'. To NODE, home starts with a basic unit, designed with the basic things a person needs: a bed, a bathroom, a water boiler, a refrigerator, a cable for a computer, a bathroom and a window with a view.

From that starting point, the user can venture into encounters with others, the idea of sharing like dining & living: commune. NODE worked on their unit-combination on the principle of home + commune. This lead to their construction principle of thinner & thinner: a room with shifting, stretching & flattening views. The design is to reinforce the thinness of naturally linear quality of the building by shifting, stretching and flattening the views of landscape on all directions, to make the building look and feel even thinner. This way we can achieve and guarantee each basic unit a different but complete view within the limited area.

In simple design NODE has tried to fit in a maximum number of units with maximum directions of views. It is a concept that is inspired by Picasso's cubism paintings.

南沙原创选择从一张草图和一首关于"家"的诗起步开始其设计。对于南沙原创来说，家的设计首先应是一个简单的房间，里面容纳了一个人所需的基本物品：一张床、一个热水器、一台冰箱、一根连接线、一个卫生间以及一个能够看到风景的窗户。

房间的使用者可在这一基础上进行与他人的接触和更亲密的交往，在进餐和生活层面进行分享。南沙原创以"家 + 亲密往来"为原则对其单元进行组合。在这一原则下诞生了他们的建筑理念"薄上加薄"：一个可平移和延伸的房间，同时具备扁平式的视野。该设计方案旨在通过平移、延伸和营造各个方向上的扁平化视野，加强整个建筑的天然线性特质所带来的薄透感。这样，在有限的区域内，我们就能为每个基本单元实现并保证其看到各不相同但却完整的景观。

基于这一简单的设计，南沙原创试图在尽可能添加更多的单元的同时实现最多样化的视角。这一灵感来源于毕加索的立体派画作。

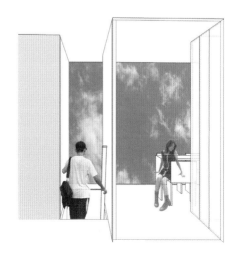

INTERNATIONAL PARTICIPATION / 国际参与

O-office Architects / 源计划

The design challenge for the lot of O-Office was how to create new high-density diversified small units of living-working and commercial space, which allow more spatial possibilities for young inhabitants, supported by the autonomy of the single building that is defined by the master plan. A certain percentage of living and retail space is required and similar section prototype of the building is defined. O-Office has split their building in the middle and divided it into two thinner slap buildings. Each 5.3m deep building contains diverse sizes of living, working, commercial or composite units and O-Office named them 'Living Cases'. The building is a collection of various living cases, a collection of various life styles. Between the two slaps, O-Office designed a landscape court which contains an artificial hill on the ground, vertical green spaces, interweaving communal platforms, stairs and corridors, which tile all "Living Cases" into a micro urban mechanism.

对于源计划的许多人来说，设计上的难度在于如何创建适用于生活、工作和商业的不同用途的高密度空间，从而在总体规划中确立的单个建筑相对独立的基础上，为入住的年轻人留出尽可能多的空间。总体规划中，对其住宅和商业空间的比例提出了要求，还为该建筑指明了类似的分隔样板。源计划将其建筑从中间分开，划分为两个较薄的扁平建筑群。每个深达5.3米的建筑群包含多种尺寸的生活、工作、商业或综合单位，源计划将其命名为"生活匣子"。整个建筑群由多个生活匣子组合而成，同时也容纳了多种生活方式。在两个建筑群之间，源计划设置了一个空中庭院，其中包含了假山、立体绿墙、纵横交错的步道、阶梯和走廊，将所有的"生活匣子"编织成了一个微型的城市体系。

HOUSING WITH A MISSION, DUTCH AND CHINESE ARCHITECTS' DESIGNS FOR THE ANTS TRIBE, THE NETHERLANDS / 住宅的使命，荷兰与中国建筑师为蚁族而设计，荷兰

Urbanus Architecture and Design inc. / 都市实践

The block designed by Urbanus is located on the north-east corner of the site. Urbanus's design focuses on re-interpreting and re-creating the young professional's community environment into "the Ant Farm." The community environment becomes a procession of public communal spaces, analogous to tunnels made by ants within an ant farm. These spaces become a unique opportunity for young entrepreneurs to not only produce and exchange ideas but to realize them.

The procession of public communal spaces in the building is linear and wraps around the building to distinguish its four fronts. It begins on the ground floor of the building's west side as retail frontage then wraps up the north side of the building as a public stair. It then travels across the second floor of the east facade as a market arcade before entering the south facade. On the second level of the west facade, the procession becomes a series of flexible spaces that step up and across each floor along the facade of the inner street until reaching the roof. With this, Urbanus has designed a flexible common space on each residential and retail level suitable for all sorts of inter-activity and collaboration.

"The Ant Farm" becomes the ultimate venue for small sporting activities, micro-farming markets, hobby clubs and many other unique entrepreneurial opportunities. By re-examining communal life for young professionals, "The Ant Farm" is a unique typology that opens new possibilities for residential life in the urban realm.

由都市实践设计的楼群位于整体建筑的东北角。都市实践的设计重在对年轻职场人士们的社区环境进行重新解读和再造，将其定位为"蚂蚁农场"。其社区被转化为一排排的公共空间，模拟蚂蚁农场中蚂蚁挖出的隧道。这些空间为年轻的创业者们提供了独特的机会，不仅可在此寻找灵感，还可交换和实现其想法。

该建筑中的公共空间呈线性结构，围绕该建筑划分其四周的外围区域。该结构从建筑一楼西边的商用区开始，一直延伸到建筑的北边，化身为公共阶梯。然后，它穿过二楼的东面外墙形成一个购物长廊，随后进入南面外墙。在西面外墙的第二层，该结构又变成了逐层爬升、形态多变的一系列空间，并穿越各楼层，沿内部街道的外墙而上，直至屋顶。都市实践借助这一手法在各个住宅和商用楼层中设计出了灵活多变的公共空间，适合人们进行各种互动与合作。

"蚂蚁农场"为小型体育活动、微型农场和集市、俱乐部和许多其他独特商机提供了大展拳脚的绝佳场所。凭借对年轻的职场人士的公共生活的重新解读，"蚂蚁农场"这种独特的形态为都市中的居民生活创造了新的可能性。

INTERNATIONAL PARTICIPATION / 国际参与

NEXT architects

For NEXT architects the biggest challenge lays in an apartment building containing units that vary from 14m² to 21m² combined with a rather thin plot to build on. In this block, NEXT started with a dense layout of the units with a double loaded corridor. By taking some space out on the upper floors, the corridor opens up and creates (communal) space. This can be used to bring more green into the project and facilitate social interaction. Although the apartments are rather minimal, they become small urban villa's on top of the building. To emphasis this, the upper floors are built up out of brick standing on its own feet. The apartments are orientated with respect for the privacy of each individual and the opposite building. On the fourth and fifth floor the owners can use the outside space to create a green atmosphere. The size of the apartments determines a very efficient organization of the interior. The idea is that there are two zones or walls. One that facilitates a bathroom, a pantry and incorporates the air conditioning unit. The other wall is a multifunctional and multiflexible unit that provides a single/double bed, a desk which can be used from two sides and storage space.

对于 NEXT architects 来说，最大的挑战来自于公寓楼的设计：必须在十分狭窄的建筑布局中安排从 14 平方米至 21 平方米的各种单元。在这个楼群内，NEXT 首先在单元的设计上采用了密集布局，走廊两侧都有房间。上层楼面拿掉部分空间之后，现出的是开放式的走廊，从而营造了（公共）空间。这样可为项目设计中的绿色景观留出更大的余地，同时也可促进社交互动。尽管这些套房面积极小，但他们算得上是整个建筑上的小型市区别墅。为了强调这一点，建筑的高层自成体系，均以砖块建成。各套房的朝向充分考虑到了每栋建筑及其对面建筑之间的私密性。在第四和第五楼层，业主可使用外部空间营造绿色景观。这些套房的大小决定了其居住者必须有效安排其内部空间。其创意在于营造了两个区域（两道墙），一道墙构成了卫生间和食品储藏室，并可容纳空调设备，另一道墙则用来满足多功能用途，包含单人/双人床、两面均可使用的桌子和储物间。

HOUSING WITH A MISSION, DUTCH AND CHINESE ARCHITECTS' DESIGNS FOR THE ANTS TRIBE, THE NETHERLANDS / 住宅的使命, 荷兰与中国建筑师为蚁族而设计, 荷兰

NL Architects

NL Architects named their proposal Maximize! The idea of their proposal is drawn out of the spatial circumstances of the site-regulations. The building height is 18m. And the streets are just 6 meters wide. The combination of the two makes it possible to see the inner street as narrow corridors. This narrowness has it's advantages. It creates protection from the sun, wind and even rain. But it also generates a feeling of togetherness.

On the first two layers there will be the vibes which come from a variety of commercial activities. These will be mixed with the liveliness of the in- and outgoing neighbors who live on the 2 floors above. The second layer is articulated by a lifted outdoor street which is generously assessable by two pairs of waterfall stairs. The extra street doubles the valued surface of exterior shop window. The requested residential unit sizes are small, only 14 and 21m2. They will give room to young urban professionals who are looking for an affordable space to live. Maximize! proposes a unit that could be named a 'hallway-house'. The dimensions of the units are stretched to the limits. The sizes are as long, as narrow and as high as possible. By raising the floor height to 4.5m daylight shines deeper into the unit. The high ceiling generates a spacious feeling. The room has a tremendous amount of wall surface. The extra height gives the opportunity to create extra horizontal surfaces. It is a unique space that challenges the user to interact: to maximize!

NL Architects 将他们的方案命名为"Maximize！"（最大化！）。该提案的创意来源于现场规程对空间环境的规定。该建筑的高度为18米，街道仅有 6 米宽，结合这两者，就可以将内部的街道视为狭窄的走廊。狭窄也有好处，这样的结构能够提供对日照、强风甚至降雨的抵御。它还能产生归属感。

各种商业活动在建筑最下方两层营造的是商业气氛，而居住在这两层楼上方的住户的穿梭来往又在其中增添了生活气息。设计师们在第二层使用了加高的户外街道进行点缀，并不惜为此在两旁修建了伸手可及的阶梯瀑布。这条额外的街道让两旁店铺橱窗的身价倍增。住宅单元的尺寸很小，被定为 14 平方米和 21 平方米。对于在承受能力范围内寻求住所的年轻的都市职场人士来说，这里将为他们提供容身之所。"Maximize！"方案推出了一种可被称为"厅房"的住宅单元。该单元的尺寸已被挖掘到了极限，在长、宽、高上都尽可能地进行延伸。通过将层高提高至 4.5 米，使得房间更深处也能获得光照。高高的顶棚让人产生开阔感。房间的墙体面积极大。多出的高度使得人们可进行更多的水平分隔。这种独特的空间结构也给居住者的动手能力提出了挑战，让他们对其空间进行"最大化"利用。

A CATALYST REACTION OF OUR CITY: "SHENZHEN AND UNIVERSIDE" SPECIAL EXHIBITION
城市触媒：大运与深圳专题展

Organization Group: RITO & Li Degeng, Shenzhen Public Art Center
策展团队：朗图 & 李德庚，深圳公共艺术中心

A CATALYST REACTION – POST CITY: 3rd SHENZHEN HONG KONG BI-CITY BIENNIAL EXHIBITION

Greetings of Exhibition Space

The exhibition space located in West Lobby of Civic Square (Area B), so the proposal brought out "sphere" as the core visual concept of the whole space base on openness, pre-fab and installationess of the background and specialty of the exhibition. The principal axis is 3 spheres of 9 meters' high and the artists' work were involved in. There're also some small spheres of 1.2m high, which are presentation area of video and introduction of the exhibition. The whole plan is three spheres would guide the audience as openness space.

Organizing the Works

After many communications, we found the material is hardly got, so we change the organizing structure and invited three authors from different fields that present their ideas via model, multi-media installation and video.

展览地点位于市民中心 B 区西礼堂，基于展览的背景及地点的特殊性，在空间设计前期提出了"开放性"、"预制性"、"装置性"等原则，并由此衍生出以"球体"作为整个空间强有力的视觉符号的概念。主轴为以三个最高达 9 米的球体展览区，将不同风格的艺术家作品纳入其中，而直径 1.2 米的小型球体则作为影像播放媒介及展览介绍媒介或聚或散地在空间中出现。整个参观动线的规划，以三个球体为主轴暗示引导，亦让人们能随意步行，形成开放式的展览布局。

经过多方沟通后，发现不能取得大部分所需资料，因此及时调整了展览的组织方案，邀请了三位来自不同领域的创作者，以现场模型装置、多媒体装置、影像等方式表达他们的立场。

Instruction of Works

Viewing (Exhibition)

Multi-media installation: In the Venue City

Artists: A Long, Qiu Meng
As citizen live in Shenzhen, the author interviewed the mayor, architects, volunteer and citizens live around to present the feeling and expectation of Post-Universide, then various perspectives are presented via video installation.

Multi-media installation: Explore - Nation Universide

Artist: Keith Lam
Architecture is the witness of generation of urban ability. The Universiade left the citizens not only the architecture, but also the emotion of citizen. So this work guide the audience to travel around all the space constructed for Universiade, to provoke the reaction and engine the emotion. All the hidden issues are recorded as data and then transformed into Post-Universiade sensation.

艺术家：阿龙、秋梦
作为深圳市民，作者以纪录片的形式采访市长、建筑师、志愿者、场馆附近居民等众多角色，获取"后大运"时期市民的真切感受和期望，以视频装置的形式呈现辩证的多元观点。

艺术家：林欣杰 (Keith Lam)
建筑对于城市的功能是诚恳的城市见证者。大运会留下来的除了是建筑物本身的实体，更多是城市人的情绪。作品引领观众重新漫游大运会场馆，触发身体反应，启动心理情绪，这些隐藏的被记录成数据，化成声音演绎漫游者的后大运感知。

Participation (Workshop)

Workshop + Installation: U Workshop
Architect: Feng Guo'an
Inviting architects and students from Guangzhou Academy of Art, South China University of Technology, Shenzhen University and Shantou University to discuss how to use the space of Universiade and present the design to the audience. And a renewal design of U Station presented Book Mall Plaza as a "Hotel for One Person".

Listening (Forum)

Forum 1: "Movement" and International Citizen
Host: Jin Minhua
Panel: Liu Gao Ming, Han Jiaying, Huang Liguang, Jin Cheng, Xiao Yu, Gu Xiaojin, Wang Shaopei.

动耳参与（工作坊）

工作坊 + 装置：《U 站工作坊》
建筑师：冯国安
以头脑风暴形式，邀请业内设计师和广州美术学院、华南理工大学、深圳大学以及汕头大学的学生讨论大运会后 U 站使用问题，在展览中把过程成果重新展示给观众。另有一个 U 站实地完成优化设计后在书城广场展示，变成"一人旅馆"作品。

听（论坛）

学术论坛一："运动"与国际公民
学术主持：金敏华
发言嘉宾：刘高鸣、韩家英、黄立光、金诚、晓昱、辜晓进、王绍培

Shenzhen has entered Post-Universiade Era since the close ceremony of this event.
Since Shenzhen submitted the application of Universiade, many groups and civic "movement" took place during this period, welcoming, re-construction and cleaning, etc... no matter positive or negative reply they received, sport is so welcomed to the city. Since six months before the opening, the Weibo Movement aroused by various phenomena is from a normal Shenzhen citizen's perspective to evaluate this "sport", many we can say which is mixed reviews.
The events includes: dangerous group of 80000 people, invented torch-relay, forbidden migrant labour payment issue, no fireworks of the opening, re-build well-done green lands, rebirth of "silkworms", living space renewal and cancel the sub-way security and re-using the U station...
In fact, such grand event and city should be interacted, the event shape up the character and appearance of a city. On the other hand, the nature and character of a city also affect the event and sensation to the audience.
Besides the urban planning architecture and hardware, what kind of fortune and spiritual heritage did the Universiade left to the city?

大运落幕已过百日，深圳业已进入后大运时代。
从开始申办到现在，七年的时间，期间发生的林林总总，从群众性的动员或者说"运动"式的欢迎、整改、清理……到不管运动会是受追捧还是遭冷落，运动本身仍然是这座城市未曾谓为风尚的一项活动，再到今年大运开幕前半年因种种现象引发的、自下而上的、以人本和公民权利为核心的"围脖运动"，从一个普通深圳市民的视角感受、审视这场声势浩大的城市"运动"，大约可以用毁誉参半、悲欣交集来形容。
这些此起彼伏的"事件"包括：八万高危人群，虚拟火炬传递；禁止民工讨薪，开幕不放焰火；绿地推倒重来，"春蚕"起死回生；住区亮灯清人，城村穿衣戴帽；取消地铁安检，统一各式招牌，U 站"节"后余生……
事实上，类似大运这样的大型事件与城市之间本来就应该是一种互动的关系，事件参与形塑城市性格、城市气质和城市形象，反过来，城市的个性和气质也会影响事件的走向、观感乃至精彩程度。
抛开城市规划、建筑等硬件设施，从媒体、社会、文化的层面看，大运到底给这座城市留下了什么样的财富或精神"遗产"？

Host: Feng Yuan
Panel: Meng Yan, Liu Heng, Feng Guochuan, Feng Guo'an, Chen Shaohua, Dai Yun

Though we all know that Chinese cities has high requirement and reason to realize speedy expansion, such requirement creates the widest urban expand phenomenon, but also unique production methodology of urban spectacular. This is the result of globalization and Chinese-tion, it is typically presented as grand event and metropolitan, from Beijing Olympics, Shanghai Expo, and Guangzhou Asian Games, then Shenzhen Universiade, which presented that grand event shape up Chinese urban and provoke media and review response.

From the perspective of media and mass reviews, such grand event and metropolitan made the GDP requirement into a imaging shaping, how such general imago established in system, how such imagination provoke urban construction practice, how such investigation on appearance construction promoted and realized? All of these are targeted to the Chinese-ness issue within the context of globalization; clearly, it is the most Chinese issue of globalization pursuit.

So, Shenzhen of Post-Universaid has to retrospect such relationship between great event and metropolitan to track the globalization and Chinese-ness issue, especially Chinese dynamical connection, which would help with understanding Chinese situation now. It is a hidden historical track and obvious modernized expectation. So our target is inviting scholars to establish common discussion, to create and analyze such phenomenon and possibilities, eventually, we hope to reveal the real Chinese impulse now.

We take such discussion and revealing as real Chinese characteristics' contribution to globalization.

学术主持：冯原
发言嘉宾：孟岩、刘珩、冯果川、冯国安、陈绍华、戴耘

中国城市的高速扩张和形象上的"增容"有着现实上的理由和需求，这种需求不仅创造了人类史上最大规模的都市化现象，而且也带来了独特的城市景观的形象生产方式。从动力学的源头来说，这是全球化和中国化合力作用的一个结果。全球化和中国化的合力方式最为典型地表达在以下这个对称性之中——大事件与大都市的相互作用——从北京奥运开始，到上海世博，广州亚运，然后到深圳大运，利用大事件塑造新的城市形象的做法构成了当下中国都市的通行做法，并在当下的传播空间中获得了巨大的媒体和舆论效应。

从媒体、公众舆论的传播效应来看，正是这种大事件和大都市的相关性，使得城市扩张和增容从一种实质性的 GDP 增长追求更多地转向到一种"想象性塑造"的层面上，城市的总体形象是在怎样的体制中得以建构出来的，关于形象的想象是如何刺激了城市的建设实践，投资于形象生产的资源耗费又是通过一种怎样的调集资源的机制得到推进和实现的？所有这些，都指向了全球化追求下的中国化问题，准确地说，是貌似全球化追求中最具有中国特色的问题。

因此，在深圳大运之后的深圳特区，重新回顾从北京奥运开启的这种大事件与大城市形象的相互关系，其目的是要在一个当下的语境中去追溯上述的那个全球化（国际化）和中国化（中国特色）的合力关系，尤其是最具有中国特色的力学关系，我们相信，追踪和分析这个双重关系中的中国特色的动力学，有助于我们更好地理解当下中国的城市状况，它是隐性的历史线索和显性的现代化欲望共同塑造的产物，我们的目标是，通过来自不同领域、不同学科的专家和学者的共同讨论，创造出分析和解构上述一系列的现象的可能性，最终是为了去揭示出中国式生产的真实动力。

我们把这种讨论和揭示看成是中国特色对全球化的一种真正的贡献。

EXHIBITIONS OF PUBLIC WELFARE OF LEADING SPONSORS
赞助企业公益文化展

CR LAND

HAND IN HAND, CHANGE OUR LIFE: Photography Exhibition by CR Land

Lifestyle of Urban Synthesis

Urban synthesis is one intensive synthetical space within office, residence, hotel, business, leisure and communicating, which integrate multi-functions of a city. The synthesis occupies the center of a city with the central preponderant resource.

CR Land introduced such urban synthesis in China which has spread over Shenzhen, Hangzhou, Shenyang, etc., and became the unprecedented but familiar and favorite new urban lifestyle everywhere along with completion of CR Centers. Due to its tense function of urban synthesis, maximizing the live value and time value has been grown up in CR Centers, and CR Centers become the luxury landmark that fulfill incomparable urban life, even the diamond of a city.

Change the city, change the life, we always devote ourselves on this project, to make a better future applying to our excellent quality.

Hope Town of CR Land

2007 is the 70th anniversary of CR LAND, president Song Lin brought forward the idea that constructing Hope Town in poor area with the ability of CR LAND that multi-cultural advantage which base on the sense of worth that gratitude and social responsibility of an enterprise. Since then, CR LAND has already constructed two Hope Towns (Baise, Xibaipo), and there are 4 are under constructing (Wanning, Shaoshan, Miyun, Gutian).

CR Land intended to realize the vision through this town that, with the notion that: Follow the guiding of Management and Supervision of State-Owned Property Committee, State Department and Party Central Committee; support from the governments in all levels, and with the notion that 'Build Our Home Together', the villagers' residential environment would be transformed, rebuilt and totally changed through unified planning. Meanwhile, basing on the advantage of industry and resource of enterprise, CR Land would help the villagers establishing cooperation and developing new rural economics, then Baise Hope Town would turn into a socialist new village with agricultural developing energy, fresh local and ethical character, which would be ecological, organic, green and gets along very well with local natural environment; and practice the the scientific concept of development with the spirit of creativity and explore a new form and new way that how an enterprise involves in construction of socialist countryside with its own resource.

华润置地

华润置地"与您携手，改变生活"主题图片展

一、都市综合体生活方式

都市综合体"，是一种融合了多种城市功能，包括商务办公、居住、酒店、休闲娱乐、交通等的高度集约的综合空间。都市综合体往往占据城市的核心位置，享有城市最核心的优势资源。

华润置地率先成功地将"都市综合体"引入中国，从深圳，到杭州，及至沈阳……一种前所未见的全新的都市生活方式，随着华润中心的落成为人们所熟悉、热爱。凭借高度集约的城市功能，华润都市综合体在各城市中心演绎出最大化的生活价值和时间价值，成为一座城市中最为繁华的地标，成就无可比拟的都市生活，成为世人景仰的都心之钻。

改变城市，改变生活，我们将始终全力以赴，以卓越的品质描摹更加美好的未来。

二、华润希望小镇

2008年，华润集团成立70周年之际，宋林董事长基于感恩回报、履行企业社会责任的价值观念，提出了利用华润多元化企业优势，到贫困地区创建华润希望小镇的构想。三年来，华润集团已建成华润希望小镇两座（百色、西柏坡），在建华润希望小镇四座（万宁、韶山、密云、古田）。

华润集团创建华润希望小镇要实现的目标愿景为：在党中央、国务院、国资委的领导下，在地方各级政府的支持下，本着"共建家园"的建设理念，通过统一规划，就地改造、重建，彻底改变农民的居住环境；利用华润自身的产业资源优势，帮助农民成立专业合作社，发展新型农村集体经济，把华润希望小镇建设成为生态、有机、绿色、和当地自然环境保持和谐一致，具有农业发展活力、鲜明地方和民族特色的社会主义新农村；用创新的精神落实科学发展观，为国家探索一条企业利用自身资源积极参与社会主义新农村建设的新模式、新道路。

EXCELLENCE GROUP | 卓越集团

SEEKING THE URBAN LANDMARK

Excellence Group will build up an open design studio through which the audience can walk, and be indulged in imaginary environment, even can piece together 8 famous landmarks with environmental materials, and they can experience the changing of intelligence, architecture, urban and the world. Individual would feel the city, participate in the city, create city, so the city could be filled with energy of humanization.

When the using the energy, dealing with abandon products, and the sustainable development of the quality of air and water became the most important challenge, we would apply the environmental material Oriented Strand Board to construct the exhibition space; and when 'green' became a word in a fever, we just keep the original color of the material to create the vitality of natural beauty; when thinking about how to quantize sustainable development of a city and architecture, we just use the plexiglass to build our own city.

寻找城市地标

通过营建开放式环保的设计工作室，让参观者在走入展场的过程中快速被带入神游畅想的境界，充满兴味地用有机玻璃棒在环保材质的欧松板上拼插出八个具有代表性的城市地标，在方寸间体验智慧、建筑、都市、世界的交融变化。让个体在这一过程中感受城市，参与城市，创造城市，正因为如此，城市才充盈着人性的力量。

当关于能源使用、废物处理、空气和水质量的可持续实践成为最大挑战时，我们利用建筑剩余材料——欧松板搭建展台；当"绿色"等词语被过度使用时，我们保持材料本身的颜色，打造自然美的生命力；当人们在考虑如何去量化城市和建筑的可持续发展时，我们利用手中的有机玻璃棒打造自己的建筑，创造自己的城市。

SKYSCRAPER AND DREAM, PRACTICE AND RESPONSIBILITY

We would express our gratitude to the city with a landmark;
We would change the history with a architecture;
We would have a conversation with the world with a height.
We're building the 8th skyscraper to express our gratitude to the city, to change the history of a village, and even talk to the world with a absolute height. Such height is not only the altitude, the technics or management...
Even more, we expect people would realize stories behind it, the transformation of the old villages, to know, to concentrate, even to involve in the transformation of the old city.
In order to improve vernacular dwelling, glorify the city, save the estate resource and advance the capability of urban competition, we are working very hard; and we're fighting for constructing the dream of architecture.

高楼与梦想·实践与责任

我们用一座地标来感恩城市；
我们用一座建筑来改变历史；
我们用一个高度来对话世界。
我们用建筑一座世界第八大高楼的梦想去感恩这座城市，去改变一个村落的历史，更用一个绝对的高度去与世界进行对话。这种高度不仅是海拔，不仅是技术，不仅是管理……
我们更期望人们去了解这栋高楼的背后，去关注那些旧村落的转变，去让更多人了解、关注并参与到对旧城的改造中来；我们为改善民居，美化城市，节约土地资源，提升城市竞争力而努力；我们正在为建筑梦想而奋斗。

| FANTASIA GROUP | 花样年集团 |

ARTISTIC LIFE PRACTITIONER: ZHI ART MUSEUM

Fantasia Group initiates the notion that 'Make the Life with Better Style', and keeps exploring the better combination of architecture, community and art, and concentrating on the development of art estate.

Zhi Art Museum is what the Fantasia Group would like to present for the 2011SZHKB, which is also the practical experience of Fantasia Group's 'Art Estate'.

Zhi Art Museum locates besides the Laojun Mountain, Xinjin, where is the southern door of Chengdu. It is 30 kilometers from Chengdu, and sit at the foot of this famous mountain.

The art museum is a design work by Japanese architect Kengo Kuma, he applies to Taoism, art and culture, links the sky and the earth with water which has connected the architecture and nature together to achieve the environment that the sky, the earth and water just integrate together closely. The implicative and quite appearance of Zhi Art Museum owns zen attitude, in which the water and Chinese traditional architecture element 'tile' are connected and endows the appearance of the architecture with smooth energy, however, the museum owns the comfort distribution of quiescent and quiescent condition.

艺术生活践行者"知·艺术馆"主题展

花样年倡导"让生活更有风格"的企业理念，持续探索建筑、社区与艺术的更好结合，践行于艺术地产的发展。

花样年在2011年建筑双年展中的参展项目以成都"知·艺术馆"为展示标本，展现花样年"艺术地产"的实践经验。

"知·艺术馆"位于成都市南大门新津老君山山侧，距成都市区30多公里，千年老君山脚下。

"知·艺术馆"由日本著名建筑设计师隈研吾设计。隈研吾在这个项目里，把握道教、艺术、文化的主脉，用水来衔接天和地，让建筑与自然有机地结合在一起，达到天、地、水相融的境界。"知·艺术馆"的外形含蓄、内敛，极有禅意，是把水的意境和中国传统的建筑元素"瓦"相联系，使得建筑的外在形态有着非常流畅的动感，但建筑本身最终呈现的是一种动、静相宜的状态。

GALAXY REAL ESTATE / 星河地产

LIMITLESS IMAGINATION, CONSTRUCTING THE FUTURE WITH HEART

During 1989 and 2011, Galaxy always exploring the truth of architecture and human life.

Galaxy Estate insists the management opinion of 'Being the pioneer of the market, exploring the brand', and 'competitive products, heartfelt service', supplies detailed, honest, professional, standard and bardian product and service, which has established its first class position. Till 2010, the whole area it has constructed over 3.5 million square meters, and only in Futian District, their work has occupied 0.6 million square meters; it stores the land over 2.4 million square meters, and the whole construction area is over 4 million square meters. In Yangtze River Delta, it has stored 187 hectares.

In 2010, Galaxy proposed the strategy of troika: Leading by the synthesis estate, supported by commercial estate and financial investment, moving forward smoothly and establishing the hundred years enterprise in economical fluctuation.

Meanwhile, Galaxy estate response the duty of enterprise citizen, and devotes into social public welfare undertaking and charity. During the last 20 years, Galaxy has donated 300 millions to education, religion and public welfare.

创想无界，心筑未来

1989~2011年，22年来，星河人不懈地探索建筑与人居生活的真谛。

星河集团秉承"做市场先锋，走品牌之路"的经营理念，坚持"星河精品，至诚服务"的服务理念，为顾客提供细致入微、诚信为本、专业化、标准化、个性化的产品和服务，奠定了深圳房地产业一流开发企业的地位。截至2010年，已累计开发的总建筑面积达350万平方米，仅深圳福田中心区的总开发面积就达60万平方米；在珠三角区域拥有土地储备面积超过240万平方米，总建筑面积超过400万平方米；在长三角区域拥有土地储备2800亩。

2010年，星河集团正式成立并明确提出"三驾马车"的战略规划：以房地产综合开发为首，以商业地产经营和金融投资为辅的三元驱动，稳步发展的同时，在经济的周期波动中打造星河百年基业。

星河集团在健康发展的同时也肩负起了企业公民的责任，积极投身社会公益事业与慈善事业，感恩社会，回馈社会。20多年来累计向教育、宗教及公益事业捐款超过3亿元。

YITIAN GROUP

益田集团

PRESENT FOR THE CITY

Every city has its own happiness, one tree, one building, one street, they all have special colors, and the combination of them is a city. Every detail mentioned belongs to the brilliant spectacular of the city, belongs to everyone in the city, and they are all the happy present.

The whole exhibition presented by Yitian Group is designed and layout accordance with the notion: Here are the presents for the city from Yitian Group, and it is a platform for everyone in city life can share their blessing. The space would be embodied with the element 'present box' to emphasize the idea that 'the present for the city and for everyone, from Yitian Group'.

Corresponding to the theme ''Architecture creates cities. Cities create architecture.'', Yitian Group creates a Platform of Love and Happiness for the city and every citizen in. It includes humanized residence and commercial products, and even marvelous public welfare events, Yitian Group is a choir of the city, and blessing for everyone in the city and the city itself. The audience would feel the blessing from Yitian Group, meanwhile, they also can communicate their well-wishing to the city and people.

献给城市的礼物

每个城市都有自己的幸福，一棵树，一栋楼，一条街都带有浓厚的独特色彩，组合在一起便是一座城市。这一切，都属于城市的精彩，都属于城市里的每一个人，这是幸福的礼物。

整个展馆，始终以"这是益田集团送给城市的礼物，又是城市生活中的每一个人互赠祝福的平台"的概念布局及设计，以"礼盒"元素贯穿空间造型，突出"这是益田送给城市的礼物，送给每个人的礼物"之意。

继承本次双年展"城市创造"的主题，益田为这座城市和城市中的每一个人创造了一个"爱与幸福的平台"。从高品质人居生活的居住及商业产品，再到各种盛大的公益文化活动，益田集团以"城市礼赞者"的角色，为我们的城市及生活在城市的每一个人送上属于益田的祝福。我们以"送给城市的礼物"为展览主题，欲使每一个参观者感受益田祝福的同时，在这个平台上，为我们的城市及共同的居住者献上属于自己的祝福和礼赞。

LVGEM, BLESSING THE CITY AS WONDERFUL AS YOU IMAGINE

Every city has its own soul: creative, passive, open and compressive... 'absolute sincerity' is not only the soul of LVGEM, but also its mission, such persistence make us witness and promote the urbanization of China. through this enterprise, we see the urban life that thousands lights are lively characters in the urban, and we treasure the faith from the society.
Since the light was invented in 1893, which has witnessed the change and development of modern city. So this exhibition will be carried by lights, apply to this usual element to design and evaluate, and then introduce the energy conservation, environment protection and creation to show our thinking on urban development.
The management philosophy of LVGEM is born in the common faith of 'absolute sincerity'. We believe that, insist such faith, the city will be as wonderful as we imagine.

绿景，祝城市和你想象的一样美好

每座城市都有她的灵魂：创新的、热情的、开放的、包容的……"精诚"，是绿景的灵魂，也是绿景坚持的使命，这种坚持使我们有幸见证并推动了中国的城市化进程。透过产业，我们看到的是城市生活，是夜幕中每个窗户亮起的盏盏明灯，是城市生活中一个个鲜活的人性。正因如此，我们对社会充满了感恩情怀，我们无比珍惜来自社会各个维度对我们的信赖。
霓虹灯自1893年诞生以来，见证了现代化都市的演变和发展。本次展览以"霓虹灯"为载体，将这个大众生活中随处可见的元素进行设计和提升，并引入节能、环保、创新等概念，表达了绿景集团对城市发展的思考。
绿景的经营哲学，从根本上说，就是源于"精诚"这一共同的信仰。我们也坚信，坚持这种信仰，城市定会和我们想象的一样美好！

HOROY — 鸿荣源

WHOLEHEARTEDLY QUALITY, URBAN WITNESS

Wholeheartedly Quality, Urban Witness! For 20 years, with advancing perspective, HOROY has promoted civic residence, update the commercial forms and life style.

As the leading member of Shenzhen estate, HOROY's project Xiyuan Estate has been nominated as 'Green Golden Certification' by Zhenshen Housing Department, and 'National 2 Starts Green Architecture Certification'. The detailed management of HOROY embodied in 6 grand projects and 45 small projects in total. HOROY would inspire residence revolution in each step they involve. Classics is the special character of HOROY project.

Since 2005, Horoy has completed 2 million squares commercial estate and its projects are the focus of Shenzhen west. Meanwhile, West Time Square occupied the center of Qianhai which take the advantage of 5 cities culture and leisure public construction, and the estate around Line 1 and 5 of subway, totally investment is 1.2 million, which has been the Shenzhen commercial estate legend. Super luxury business center which has been the landmark of all the Shenzhen urban synthesis which contains super shopping centers, local business street, grade A office building and business apartment, and it would bring the residences new experience of shopping, leisure, humanity and working, even housing experience, and it will bring Qianhai Center the new business form and take the lead of new lifestyle. 'Architecture creates cities. Cities create architecture' makes everything beyond the imagination!

品质见心，城市见证

品质见心，城市见证！鸿荣源20年优越筑城，以运营城市的前瞻眼光，推动优越的人居建设，推动商业模式和生活方式的升级。

作为深圳地产开发第一方队的成员，鸿荣源开发的熙园山院项目荣获深圳住建局颁发的"绿色建筑金级认证"，住建部"国家二星级绿色建筑设计评价标识"。鸿荣源的"精细化管理"主要体现在工程开发六大项共45小项的层层精确的控制、落实上。鸿荣源每到一处，都引领区域的人居革命。给城市留下经典，是我们的特质。作为深圳西岸的建设先锋，我们推动城市进程。

自2005年起，鸿荣源仅在深圳西岸就打造了近200万平方米商住建面，"鸿荣源深圳西岸项目群"成为了市场中最闪耀的焦点。同时，鸿荣源正在缔造深圳商业地产新锐传奇——西岸时代广场，项目雄踞前海中心，享五大市级文化娱乐公建，地铁1号和5号线上盖物业，计划总投资逾120亿元。鸿荣源将打造超级豪华的商业航母，成就深圳都市综合体标杆力作。届时将建成集大型购物中心、风情商业街、超甲级写字楼、商务公寓和高端住宅于一体的超大型城市综合体，为前海中心带来新锐的商业模式，引领全新的生活方式。

城市创造，让一切尽可超越想象！

OCT Properties | 华侨城地产

QUALITY OF OCT, HAPPINESS OF HOMES

OCT is a long term achievement! Since 1985, OCT has became national civilization of tourist area, demonstration plot of human residence and national cultural enterprise demonstration park. Now it maintains international reputation, and a tourist city, ecological city and cultural city, we are the excellent imaginer.

We insist the pursuit of high quality, and insist developing the mode that modern service synthesis and integrating the cultural-artistic connotation, green-LC, ecological environment idea, and practice many competitive products.

We denote ourselves in the perfect combination of humanity and poetic: cultural themed community (Portofino, Xi Citym cultural-artistic area OCT-LOFT; urban cultural synthesis: Su Bend, OCT Bay, etc. They are all the fruits of OCT's long march.)

During the 2 decades, we practice our idea "Quality of OCT, Happiness of Homes" with the spirit of creation, perspective of humanity, thinking of advance, attitude of honest.

品质华侨城，幸福千万家

华侨城不是一天建成的。从 1985 年至今，这里已成为了全国文明风景旅游区、人居环境示范区和国家级文化产业示范园区，成为了举世闻名、独具特色的旅游城、生态城、文化城。我们是优质生活的创想家。

我们坚守对品质的追求，坚守现代服务业成片综合开发与运营模式，坚守融入文化艺术内涵，坚守绿色低碳、生态环保理念，为城市铸造出一个又一个的精品。

我们致力于人文精神与诗意栖居的完美结合：文化主题社区——波托菲诺、曦城，文化艺术高地——OCT-LOFT 华侨城创意文化园，都市文化综合体——苏河湾、欢乐海岸等作品，都是华侨城创想征程上取得的累累硕果……

20 多年来，我们一直以创想的精神，人文的视角，前瞻的思维，真诚的态度，不断诠释和践行着华侨城人的理念——"品质华侨城，幸福千万家"。

SATELLITE EXHIBITIONS AND EVENTS
外围展及活动

SATELLITE EXHIBITIONS AND EVENTS / 外围展及活动
FRINGE X ARCHITECTURE: ART CREATES URBANISM. URBANISM CREATES ART. / 艺穗 x 建筑：艺术创造

Co-organized by Shenzhen International Fringe Festival Organizing Committee, Shenzhen ATU Architectural Development Communication Center, and Shenzhen Fatbird Theater, this project is to explore the relationship between public art and urbanism.

A temporary pavilion installation is created to function as public space for art performances and various educational public events, as well as experiments on urban planting, urban transport and mobile performing art. All these activities further activate the vitality and social function of the public space. Under the city urbanization context, public art is like intangible weapons to inspire the city and boost sustainable development, while architecture often facilitates as tangible weapons to build up the city and the urban culture. It would be an inspiring social practice about public art and urbanism in such a rapidly urbanizing city like Shenzhen.

This project is also one of the Special Project of 2011 Shenzhen International Fringe Festival.

由深圳湾国际艺穗节组委会、深圳观筑建筑发展交流中心、深圳胖鸟剧团共同合作，借助建筑空间装置的创意和实践，为一系列艺术表演及文化交流活动度身创造临时公共空间。

同时，这些文化活动以空间流动表演艺术、城市空间探索、城市综合用途开发、绿色低碳出行、公众建筑教育等多种元素来激发城市空间的活力和公共功能，形成艺术与建筑共生互创的城市空间舞台，推动城市空间和公共艺术的探索与实践，为深圳城市文化建设注入新的活力。

Venue:
The east side of Hou Hai Bin road (Take the Metro Line 2 "Shekou Line" to HouHai Station, Exit A
Duration:
from Dec 8th, 2011 to Jan 8th, 2012
Co-Curators:
Eric ZHU, BAI Xiaoci, CHEN Weihang, Carol MOK
Organizer:
Shenzhen International Fringe Festival Organizing Committee
Co-organizers:
Shenzhen ATU Architectural Development Communication Center, Shenzhen Fatbird Theater

地点：
深圳市南山区后海滨路海德二道交界东端（后海滨路以东，地铁二号线蛇口线后海站A出口处空置草地）
时间：
2011年12月8日 – 2012年1月8日
策划：
朱德才、白小刺、陈伟航、莫舒敏
主办：
深圳湾国际艺穗节组委会
协办：
深圳观筑建筑发展交流中心
深圳胖鸟剧团

SATELLITE EXHIBITIONS AND EVENTS / 外围展及活动
THE FLYING GRASS CARPET / 来自风车王国: 荷兰的绿色飞毯

Hop on board the Flying Grass Carpet and join it in its quest for adventure and urban leisure. Studio ID Eddy and HUNK-design have designed the Flying Grass Carpet; it's an instant park that can be unfolded in any city. This way bringing a unique experience of fun and relaxation that can be shared worldwide.

The Flying Grass Carpet is designed to look like an immense Persian rug with its pattern executed in different types of artificial grass, giving it a typical look and touch. The design consists of different parts that can be adjusted to any location. The size of the Flying Grass Carpet can be altered from 18 by 22 meters to 25 by 36 meters.

When landed in a city the Flying Grass Carpet is ideal for all kinds of events. Its attractive appearance is also perfect for special events such as a city picnic and all sorts of contests and performances.

The Flying Grass Carpet delivers a beautiful alternative for city dwellers to enjoy the city. It brings instant cosines and a green leisure feeling to any city where it lands. The combination of green, beauty and activity makes the Flying Grass Carpet irresistible!

乘上会飞的绿毯,一起探险,一起寻找都市忙碌生活中的久违闲暇。来自荷兰的 Studio ID Eddy 和 HUNK-design 为您呈现了绿色飞毯: 一个可以简单打包便行走在城市间的便利公园,一种全世界共享的独一无二的快乐轻松。

绿色飞毯的创意来自于古老的波斯地毯。根据图案的组成,它由相应质感的人工草坪拼接而成。灵活的设计使得这块巨大的绿色飞毯可以在任何地方快速铺展。我们设计的飞毯尺寸从宽 18 米、长 22 米到宽 25 米、长 36 米不等。

当绿色飞毯进入我们的生活,它是各种场合的理想选择。正如我们印象中的公园,游人们可以躺在草坪上休憩,朋友们可以互相邀约来踢球,不过绿色飞毯还提供了更多的选择: 飞盘锦标,城市野餐以及各种各样的竞赛,展示和表演活动。

如今,世界上每一个城市中心都被快速地私有化,公共区域正在压力中艰难地生存,环境质量每况愈下。来自荷兰的绿色飞毯将给城市居民带来一个崭新的空间,一块"飞"来的既成的绿地,一片高楼大厦缝隙间的绿色惬意,一种不可抗拒的闲适享受。

Venue:
West square of Yitian Holiday Plaza, Nanshan District, Shenzhen
Duration:
Jan. 7th, 2011 – Feb. 10th, 2012
Organizer:
Shenzhen Biennale of Urbanism\Architecture Organizer Committee
Sponsor:
Consulate-General of the Kingdom of the Netherlands in Guangzhou

地点:
深圳市南山区益田假日广场西下沉广场
时间:
2012 年 1 月 7 日 – 2012 年 2 月 10 日
展览策划:
Studio ID Eddy and HUNK-design
展览主办:
深圳城市\建筑双年展组织委员会
展览支持:
荷兰王国驻广州领事馆

REVIEWING THE PAST AND ENVISIONING THE FUTURE: 20 YEARS OF SHENZHEN PUBLIC ART CENTER / 回望与前瞻—深圳市公共艺术中心(深圳雕塑院)二十年

Shenzhen Public Art Center, formerly known as Shenzhen Sculpture Academy, was established in 1991 as a directly affiliated undertaking institution under Shenzhen Urban Planning and Land Resources Committee, which was originally an office of Shenzhen Urban Sculpture established in 1982. Shenzhen Sculpture Academy was officially renamed as Shenzhen Public Art Center at 17th March, 2009, which is the first undertaking institution with "Public Art" in the title and which carry out the following responsibilities: researching, planning and promoting the urban public space art; creating public space arts (urban sculptures); carrying out other projects committed by Shenzhen government, such as researching on public art system and organizing big events on public art. Two decades since the establishment of Shenzhen Public Art Center (Shenzhen Sculpture Academy), which is not a short period of time. It is necessary for us to summarize, research, discuss and present the works of Shenzhen Public Art Center (Shenzhen Sculpture Academy) during the 20 years and to show them on an exhibition for more discussion and attention from the public.

深圳市公共艺术中心（深圳雕塑院）为深圳市规划和国土资源委员会直属事业单位，成立于1991年，其前身为深圳市城市雕塑领导小组办公室，成立于1982年。深圳雕塑院于2009年3月17日正式更名为"深圳市公共艺术中心"，是国内第一家以"公共艺术"命名的国家事业单位，职责是承担城市公共空间艺术的研究、策划和推广工作，开展公共空间艺术（城市雕塑）的创作工作，承担政府相关部门委托的其他项目，如公共艺术制度的调研工作和大型公共艺术活动的组织工作。 深圳市公共艺术中心（深圳雕塑院）从成立之初到今天整整走过了20年的时间，这是一个不短的时间，因此有必要对深圳市公共艺术中心（深圳雕塑院）这些年的工作进行梳理、研究、探讨、展示，并呈献给大家进行交流，以得到业界的关注。

Venue:
Contemporary Art Gallery of Shenzhen Public Art Center,
the first floor, Block A , Shenzhen Sculpture Academy, No.8 Zhongkang Rd, Shangmeilin, Futian District, Shenzhen
Duration:
Dec.15th, 2011 – Jan. 16th, 2012
Organizer:
Shenzhen Public Art Center
Chief Director:
Huang Weiwen
Artistic Director:
Sun Zhenhua
Curator: Dai Yun

地点：
深圳市福田区上梅林中康路8号
雕塑院A座一层
深圳市公共艺术中心当代艺术馆
时间：
2011年12月15日 –
2012年1月16日
主办：
深圳市公共艺术中心
总策划：
黄伟文
学术主持：孙振华
策展人：戴耘

SATELLITE EXHIBITIONS AND EVENTS / 外围展及活动
QIANHAN WORKSHOP / 前海工作坊

This exhibition aims at discussing the development direction of the Qianhai District in Shenzhen. In May 2011, Urban Planning, Land and Resources Commission of Shenzhen Municipality organized the workshop, 'The complexity and Contradiction of Future City Urban Form', and invited many international and national design and planning experts to contribute their ideas about the Qianhai development. Other than showing outcomes of the workshop and previous international design competition, the exhibition also creates an open platform for the public to engage. We wish to include the voice of the public, so that we can project the future of our city together.

Thus, the exhibition emphasizes public general participation. We provide many public interaction installations, so that people can voice out their comments through voting, playing games and drawing. At the same time, they can receive information and knowledge about the development.

本次展览主题是探讨深圳前海地区的发展。在2011年5月，规土委策划了"前海 —— 未来城市形态的复杂性与矛盾性研究"，并邀请多家国内外不同专长的设计与规划单位为前海这片属于未来的土地出谋献策。这次展览的主要内容，除了展示工作坊及设计竞赛的成果外，我们更希望能打造一个向公众开放的平台。我们希望公众能在本次展览中积极发表自己的意见，为我们的未来共同制定目标。

故此，"前海"展览强调参与性。我们提供了多个公共互动装置，令公众透过投票、游戏及涂画等肢体体验，向我们诉说大家的意见，与此同时，获得我们对前海研究的认识。

Venue:
B-10, OCT-Loft, Shenzhen
Duration:
Dec. 8th, 2011 – Feb. 18th, 2012
Organizer:
Urban Planning, Land & Resources Commission of Shenzhen Municipality Urban Design Division
Curator:
Urban Planning, Land & Resources Commission of Shenzhen Municipality Urban Design Division, Shenzhen Urbanus Architecture and Design
Agent:
URBANUS

地点：
深圳华侨城创意文化园北区 B10 展厅
时间：
2011年12月8日-
2012年2月18日
主办：
深圳市规划和国土资源委员会城市设计处
策划：
深圳市规划和国土资源委员会城市设计处
承办：
都市实践建筑事务所

SPACESHIP HEART-EXPERIENCING S, M, L, XL FOR ALL / 太空船之心——让所有人经历由小到大的美妙旅程

Collective Paper Aesthetics is a play in design models. The work is expressing the Dutch notion of architecture as a framework. The project Initiated in a site specific installation made for London Festival of Architecture 2008. Through a series of explorations in objects, space and experience Collective Paper Aesthetics are reloading Buckminster Fuller octet truss patent to be merged with the idea of the world game.

At Shenzhen & Hong Kong Bi-City Biennale of Urbanism \ Architecture, Collective Paper Aesthetics will present a new work titled: Spaceship Heart - Experiencing S,M,L,XL for all.

The work is examining the question how can we make together a place people truly loves?

During the biennale, visitors in the exhibition will construct Spaceship Heart, using modular cardboard building block. Each element in the installation consists of eight similar faces. Each face is representing a program in hypothetically mixed-use structure.

The end result will embody, apart from the symbolic heart shape a self-planned programmatic organization for a new place people truly loves.

集纸美学在模型设计中得以展示。作品作为一个框架，表现了荷兰的建筑概念。该项目特别为 2008 年伦敦建筑节安置，并从此开始了集纸美学的旅程。经过对物体、空间和经验的一系列探索，集纸美学重新载上了巴克敏斯特·富勒的澳克太特桁架专利（也就是 Tetraphedron，是指以自然为面的四个立方面，也就是如金字塔般的立方体；Octet Truss 是 Tetraphedron 的繁衍体，可形成更有力的组织架构），并且与世界游戏的概念相融合。

深圳·香港城市\建筑双城双年展中，集纸美学将会向所有人展示最新作品，名为：太空船之心——让所有人经历由小到大的美妙旅程。

这个艺术作品可以帮我们解决疑惑：我们如何才能共同建造一个使每个人都由衷地热爱的地方呢？

双年展期间，参观者可以用纸板模块来构建一个属于他的太空船之心。每一个安装的部件都包括 8 个相似的面，每个面都代表一个假设的复合型使用结构的项目。

在这里体现出的最终成果将是一个标志性的心形，一个自创的结构，一个让人们由衷热爱的地方。

Venues:
2 atrium B2 Floorm,Yitian Plaza Holiday, Nanshan District, Shenzhen
Duration:
30th December, 2011 -30th January, 2012
Curator:
Noa Haim/ Collective Paper Aesthetics

地点：
深圳市南山区益田假日广场 B2 层 2 号中庭
时间：
2011 年 12 月 30 日 - 2012 年 1 月 30 日
策划：
Noa Haim/ 集纸美学

SATELLITE EXHIBITIONS AND EVENTS / 外围展及活动
LIVING-ROOM CULTURE / 客厅文化

Using living-room as a cultural set, the exhibition presents the transition of our living space, the interaction between people and living-room, the connections between social-cultural contexts and living-room. Bringing a space from our daily life under social, cultural and historical scrutiny, this is the motif and purpose.

With texts, photos and a variety of art works, the exhibition presents the motif with three components:

1\Chinese Living-room in the eyes of three photographers
2\Graphic design on living-room culture
3\Community Design perspective on living-room culture

The interior space is configured according to a stardard residence this provides the city with a new possibility.

"客厅文化"展览从客厅文化入手，考察我们日常居住空间的演变，人与空间的互动，当下生活的主动性和被动性，社会文化背景和客厅之间的关系。

展览由三个部分组成：
1. 三位摄影师眼中的中国客厅
2. 客厅文化的多元表达
3. 客厅文化《住区》视觉

室内按标准住户配置——这给城市提供了一种新的可能性。

Venue:
Zai Gallery, Shekou, Shenzhen
Duration:
Dec.10th, 2011—Dec. 30Th, 2011
Organizer: Community Design Magazine of Tsinghua University
Curator:
DAI Jing & LI Wenhai

地点：
深圳蛇口自在空间画廊
时间：
2011年12月10日-
2011年12月30日
主办：
清华大学《住区》杂志
策展人：
戴静，李文海

COASTER RAID – SHENZHEN IS NOT TOMATO FIELD / "寻找深圳"成果展: 深圳不是番茄田

For many years people have not thought of Shenzhen as being a city with traditions, but rather as a city expanding in all directions, a city at the intersection of an endless network of roads. This is a place where rapid transformation has been maintained for years; the city continuously changes.

"Coaster Raid" brings together nine artists, intellectuals, and creative designers (or groups) to explore, discuss and ultimately produce a piece of work about a place in the city. We hope that through this process, those of us who live and work in Shenzhen can define and re-evaluate the city in our own terms.

An exploration consists of a two day workshop, a creative period, and a presentation of the works to a local audience. During the workshop, the nine participants explore an area and speak with local residents. In the evening, there is a dedicated time to the exchange ideas and creative brainstorming. Following the workshop is the creative period. During this time, participants will have a month to create an original work. Finally, these works are presented to a local audience to encourage deeper conversation and debate about the meaning of the city.

In the past year, "Coaster Raid" discovered Dongmen, Meilin, Nantou, Xixiang and Wutong Mountain. We would like to invite you to explore these places of Shenzhen with us in this exhibition.

很长时间里，深圳都被认为是一个没有传统的都市，触手向四面伸展，路网无远弗届。这是一个高速运转而能量持续涌出、不断变动的地方。

"寻找深圳"是一场由九位（组）不同领域的艺术工作者，知识分子以及创意人士围绕一个区域进行探索、交流以及以该地区为主题完成一项作品的城市创意探索活动。我们希望通过这个活动让我们这些居住和工作在深圳的人，可以用自己的想法来定义或是重新审视这座城市。

活动包括一个为期两天的的工作营，由这九位（组）人士对目标区域进行自由探索，与当地居民交流，其中有一个晚上与其他参与者进行思想交流、创意碰撞。然后，参与者将有一个多月的时间进行独立创作。最后，我们会将九部作品汇集起来，举行一个大型的发布会向公众展示，并期望与公众展开更多的讨论。

过去一年多的时间里，"寻找深圳"走过罗湖东门，福田梅林，南山南头，宝安西乡以及梧桐山片区，诚邀您参与展览，一起发现深圳。

Venue: East of Multifunctional Hall, Civil Center, Shenzhen
Time: Jan. 8th, 2012 - Feb. 18th
Curator: RIPTIDE
Organizer: RIPTIDE

地点：深圳市民中心 多功能厅东侧展厅
时间：2012 年 1 月 18 日 — 2 月 18 日
策划：锐态
主办：锐态

SATELLITE EXHIBITIONS AND EVENTS / 外围展及活动
LISTEN WITH RESPECT: DOWNTOWN EN NING / 聆听恩宁

Enning Road
A Round-table Conference in the Ruins.

As one district of the old city, Enning Road is the first refurbishment program of Canton. Since 2006, the city plan of it has been changed four times and haven't finished yet. There is some discuss about the legality of removing, the protection of the culture and rehousing the residents. We Enning Road Concerning Group have been studiing such topics for one year and a half.

We exhibit the different roles who connecting to Enning Road. Their aimless, helpless ness will be revealed. Instead of just criticizing or sinking into the memory of past times, we will sit down at the round table in the ruins to think of our age, to bring about a solution of mutiple balance city planning.

恩宁路——废墟上的圆桌会议

恩宁路是广州的一处老城区，也是广州第一个旧城改造项目。但是恩宁路规划自2006年至今，已四易其稿，由于拆迁合法性、文化历史保护、居民安置等存在一些不完善之处，恩宁路学术关注组进行了一年半的研究。

此展览呈现了与恩宁路变迁相关的不同角色的声音和处境，展现了他们各有的盲目、无奈与无助之处。在废墟上设圆桌，我们几方代表坐下来，为的不是批判彼此或是悲情怀旧，而是希望反思我们所共处的时代，并寻求如何在现实约束之下实现多元平衡的城市规划。

Project Team:
XING Xiaowen, ZHONG Weijun,
WU Weipeng, WU Yuan'er,
WANG Wen, LI Liben,
LIN Changrong, LI Xue,
ZHAO Siyu,
CHEN Xulu, XIONG Yulin,
JIANG Chao, LIU Jun, LIU Chaoqun

项目团队：（排名不分先后）
邢晓雯、钟伟君、吴伟鹏、吴元儿、王文、李本立、林常荣、李雪、赵斯羽、陈旭路、熊毓麟、姜超、刘珺、刘超群等

THE MEANING OF TIME: THEME PICTURE EXHIBITION/ "时间的意义"主题图片展

The ultimate meaning of a city is to create the value of time.

The growth and the creative process of the city are enriching the value of time constantly, and what we are looking for are such general characters or differentiations. Through displaying pictures of metropolitan life and taking Gangxia's cases of creating practices for example, the exhibition is meant to unscramble a city's past, present and future, to discuss the path of creating metropolitan life-mode of Shenzhen, as well as to search for the locations of Shenzhen in internationalization.

城市的终极意义是创造时间价值。

城市的生长和创造过程就是不断赋予时间更丰富的价值。我们要寻找的正是这种共性或者差异性。展览将通过世界大都会生活场景图片和以岗厦为例的城市创造实践案例展示，解读一座城市的前世今生和未来，探讨深圳创造国际大都会的生活模式路径，寻找深圳在国际化中的城市坐标。

Venue:
Courtyard of second floor in Central Book Mall, Shenzhen
Time:
December 11th-25th, 2011
Organizer:
Urban Research Institute of Shenzhen
Curator:
Gao Haiyan

场地：
深圳中心书城二楼中庭
时间：
2011年12月11－25日
主办：
深圳市都会城市研究院
策展人：
高海燕

SATELLITE EXHIBITIONS AND EVENTS / 外围展及活动
ENDEMONIC, GENESIS, EUDEMONIC, DECADE / 幸福·造物 幸福·十年

It is the second summary Mi Qiu's architecture and art for last ten years. The "Eudemonic· Genesis, Eudemonic · Decade" created this time is a phased critical of Mi Qiu's previous works. They are both discarding and integrating, which producing a new possibility. The identity of "feathered man" began in 2000, and now it is a new start of the whole project ten years later. He resequenced the typical figures and then start the creation, which has . The previous ideas of "happiness-survival" has been overthrown. Everything is updated. It took Mi Qiu ten years to rethink about this subject. He even demonstrated it repetitively by establishing a themed magazine. From two dimension to three dimension, he practiced non-typical collection of "happiness-survival" on a continuous basis. We can get some inkling of Mi Qiu's artistic creation for ten years from what we see today.

米丘建筑及艺术的十年总结作品之二。本次创作"幸福·造物 幸福·十年"是米丘对其以往作品的一次阶段性判断，舍弃及融合，产生一种新的可能性。"羽翼人"形象开始于2000年，经历十年时间后，新的方案开始启动，将典型形象进行排序后重新创作，颠覆曾经建立的对于"幸福-生存"的看法，一切都是更新。米丘花了十年时间重新思考这一命题，甚至通过建立主题杂志来反复论证，从平面到空间，持续地实践对"幸福-生存"的非典型性采集工作。今日所见，为米丘艺术创作十年之一斑。

Venue:
Shenzhen Civic Square
Time:
Dec.8th, 2011- Feb. 18th, 2012
Curator:
Mi Qiu Modern Art Workshop

展览场地：
深圳市市民中心
展出时间：
2011年12月8日-
2012年2月18日
展览策划：
米丘现代艺术工作室

CONCRETE & POSSIBILITY / 混凝土的可能

As the pacemaker in the process of residential industrialization, Vanke always keeps contributing strength to research and develop sustainable products and ideas. Possibilities of Concrete, which is planned by Vanke and INOUT magazine of Lemon Media, offers chances to designers to show their talents. The design event encourages designers to explore different and amazing possibilities of using concrete. Their excellent creations with concrete give us a fresh view of the connection of architecture, art, people and society.

The design event will keep inviting known designers from the areas of architecture, interior design, products design and graphic design. With all their wisdom, the design event will achieve more ideas and more creations.

万科作为全国住宅产业化的领跑者，一直致力于可持续发展理念的产品研发和深入实践。由万科集团建筑研究中心和柠檬传媒——《里外》杂志共同发起的"混凝土的可能"设计计划，拟为设计师提供充分的创作平台，发掘清水混凝土的N种可能，展现建筑、艺术、人群与社会的全新关系。

此计划将持续邀请建筑、室内、产品、平面等领域中具影响力的设计师参与，寄望多角度地思考和实践，创造更多的可能。

Venue:
Shenzhen Vanke Center
Duration:
Dec. 9th, 2011 – Jan. 9th, 2012
Curator:
IN·OUT GROUP, RITO
Organizer:
Vanke Construction Research Center, IN·OUT Magazine
Undertaker:
Shenzhen Public Art Center

地点：
深圳市盐田区万科中心
时间：
2011年12月9日 - 2012年1月9日
策划：
朗图——里外团队
主办：
万科集团建筑研究中心、柠檬传媒——《里外》杂志
承办：
深圳市公共艺术中心

SATELLITE EXHIBITIONS AND EVENTS / 外围展及活动
DETERRITORIALIZATION & EMERGING CENTRALITIES FOR PRD / 珠三角城市化形体研究

Traditionally the streets as the city's main communication space has been sentenced to death in motorization and urbanization, the streets became pure traffic infrastructure; The urban is sprawling, the strategy points that benefit the system of fast track become the new centrality, the social interactions and activities are polarized in those emerging centralities, they are the substitute of traditional communication places. The Pearl River Delta city, especially Guangzhou and Shenzhen, the feature is highlighted particularly.

Metaphorically, La Ville des Lieux (centrality) acting like a vitality oasis floating in La Ville des Liens (Deterritoriazation). This exhibition attempts to use an installation and a movie to probe into the Pearl River Delta urbanization model: a scene of a city spread and social intercourse space polarization coexists, let the audience experience and introspection on the current city mode.

传统街道作为城市的主要交往空间，在机动化及城市化中被宣判死刑，街道变成纯粹的交通空间；另一方面，进入快速交通系统的点成为城市蔓延过程中的中心点，大量社会交往被极化在这些中心点上，例如大型购物中心，凸显中心性成为传统交往空间的形态。珠三角城市，特别是广州及深圳，该特征尤为明显。

隐喻上，场所之城（中心性场所）像一个充满生机的绿洲漂浮于联系之城（快速交通空间）。本展览试图通过一个装置及影像探讨珠三角城市化模式：一个城市蔓延及社会交往空间极化并存的景象，让观众体验及反思当前的城市模式。

Venues:
A Building, OCT Ecology Plaza, Nanshan District, Shenzhen
Duration:
Dec. 8th, 2011 – Feb. 8th, 2012
Curator:
Linshou Wu & Xiangying Zhao
Organizer:
WAU Studio

地点：
深圳华侨城生态广场 A 座首层
时间：
2011 年 12 月 8 日 -
2012 年 2 月 18 日
策划：
吴林寿，赵向莹
主办：
深圳市原点上建筑设计有限公司

ECOTOPIA: URBAN DESIGN & ART WORK -PERSONAL EXHIBITION OF YE CHENG / 《生态乌托邦》 城市设计 & 装置艺术 – 野城个人作品展

Eco-architecture, a curative architecture

The eco-architectures should be curative, a kind of living machine. As an important link of urban ecosystem, they will be inserted into cities, as if transplanting new artificial organs to a withering ill body. ECOTOPIA brings the eco-agriculture into the city as a treatment measure, in another words, to insert the eco-agricultural system into the urban ecosystem as a restoration therapy in order to accelerate the system's recycling and regeneration with the help of eco-technologies so as to restore the unbalanced ecosystem.

Urban Farmer, Restructure of Society Frame

This restoration system consists of a series of sky-farms which are not isolated from each other but are making up an "eco-tribe" which perfects the urban biological chain and supplying predominant food. It results in new social type called "Urban farmer". The community residents can employ a certain number of "Urban farmers". A new rationing and consumption system will come into being based on the eco-community – "The collective ownership of the urban agricultural community", producing independently and self-sufficiently in the community. Indeed, the community residents must assume their responsibility for the agricultural product quality, attaching great importance to its supervision and management.

Civil Society

ECOTOPIA aims to narrow the gap between rich and poor, eliminates the segregation between people, enhances the exchange and understanding among the residents, endows the citizens with more participation rights and thus promotes the democratization reform. It is a reform to the existing mode of urban communities as well as a new approach to revamp the community without large-scale building demolition and resident moving.

生态建筑，一种治疗性建筑

生态建筑的本质是治疗性，犹如一种活的机器，作为城市生态系统的重要环节，被植入城市内部，如同给生病的躯体更换新的人造器官。"生态乌托邦"把生态农业作为一种治疗方法引入城市内部，生态农业系统作为一个修复性环节被植入城市生态系统，并利用生态技术加速系统的循环再生，修复失衡的生态系统。

城市农民，社会结构重组

该修复系统由一系列垂直农场组成，并在空中相互连接形成生态群落，完善了城市生物链和食物供给循环。一个新的社会阶层"城市农民"由此产生。部分社区居民被邀请成为城市农民，负责农业生产，控制食品质量，并通过"城市农业社区集体所有制"这一新的消费配给制度进行社区内相对独立而自给自足的生产管理和分配。

公民化社会

"生态乌托邦"旨在降低贫富差距，减少人的隔阂，增进居民的交流，赋予市民更多社会参与权，促进民主化改革。这也提供了一条避免老社区大面积拆除搬迁的新道路。

Venues:
101, Unit A, Shenzhen OCT Ecological Square, Shenzhen

Duration:
Dec. 8th, 2011 – Feb. 18th, 2012

Curator:
Ye Cheng

地点：
深圳市华侨城生态广场 A 座 101

时间：
2011 年 12 月 8 日 – 2012 年 2 月 18 日

策划：
野城

SATELLITE EXHIBITIONS AND EVENTS / 外围展及活动
City, People, Nature: Mangrove Eco-Arts Festival / 城市·人·自然: 红树林生态艺术节

The Mangrove Eco-Arts Festival is a carring environmentally friendly and humane care of the eco-show event. The exhibition takes "City·People·Nature" as the theme, the main form is eco-art show and concept spread, based on strengthening the city of Shenzhen, the local ecological problems of research and discussion, continuation and strengthening the idea of " City creates" direction, extracting a kind of slow living lifestyle, continuing to create a green eco-homes.

契合本届深港双城双年展主题精神举办的双年展外围展暨红树林生态艺术节，是一场承载环保理念与人文关怀的生态展示盛会。该展览以"城市·人·自然"为主题，以生态艺术品展示与理念传播为主要形式，立足于加强对深圳本土城市生态问题的研究和讨论，延续和强化"城市创造"这一主题方向，提炼出一种慢生活的生活方式，持续创造绿色的生态家园。

Venue:
Shenzhen Bay Park, Mangrove, Shenzhen
Duration:
Dec. 8th, 2011- Feb. 10th, 2012

地点：
深圳市红树林深圳湾公园
时间：
2011年12月8日-
2012年2月10日

FUTURE RELEVANCE: THE 6TH OCAT INTERNATIONAL ART RESIDENCY / 明天，谁说了算？—第六期 OCAT 国际艺术工作室汇报展

'Future Relevance' is an exhibition of four international artists from Europe. The Sixth OCAT International Residency Programme has selected artists Thomas Adebahr (DE), Frank Havermans (NL) and Nika Oblak & Primoz Novak (SI) and curator Paula Orrell (UK). This exhibition includes the residual ideas developed on the residency.

Contemporary art now illuminates the present and explores the world around us. Through observation, research and process, the artists consider context and histories, both real and imagined. The exhibition title 'Future Relevance' explores the potential, to look to the past to inform the future, understanding what are the conditions we need to establish to have a global discussion about the future of contemporary art.

"明天，谁说了算？"是第六期 OCAT 国际艺术工作室交流计划的汇报展，本期交流计划邀请了来自德国的 Thomas Adebahr，来自荷兰的 Frank Havermans，来自斯洛文尼亚的艺术家组合 Nika Oblak & Primoz Novak 以及来自英国的策划人 Paula Orrell。

当代艺术丰富了这个世界，同时也在不断探寻着这个世界。艺术家们通过观察和研究，来探查现实、想象中的上下语境和历史遗存。这个展览提出了这样的问题：如果我们要一起谈当代艺术的国际未来，那么我们必须建设什么样的条件才能够跨过不同文化和历史的差异？

Venue:
Northeast of B10, Northern OCT-LOFT, Shenzhen, Nanshan District, Shenzhen, China
Duration:
Nov. 11th, 2011 - Jan. 8th, 2012
Curator:
Paula Orrell (UK)
Organizer:
OCT Contemporary Art Terminal of He Xiangning Art Museum

地点：
深圳南山区华侨城创意园北区 B10 栋内一楼东北侧展厅
时间：
2011 年 11 月 11 日 - 2012 年 1 月 8 日
策划：
Paula Orrell（英国）
主办：
何香凝美术馆 OCT 当代艺术中心

Bi-City Biennale:
Conveying the Urban Creativity Spirit

On the evening of 19 February 2012, 2011 Shenzhen & Hong Kong Bi-city Biennale of Urbanism Architecture was closed, and it has been taken two years of thoughtful planning and preparing and achieved a very good performance.

The Biennale brought a new international cultural feast for the city with support from various areas. This will be a sound foundation for the "Bi-city Biennale" to earn her fame in the international Biennale circle. We invited renowned curator Mr. Terence Riley as the chief curator, the of exhibition's theme is "Architecture creates cities. Cities create architecture", more than 60 project included in the show, all of them are excellent work which studied city issues with global view, among them were the Netherlands, Austria, Chile, Finland, Bahrain national Museum project. In addition, the Biennale has also successfully implemented 30 Academic Forums, "Shenzhen and Universiade" Special Exhibition, 15 Satellite Exhibitions as well as street theater and many other public activities, greatly enriched the exhibition content. Interaction and display during the past three months, the Biennale has attracted about 150,000 spectators attending the event, its strong architectural feature and highly professional and international performance has won acclaim in and out the industry.

With the closing of the Shenzhen Biennale, plenty of ideas and phenomena have found extension from it, which still worth our sustained attention. The function of Shenzhen Biennale in activating urban public space has got better chances of expression. At its opening ceremony, the 10,000-flower maze on Citizen's Plaza, together with lighting, music and skating performances, has greatly stunned the audience. This project took advantage of thousand orange safety cones to vivid the once empty and quiet Government Plaza, and it welcomed the public to enter for personal experience and re-creation. And safety cones for the work 10,000 Flower Maze in the opening became the nice entertainment for the citizens, the recreated new shapes which enlarge and promote the value of public art. This year's exhibition chose the vacant and ready-to-transition OCT industrial plants as its site – the same as the first Biennale in 2005 and second one in 2007. From OCAT to B10, the Biennale continuously activated the upgrading and transformation of the industrial zone, altering the former old plant into a colorful OCT LOFT where is bristled with designs and creations and flourishes cultural and artistic activities. As the closing ceremony speech delivered by the executive vice Mayor Mr. Lv has concluded, "Being a significant public and civic event, the Biennale plays a prominent role in shaping the brand of our city and enhance its charm." An attentive eye will discover that the current exhibition site, i.e. the Citizen's Plaza (including underground exhibition halls) and the OCT LOFT (the main and vice halls in southern district; B10 in north district), happens to be the comprehensive presentation of three previous exhibition places.

As Biennale itself has always been a platform to witness the growing of cutting-edge architects, the chief curator of this biennale organized "The Street" & "Ultra-light Village" to pay tribute to the first Venice Architecture Biennale in 1980. In these two projects, all the 18 architects are recommended by critics and curators all over the world, thus Shenzhen Biennale became a stage for the collective show of cutting-edge architects. Among them accounted to one-third are Chinese Architects, their performance were eye-catching, also they acted on behalf of Chinese architects, now they play under the same rule, in the same arena with their international colleague, without bias or ism, compete on an equal footing. The housing problem, especially the three thematic exhibitions on affordable housing / social housing, including "10 Million Units Housing an Affordable City", "Housing with a Mission Dutch and Chinese architects' designs for the ants tribe, The Netherlands" and "Housing in Vienna. Innovative, Social and Ecological" were one of the highlights the Biennale with much public concern. In addition to the possible strategies for a variety of design innovation, it is also very important that affordable housing topics convey to a new residence values.

There were around 30 academic forums were held during the opening, international experts and scholars of urbanism and architecture mentioned that Shenzhen as new urban in the discussions of international context. Besides, they criticized Shenzhen's urban development from the points of environment protection, social anthropology, meanwhile, no one denied that Shenzhen owns the young, dynamic and flexible creativity. In the last forum, "Shenzhen: became a city... future metropolis of the world", all the experts expressed their expectation of Shenzhen. Prof. Lee Ou Fan said, "Shenzhen is very energetic, when I come to Shenzhen, it feels like the west of America in the 19th century, both the spirit and the hard & soft power." And Michelle Provoost, (International New Town Institute) said, "We can say, Shenzhen is the most successful new town in the world so far. It's not the nicest, but its energy impressed us. It's a place which is full of experiments, and you're free to do quite a lot of things. You can venture, experiment your hypothesis." Then how to explain the normal failure and the exception of Shenzhen? Michelle thought "the urban planning of Shenzhen might different from the expectation 30 years ago." Her sensible judgement has expressed the un-planning, informal which support the whole developing process of Shenzhen. In such informal area which requires more investigation.

Besides, the curatorial structure of 2011 biennial which innovated Invitational Exhibition, there are Housing in Vienna, Innovative, Social and Ecological (Austria), Gimme Shelter! (Chile), Newly Drawn – Emerging Finish architects (Finland), Solution Finland: The Welfare Game (Solution 239-246), and Reclaim, by the Kingdom of Bahrain. These 5 nation pavilions have brought the biennale to another phase of international communication (total 6 countries, the Switzerland had joined the 2009 Shenzhen biennale, "Arch/Scapes: Swiss Architecture"). And it has attracted more consulates and cultural institutes, such as the consulate general of Belgian and promotion institute WBA would love to join the biennale in 2013. We all witness the increasing attraction and cultural influence of Shenzhen. And those international cities brought their experience here, and we also hope the biennale also can present Shenzhen experience to the world.

For the interplay between Shenzhen and Hong Kong, the projects Counterpart Cities: Climate Change and Co-Operative Action in Hong Kong and Shenzhen, 100 Million Units: Housing an Affordable City, and satellites exhibitions Coaster Raid: Shenzhen is Not a Tomato Field and Qianhai Workshop represent the cooperation between SZ and HK, housing in Shenzhen, and research on present practice and about future, they may move to the Kowloon Park in Hong Kong as Shenzhen Pavilion. The biennial in HK opened on Feb. 15th 2012, and will last until Apr. 22nd, 2012, chief curators Gene K. KING and Anderson LEE would closely cooperate with Shenzhen, and the Hong Kong & Shenzhen Bi-City Biennale of Urbanism \ Architecture will be a consentaneous one with Shenzhen part. During the presentation for 3 months, the biennale received many audiences from different fields, meanwhile, feedback and suggestion from the mass and experts. Mayor Xu Qin visited the exhibition and said, "The special area should take the leading position which means we take the opportunity and own the ability of competence. On this platform, we should think about how to transform, how to practice the scientific development, how to obtain new comparative ability, how to tend towards modernization and international."

Vice mayor of Shenzhen, Mr. Lv Ruifeng was invited giving a speech in the closing ceremony, he pointed out that, "Bi-city Biennale is aiming for urban developing and regional cooperation, which has created a new interactive model of cultural and urban development. And it has taken significant position with its unique expression and luxuriant content, from which we can observe the latest cultural vision and expressive modes. The event has pulled and promoted the urban creative industry, activated the public space, even drove the transformation of certain regions. As an important public event and urban issue, SZHKB contributes a lot for the city brand and charming." In the end, Mr. Lv Ruifeng express the appreciation to all the sponsors, media friends, volunteers and our citizens who participated in and showed great support for the biennale.

2011 Shenzhen & Hong Kong Bi-City Biennale of Urbanism\ Architecture also awards some projects as before, one Organizer Committee Award nominated by curator, chose by the committee; 3 Academic Committee Awards, by the Academic Committee; 4 Public Choice Awards: voted by the audience. Mayor Xu Qin has awarded the projects.

There were over 70 academic forums and 360 works for the 4 biennales since it established, the urban spirit within have widened our view and transformed our knowledge on urban and architecture, which close the notion and practice. As the citizens and media suggested, some of the works would stay in Shenzhen longer, in the plazas, parks, public architecture and art institutes, and the works and forums would be published, which would be transformed into the culture and urban spirit of Shenzhen.

As Mr. Lvs Ruifeng mentioned in his speech, he appraised such culture development strategy and modes, the government would further more support the biennale in the spirit of urban development, and we hope the whole society could join us and support the biennale to promote it as a namable international brand, a platform of cooperation for Shenzhen & Hong Kong, and a powerful engine of Shenzhen development. Since the 2009 biennale, Shenzhen Public Center is the standing institution for running the biennial brand in charge of the function of the event. With the assisting of members the organizing committee, they would evaluate the brand by social investigation, and they would try to cooperate with more local enterprise, cultural institution and consulate all around the world. They would concentrate on long-term development and has attracted a hundred domestic and international, mass media and professional media's reports. Meanwhile, the biennale's creativity has attracted social new power to join in our volunteer team. They're creative and active, they are the hope and future, they're the link between the urban creativity spirit and the citizens.

Shenzhen Biennale of Urbanism\Architecture Organizing Committee

EPILOGUE / 跋语

双城双年展，
持递城市创造精神

历经两年的精心筹划和近 3 个月的精彩展示，2011 深圳·香港城市\建筑双城双年展于 2012 年 2 月 19 日落下帷幕。

在社会各界的鼎力支持下，本届双城双年展给深圳带来了一场全新的国际文化盛宴，为"双城双年展"品牌迈向国际双年展舞台奠定了良好的基础。本届展览邀请国际著名策展人泰伦斯·瑞莱先生担任总策展人，以"城市创造"为主题，策划了 60 件以全球视角研究城市问题的优秀作品，其中包括荷兰、奥地利、智利、芬兰、巴林的 5 个国家馆项目。除此之外，双年展还成功实施了约 30 场学术论坛、大运与城市专题展、15 个外围展以及街道剧场等多场公众活动，极大地丰富了展览内容。通过近 3 个月的互动和展示，本届双年展吸引了约 15 万观众到场参观，其强烈的专业特点及国际化程度赢得了业内外的一致好评。

随着双年展降下帷幕，诸多通过展览延伸出来的观点和现象仍值得我们持续关注。深圳双年展对城市公共空间的激活功能在本届得到了更大的体现。市民广场上的《万花阵》在开幕式上结合灯光、音乐及轮滑表演惊艳全场，该项目利用数千个橙色的安全锥激活了市民广场，并欢迎市民进入其中体验，进行再创造。《万花阵》的上千个雪糕筒在展览期间成了市民在市民广场上最喜爱的玩具，他们重新创作了各种图案，延续和提高了这件广场公共艺术的公共价值，令人印象深刻。本届展览又回到了首届双年展及第二届双年展主场馆，持续激活着产业升级转型的工业区，将区域内的旧厂房催生转化为设计和创意集中地、文化与艺术活动精彩纷呈的华侨城创意文化园。吕锐锋常务副市长在闭幕式致辞中总结道"双年展作为重要的城市公共活动和城市事件，对塑造城市品牌、提升城市魅力发挥了突出作用。"细心的观众会发现，本届双年展所选择的展场——市民广场（含地下室内展厅）及华侨城创意文化园（南区 OCAT 主副厅、北区 B10），恰好是前三届展览场地的综合呈现。

双年展历来是见证新锐建筑师成长的平台，本届总策展人通过《街道》、《超轻村》两个项目，向 1980 年的首届威尼斯建筑双年展致敬。通过全球评论家与策展人的推荐而来的两个项目的 18 位建筑师，共同筑起深圳双年展这个国际新锐建筑师集中亮相的舞台。这中间约占三分之一的中国建筑师的不遑多让的醒目表现，也代表着中国建筑师在同一国际舞台和游戏规则下的不带偏向或主义的、平起平坐式的崛起。而居住问题，特别是关于保障房 / 社会住宅的三个专题展《广厦千万·居者之城》、《住宅的使命——荷兰与中国建筑师为蚁族而设计》、《维也纳住房——革新的，社会的以及生态的》，是本届双年展备受公共关注的亮点之一，除了各种设计创新的可能策略，保障房专题最重要的是着重传达了一种建设热潮中的冷静思考、更加可持续的居住价值观和人居关怀。

来自全球各地城市与建筑研究设计领域的专家、学者、评论家共聚一堂，在近 30 场学术论坛中间关于城市讨论的国际语境下，深圳作为典型的新城被频频提及。专业嘉宾们除了从城市非人尺度、环境保护、社会人类学等角度对深圳的城市发展进行了评论外，对深圳的年轻活力和灵活的创意能力都给予了普遍认可。如在开幕周最后一场研讨会"深圳，成为一座城，未来的世界都会"上，嘉宾演讲者们纷纷表达了对深圳未来的期许。文化学者李欧梵在研讨会上这样评价："我每次到深圳来，我就感觉到深圳在各方面就像 19 世纪美国的西部一样，西部的猛冲的精神，牛仔精神，包括硬体和软体的发展。"与会另一嘉宾 Michelle Provoost（INTI International New Town Institute 国际新城研究院总监）也高度评价深圳："我们可以这么说，深圳是世界上最成功的一个新城……它可能并不是最美丽的，但是很有活力。另外这也是一个充满试验的地方，给予你很多的自由，你可以尝试做很多事情。这个城市里面，你可以冒风险，验证你的假说。"那么如何解释新城城市规划普遍的失败和深圳的某种例外成功呢？专家讨论和部分展览作品指向了深圳非规划的、自发的（Informal）的力量对深圳规划的修正、对快速发展起到的支持。这一巨大的自发的领域，还有待更多的专业工作者来挖掘。

此外，本届新增的"深圳邀请展"荟萃了奥地利的《维也纳住房——革新的、社会的以及生态的》，荷兰的《住宅的使命——荷兰语中国建筑师为蚁族而设计》、智利的《给我避难所！》、芬兰的《新绘图——芬兰新生代建筑师》、《芬兰方案：福利博弈（解决方案 239-246）》、巴林王国的《再生》五国国家馆，令双年展的国际文化交流活动步入新的台阶。这两届已有 6 个国家（包括 2009 届来自瑞士的《建筑风景：瑞士建筑展》）主动参加了我们的展览，此种国家馆的形式成功吸引了更多的国家有意向参加今后的双年展，如比利时王国驻广州总领馆总领事达乐文先生及建筑推广机构 WBA 已与组委会接洽，明确提出参加 2013 年双年展国家馆的合作意向。这充分显示了深圳双年展不断增强的国际吸引力和文化影响力，成功起到了把国际城市经验带到深圳，也把深圳带进全球视野的交流平台作用。

本届双城双年展，在深港互动方面，深圳展中《对应双城：香港与深圳的气候变迁及合作行动》、《广厦千万·居者之城》及《寻找深圳——创意的都市实践》、《前海工作坊》这四个代表着深港研究合作、深圳住房领域研究、当下实践和未来展望的专题展览，获邀移师到香港九龙公园内香港展的"深圳馆"展出。本届双城双年展香港展已于2012年2月15日开幕，展期至4月22日，总策展人金光裕与李亮聪先生与深圳展紧密合作，共同创造了一个非常统一的双城双年展。

在近3个月的展期中，双年展展场迎来了来自社会各界的观众，接收到了来自大众与专业人士的回馈与建议。许勤市长在参观双年展后发表讲话："双年展这一创新展览成功激活展场地区，是促进产业转型的成功案例。这充分说明，任何一件事，特区都得走在全国的前面。走在前面，就意味着我们抓住先机，就意味着我们更早地构建竞争力。大家应该从双年展这个平台上思考更多的问题，更多地思考如何转型、如何落实科学发展观、如何构建新的竞争力、如何使我们的城市走向现代化、国际化。"此外，来自南京市委、江苏省住建厅、澳门文化局、香港建筑师学会、香港规划署等的重要客人参观了双年展，均给予了较高的评价。闭幕式邀请到吕锐锋常务副市长致辞，他高屋建瓴地指出："双年展以文化发展促进城市发展和区域合作，创造了一种文化发展与城市发展互动的新模式。双年展独特的表现形式和丰富的文化内涵已经成为深港两地文化发展的重要内容，充分展示了当代最新的文化发展动态和艺术表现形式，极大地带动和促进了城市文化创意产业的发展，促进了产业升级，激活了城市公共空间，带动了城市中一些特定区域的发展转型。"最后，吕锐锋常务副市长代表深圳市委、市政府，向参与、支持和关注双年展的赞助机构、新闻媒体、双年展志愿者和市民朋友表示了衷心的感谢！

秉承往届传统，本届针对参展作品设置了组委会奖（共1名）：由策展人提名，组委会选定，学术委员会奖（共3名）：由学术委员会评定，公众奖（共4名）：由公众投票决定，并在展览闭幕式上由许勤市长主礼颁发各个双年展奖项。

四届双年展的成功举办，在深圳展示了总计360多件来自全球的优秀作品，举办了70多场论坛。这些作品以及论坛中的城市思想，开阔了我们的视野，丰富了城市的文化内涵，也潜移默化地改变着我们关于城市与建筑的认识，拉近了深圳与全球城市先进观念与实践的距离。应公众及媒体的不断呼吁和要求，部分作品将不会随着展览结束而消失，而是作为深圳的文化艺术财富，继续留在深圳的广场、公园、公共建筑大厅以及艺术机构的收藏空间中展出，展览作品及论坛思想也会进一步整理出版，使得双年展这场文化盛宴能被更多地分享、体验和消化，转化为我们深圳的文化基因和城市的精神财富。

正如吕锐锋常务副市长在闭幕式致辞中高度肯定的双年展的文化发展策略及模式，市委市政府将从城市发展战略的高度，进一步加大对双年展的扶持力度，也希望社会各界积极参与支持和赞助双年展，共同将其打造成为知名的国际文化品牌，使之成为深港城市合作的重要平台和深圳城市发展的重要引擎。为了长期经营"双城双年展"的品牌，自上届始深圳市公共艺术中心作为常设机构，统筹执行双年展的机制运作，在组委会各个成员单位的协助下，通过社会调查来考察评估双年展品牌，尝试与更多本土企业、文化机构及各国领事机构等建立合作关系，来探索长期发展的道路，并成功吸引了上百家国内外大众及专业媒体对双年展进行了大量深度专题报道。同时，双年展的锐意创新特性也吸引了一批热心社会公益事业的新生力量加入到双年展志愿者的队伍中。他们充满创意、永不停滞，代表着城市希望与未来，将双年展所倡导的城市创造精神传递到在城市生活的广大人民中，使其共同参与到城市建设中来。

深圳城市\建筑双年展组织委员会

AWARDS LIST / 获奖名单

Organizer Committee Award / 组委会奖（共1名）：

SZ Invitational: Netherlands Architecture Institute
"Housing with a Mission, Dutch and Chinese Architects' Designs for the Ants Tribe"
Curator: Ole Bouman, Co-curator: Korn Konijn.
深圳邀请展之荷兰馆《住宅的使命——荷兰与中国建筑师为蚁族而设计》
策展人：Ole Bouman 联合策展：Jorn Konijn

Academic Committee Awards / 学术委员会奖（共3名）：

Counterpart Cities: Climate Change and Co-Operative Action in Hong Kong and Shenzhen,
Curators: Jonathan Soloman, Dorothy Tang
对应双城—气候变化 & 香港和深圳的联合行动
策展人：Jonathan Solomon，邓信惠

Rebirth Brick Development 2011
Curator: Liu Jiakun
再生砖进展 2011
策展人：刘家琨

Special Plaza Project: 10,000 Flower Maze
Designers: John Bennett, Gustavo Boneva
特别广场项目：万花阵
设计师：John Bennett 与 Gustavo Bonevardi

Audience Awards / 公众奖（共4名）：

A Catalyst Reaction of Our City:
Shenzhen and Universide Special Exhibition
城市触媒—大运与深圳专题展

Regional Construction Studio (College of Architecture Hunan University),
YanPaiXi Village
岩排溪村　地方营造工作室（湖南大学建筑学院）

Clavel Arquitectos (Murcia, Spain),
Centrifugal Village
离心村　Clavel Arquitectos（西班牙穆西亚）

Gimme Shelter!
Chilean Davillion
给我避难所！　智利国家馆

At the Closing Ceremony,
Xu Qin, Deputy Secretary of Leading Party Committee of Shenzhen Municipal
Government, Mayor of the Shenzhen, awarded the winning projects
闭幕式上，深圳市委副书记、市政府市长许勤为获奖项目颁奖

ABOUT TERENCE RILEY / 主策展人泰伦斯·瑞莱简介

Terence Riley is an internationally – recognized leader in the design and development of cultural facilities and programs with great architectural significance worldwide.

He has played a lead role in the architect selection and design processes for the renovation and expansion of the Museum of Modern Art (New York), the Miami Art Museum, and the Museum of Art, Design and the Environment (Murcia, Spain). In 1991, Riley joined the curatorial staff of the Museum of Modern Art, New York (MoMA). In 2002 he was given the title of Philip Johnson Chief Curator for Architecture and Design, in recognition of his accomplishments over 10 years.

In addition, Riley has served on international juries for numerous important projects, including the Reina Sofia National Museum (Madrid), the 9/11 Memorial at the Pentagon (Washington D.C), and the Museum of Latin American Art (Buenos Aires). He has also served as an advisor to both the public and private sectors with regards to such projects as the Taiwan Tourism Infrastructure Development, the World Trade Center site redevelopment in New York City, and the West Kowloon Cultural District in Hong Kong. During his tenure at MoMA, Mr. Riley played a key role in the successful development and launch of MoMA's expanded and renovated facility, which opened to international acclaim in 2004. He was also responsible for the design and installation of the new Architecture and Design Galleries, housing the pre-eminent collection of its kind in the world.

As Chief Curator at MoMA he organized exhibitions of well-known figures, including Rem Koolhaas and Bernard Tschumi, as well as presenting emerging voices. In addition to international architectural surveys, such as Light Construction, Tall Buildings and On-Site: New Architecture in Spain, Riley organized two major scholarly retrospectives: Frank Lloyd Wright: Architect and Mies in Berlin.

Riley joined the Miami Art Museum (MAM) as its director in March 2006. In this role, Mr. Riley led the Museum through the design phase of a major expansion. He lead the museum's Architect Selection Committee, which selected Herzog & de Meuron to design an innovative new waterfront home in Miami's Museum Park. Subsequently, he oversaw a coordinated design process that involved architects, engineers and consultants in Miami, Basel, New York, London and Frankfurt.

Riley studied architecture at the University of Notre Dame and Columbia University and is a licensed architect, certified nationally by the National Council of Architectural Registration Boards. Riley is a founding partner of K/R (Keenen/Riley, 1984), an architectural studio well – known for its designs for art museums, galleries, artists and collectors. An acclaimed author and contributor to journals and other publications on design, he lectures frequently and has taught at the Harvard Graduate School of Design and at numerous other architecture schools. K/R is currently designing a Master Plan for a 100-acre sculpture park in Murcia, Spain and advising on the development of a Museum District in Hangzhou, China.

泰伦斯·瑞莱，美国人，生于 1954 年。作为国际著名建筑主题展策展人、建筑师、博物馆专家、教授及评论家，瑞莱先生是享誉国际的文化艺术领袖。

瑞莱先生拥有丰富的国际策展经验，他历年来多次为世界权威双年展——威尼斯建筑双年展担任评委会主席及策展人。1991 年，瑞莱受邀加入纽约现代艺术博物馆（MoMA, The Museum of Modern Art, New York）的策展队伍。在之后的 14 年间，瑞莱于 MoMA 先后就任菲利普·约翰逊建筑设计部总监、高级策展人，并于 2002 年被授予建筑及设计总策展人的称号。其间，瑞莱成功完成了由设计到执行指导备受瞩目的 MoMA 新馆革新扩张计划。

在担任纽约现代艺术博物馆总策展人期间，瑞莱主要关注当代建筑、设计趋势以及大众所关注的全球问题。他策划了许多知名建筑师的展览，包括 Rem Koolhaas、Bernard Tschumi 以及大量建筑界后起之秀。除了大量开展国际性建筑调研，如 Light Construction, Tall Buildings and On-Site: New Architecture in Spain，他还组织了两项重大的学术回顾研究：Frank Lloyd Wright: Architect 是全面总结设计师 Frank Lloyd Wright 64 年职业生涯的建筑回顾展；Mies in Berlin 则是首个深入研究 Ludwig Mies van der Rohe 早期作品的项目。瑞莱策划的展览一直以深切洞悉博物馆历史著称。

他在纽约现代艺术博物馆任职期间，展馆设计收藏一直稳步增加，包括时间跨度为从 18 世纪到 20 世纪不等的著名设计物件、建筑设计图纸以及来自世界各地的模型。

随后，于 2006 年 3 月，瑞莱加入了迈阿密美术馆（MAM, Miami Art Museum）并担任馆长职务。在其带领下，博物馆经历了展馆的设计、扩张、改革阶段。他领导建筑甄选组委会，为迈阿密博物馆公园挑选出 Herzog & de Meuron 所设计的极具创意的湖畔小屋。随后，他督导了融合建筑师、工程师及来自迈阿密、巴塞尔、纽约、伦敦和法兰克福多个地区顾问的综合设计项目。此项目获得了美国博物馆理事会及迈阿密政府的一致认可，且现阶段已获得 1 亿美元公共基金资助，并于 2010 年正式动工。瑞莱已于 2010 年离任 MAM。

另外，在西班牙艺术设计及环境博物馆的改建和扩张项目中，瑞莱担任过建筑作品甄选以及整体设计过程中不可或缺的领导性角色。

瑞莱毕业于美国圣母大学及哥伦比亚大学建筑学系，是美国注册建筑师。作为 K/R(Keenen/Riley, 1984) 的创始人之一，他建立了知名的建筑工作坊，专为美术博物馆、画廊、艺术家及收藏家设计建筑作品。瑞莱曾参与多项国际重要项目，包括西班牙马德里瑞内索菲亚美术馆、9·11 五角大楼纪念碑和布宜诺斯艾利斯拉丁美洲艺术博物馆的设计及建设工作。同时，以顾问的身份，他参与了多项公共及私人项目，包括台湾旅游基础设施建设、纽约世贸中心重建及香港西九龙文化区建设。另外，作为大获评论界赞颂的作家及撰稿人，瑞莱还经常于著名建筑学院哈佛设计研究所等学府授课。瑞莱先生于 2010 年通过深港双城双年展全球策展方案的甄选，被选定为 2011 第四届深港双城双年展的总策展人。

SHENZHEN BIENNALE OF URBANISM\ARCHITECTURE ORGANIZING COMMITTEE

Director: Lv Ruifeng (Vice-Mayor of Shenzhen Municipal Government)
Vice Director: Wang Jingsheng (Director of Shenzhen Municipal Bureau of Propaganda)
Wang Peng (Director of Shenzhen Municipal Planning and Land Resource Commission)
Chen Wei (Director of Shenzhen Municipal Bureau of Culture, Sport and Tourism)
Secretary General: Xue Feng (Vice Director of Shenzhen Municipal Planning and Land Resource Commission)
Chen Xinliang (Vice-Director of Shenzhen Municipal Bureau of Culture, Sport and Tourism)

SHENZHEN BIENNALE OF URBANISM\ARCHITECTURE ORGANIZER COMMITTEE OFFICE (2011-2012)

Executive Director: Huang Weiwen (Director, Shenzhen Biennale Organizers Committee/ Shenzhen Center for Design)
Director of Committee Office: Sun Zhenhua (Art Director of Shenzhen Public Art Center)
Project Director: Liang Tian
Branding & Resource Management: Dong Chaomei, Liang Jiehua, Zhou Tianqi, Fan Zhen (Internship), Hu Ting (Internship)
Exhibition Coordination: Mi Lan, Huang Zihui, Mei Zhen, Liu He (Internship)
Administration: Chen Yuling, Jessica Sun, Li Wen, Zhang Lu, Xie Kun
Finance: Jiang Peng, Xie Tian, Yang Li, Chen Bin
Opening Events Coordinator: Shenzhen Center for Design
Web Support: Bai Xiaoci, Chenjia

Shenzhen Biennale of Urbanism\Architecture Organizing Committee Associated Members (2011-2012)

Finance Commission of Shenzhen Municipality | Urban Planning, Land and Resources Commission of Shenzhen Municipality | Transport Commission of Shenzhen Municipality | Health, Population and Family Planning Commission of Shenzhen Municipality | Education Bureau of Shenzhen Municipality | Public Security Bureau of Shenzhen Municipality | Civil Affairs Bureau of Shenzhen Municipality | Culture, Sports and Tourism Administration of Shenzhen Municipality | Housing and Construction Bureau of Shenzhen Municipality | Local Taxation Bureau of Shenzhen Municipality | Market Supervision Administration of Shenzhen Municipality | Urban Management Bureau of Shenzhen Municipality | Port Office of Shenzhen Municipal People's Government | Foreign Affairs Office of Shenzhen Municipal People's Government | Taiwan Affairs Office of Shenzhen Municipal People's Government | Shenzhen Public Security Bureau | Fire Services Department | Traffic Police Department of Shenzhen Public Security Bureau | Government Offices Administration of Shenzhen Municipality | Shenzhen Publicity Ministry for Culture Foundation Office | Reception Office of Shenzhen Municipality | Shenzhen Municipal Committee of the Communist Youth League of China | Shenzhen Customs District of People's Republic of China | Shenzhen Entry-Exit Inspection and Quarantine Bureau | General Station of Exit & Entry Frontier Inspection | Shenzhen Municipal Literature & Arts Association | Shenzhen Media Group | Shenzhen Press Group | Shenzhen University

Academic Committee Members (the 4th edition)

Curators of previous editions: Yung Ho Chang, Ma Qingyun, Ou Ning
Committee members: Sun Zhenhua, Liu Xiaodu, Lu Hong, Han Jiaying, Bi Xuefeng, Doreen Liu Heng, Feng Yueqiang, Wang Mingxian, Yin Shuangxi, Meng Jianmin, Fenf Yuan, Yang Xiaoyan, Wang Wei Jen, Fu Chin Shing Ivan, Lim Wan Fung Bernard, Yip Alvin, Huang Zhuan, Zhu Rongyuan, Hans Ulrich Obrist, Rem Koolhaas, Steven Holl, Ole Bouman, Toyo Ito

CURATORIAL TEAM

Chief Curator: Terence Riley
Assistant to the Chief Curator: Luna Bernfest
Project Manager: Liu Lei
Assistant Project Manager: Du Liang

Project Curators: Aaron Betsky, Du Juan, Jeffrey Johnson, Jiang Jun, David van der Leer, Li Xiangning, Mary Ann O'Donnell, Jonathan Solomon, Rochelle Steiner, Dorothy Tang, Tang Keyang, Su Yunsheng
Invitation Exhibits Curators: Dietmar Steiner, Cristóbal Molina Baeza, Sebastián Irarrázaval, Hugo Mondragón, Martta Louekari, Sh. Mai Al Khalifa, Noura Al Sayeh, Fuad Al Ansari, Ole Bouman
Special Project Curator: RITO·In&Out & Li Degeng

Participants

Stefan Al, Amateur Architecture Studio, Aranda Lasch, Atelier Deshaus, John Bennett, Gustavo Bonevard, Astrid Bussink, CCDI, Clavel Arquitectos, Coop Himmelb(l)au, David Chipperfield Architects, ELEMENTAL, John Ewing, Fake Industries Architectural Agonism, Fang Zhenning, FCJZ, Field Operations, Scott Fisher, Sam Green, Cao Guimarães, Hashim Sarkis Studios, Michiel Hulshof, J. Mayer H., Johnston Marklee, Jeroen Koolhaas, Liu Jiakun, Ma Qingyun, MAD Architecture, Vincci Mak, Stefano di Martino, Mass Studies, Massimiliano Fuksas Architetto, Carmen Montoya, MOS, NODE, OBRA Architects, OMA, OPEN Architecture, Miki Redelinghuys, Christopher Robbins, Daan Roggeveen, Regional Construction Studio, Shanghai Tongji Urban Planning & Design Institute, Surabhi Sharma, SO-IL, Spbr, Steven Holl Architects, Jennifer Stein, Studio Up, Urbanus, Dre Urhahn, Tom Verebes, Wang Gongxin, Woods Bagot Asia, WSP, Zhubo Design

VOLUNTEERS

Exhibition:
LIN Lin, WANG Long, XU Li, ZHAO Yu, YANG Xinxin, WANG Shaonan, SHI Chunyang, ZHAO Yu, LIANG Feng, SUN Ying, YANG Tianyi, ZHENG Huiyi, DONG Zhijian, XU Liang, BING Yingjiao

Project:
HE Yuxin, LIAN Xin, SU Yahui, LIU Huiqun, ZHANG Chao, LIN Yulin, ZHANG Yafang

Venues: (total working hours over 8 days)
CHEN Zhipeng, WEI Weilin, HU Tingting, LAI Xiaowen, ZHU Anqi, LUO Kaidan, LIAO Zhaobin, ZHANG Yun, HUANG Xiaojun, XING Tao, ZENG Yuanlun, LI Jiaqi, LI Futing, SU Yuening, ZHONG Xiaoyang, PENG Liping, WU Qiuting, LI Jin, ZHOU Meiting, ZHU Bin, LI Yiting, HUANG Lijuan, HUANG Huisi, WU Qianhui, ZENG Xiaoman, ZHEN Meiting, LI Kunming, LIU Huiqun, SITU Dongliang, YU Mingyan, ZHANG Jinyu, HUANG Lei, GUO Peishan

Office:
WU Juntong, FAN Yujian, YANG Jia guo, WANG Yuanfang, Li yang, CHEN Hao, XIONG Jiangchao, LI Qiuna, ZHEN Xiaoxue, YU Lu, CHEN Nan, XU Ningman, LIN Julian, CHEN Huiling

Honored Volunteers:
CHEN Zhipeng, WEI Weiling, HU Tingting, ZHU Anqi, LUO Kaidan, LIAO Zhaobin, ZHANG Yafang, LIN Julian, ZHANG Chao, LIN Ling, WANG Long, XU Li, ZHAO Yu, YANG Xinxin, WANG Shaonan, SI Chunyang, LIANG Feng, SUN Ying, YANG Tianyi, ZHEN Huiyi, DONG Zhijian, XU Liang, BING Yingjiao

EDITORIAL TEAM

Editorial Board: Zhang Yuxing, Zhou Hongmei, Huang Weiwen, Sun Zhenhua
Coordinator: Luna Bernfest, Dong Chaomei
Contributed Editor: Liu Lei, Liu Xiao (Introdution of 6 films of 6<60 are contributed by Alan Zhang)
Print Coordinator: Xie Qiongzhi, Sun Cui
Photographer: Chen Rao, Chen Nan, Miljenko Bernfest, Royce Chen
Translator: He Cong, Gong Linlin, Luo Yuan, Zhang Jing, Yuan Wenshan, Han Li
Graphic Design: wx-design

We apologize for any errors that may have slipped through the net. We would like to also extend our sincerest thanks to the many friends and colleagues who lent us their support.

深圳城市\建筑双年展组织委员会

主任： 吕锐锋（深圳市常务副市长）

副主任：

王京生（深圳市市委宣传部部长）

王芃（深圳市规划和国土资源委员会主任）

陈威（深圳市文体旅游局局长）

秘书长：

薛峰（深圳市规划和国土资源委员会副主任）

陈新亮（深圳市文体旅游局副局长）

深圳城市\建筑双年展组委会办公室（2011-2012 年度）

秘书处执行主任： 黄伟文（深圳市公共艺术中心主任、深圳市城市设计促进中心负责人）

办公室主任： 孙振华（深圳市公共艺术中心艺术总监）

项目总监： 梁田

品牌推广与资源整合部：

董超媚、梁杰华、周天琦、樊真（实习生）、胡婷（实习生）

展务部： 米兰、黄子汇、梅臻、刘赫（实习生）

行政部： 陈玉玲、孙粹、李雯、张路、谢坤

财务部： 江鹏、谢田、杨丽、陈斌

开幕活动协调： 深圳市城市设计促进中心

网站支持： 白小刺，陈嘉

深圳城市\建筑双年展组织委员会成员单位（2011-2012）

深圳市财政委员会，深圳市规划和国土资源委员会，深圳市交通运输委员会，深圳市卫生和人口计划生育委员会，深圳市教育局，深圳市公安局，深圳市民政局，深圳市文体旅游局，深圳市住房和建设局，深圳市地方税务局，深圳市市场监督管理局，深圳市城市管理局，深圳市口岸办公室，深圳市人民政府外事办公室，深圳市人民政府台湾事务办公室，深圳市公安消防局，深圳市公安局交通警察局，深圳市机关事务管理局，深圳市宣传文化事业发展专项基金领导小组办公室，深圳市接待办公室，深圳市团市委，中华人民共和国深圳海关，深圳出入境检验检疫局，深圳出入境边防检查总站，深圳市文学艺术界联合会，深圳市报业集团，深圳市广电集团，深圳大学

学术委员会（第四届）

历届策展人： 张永和、马清运、欧宁

委员： 孙振华、刘晓都、鲁虹、韩家英、毕学锋、刘珩、冯越强、王明贤、殷双喜、孟建民、冯原、杨小彦、王维仁、符展成、林云峰、叶长安、黄专、朱荣远、汉斯·尤里斯·奥布里斯特、雷姆·库哈斯、斯蒂文·霍尔、奥雷·伯曼、伊东丰雄

策展团队

总策展人： 泰伦斯·瑞莱

总策展助理： Luna Bernfest

项目经理： 刘磊

项目经理助理： 杜量

项目策展人： Aaron Betsky，杜鹃，Jeffrey Johnson，姜珺，David van der Leer，李翔宁，Mary Ann O'Donnell，Jonathan Solomon，Rochelle Steiner，邓信惠，唐克扬，苏运升

国际邀请展策展人： Dietmar Steiner, Cristóbal Molina Baeza, Sebastián Irarrázaval, Hugo Mondragón, Martta Louekari, Sh. Mai Al Khalifa, Noura Al Sayeh, Fuad Al Ansari, Ole Bouman, Jorn Konijn

"大运与深圳"特别项目策展人： 朗图里外 & 李德庚

参展人

Stefan Al，业余建筑工作室，Aranda Lasch，大舍建筑，John Bennett，Gustavo Bonevard，Astrid Bussink，中建国际设计，Clavel Arquitectos，Coop Himmelb（l）au，David Chipperfield Architects，ELEMENTAL，John Ewing，Fake Industries Architectural Agonism，方振宁，非常建筑，Field Operations，Scott Fisher，Sam Green，Cao Guimarães，Hashim Sarkis Studios，Michiel Hulshof，J. Mayer H.，Johnston Marklee，Jeroen Koolhaas，刘家琨，马清运，MAD Architecture，Vincci Mak，Stefano di Martino，Mass Studies，Massimiliano Fuksas Architetto，Carmen Montoya，MOS，南沙原创，OBRA Architects，OMA，开放建筑，Miki Redelinghuys，Christopher Robbins，Daan Roggeveen，地方营造工作室，上海同济城市规划设计研究院，Surabhi Sharma，SO-IL，Spbr，Steven Holl Architects，Jennifer Stein，Studio Up，都市实践，Dre Urhahn，Tom Verebes，王功新，Woods Bagot Asia，维思平，筑博设计

志愿者团队

参展人志愿者：
林琳，王龙，徐丽，赵宇，杨欣欣，王少南，施春阳，赵宇，梁峰，孙瑛，杨天翼，郑慧怡，董治坚，徐亮，邴莹娇

项目志愿者：
何宇馨，连欣，苏亚辉，刘慧群，张超，林玉琳，张雅芳

场馆志愿者（参与时间超过八天）：
陈志鹏，魏炜琳，胡婷婷，赖筱雯，朱安琪，罗凯丹，廖兆彬，张韵，黄晓俊，幸涛，曾苑伦，李嘉琪，李福婷，苏粤宁，钟晓阳，彭丽萍，邬秋婷，李瑾，周美婷，陈婉虹，章敬义，陈若虹，范君健，陈颖怡，何晓君，罗凯耀，周晓娟，林婉珊，朱斌，李艺挺，黄丽娟，黄慧思，伍千慧，曾肖蔓，郑美婷，李坤铭，刘慧群，司徒栋梁，余明燕，张瑾瑜，黄磊，郭佩珊

办公室志愿者：
吴俊通，范宇健，杨佳国，王苑芳，李杨，陈浩，熊江朝，黎秋娜，郑晓雪，于露，陈楠，徐宁曼，林菊莲，陈慧玲

优秀志愿者名单：
陈志鹏，魏炜琳，胡婷婷，赖筱雯，朱安琪，罗凯丹，廖兆彬，张雅芳，林菊莲，张超，林凌，王龙，徐丽，赵宇，杨欣欣，王少南，施春阳，梁峰，孙瑛，杨天翼，郑慧怡，董治坚，徐亮，邴莹娇

编辑团队

编委会： 张宇星，周红玫，黄伟文，孙振华
内容统筹： Luna Bernfest，董超媚
特约编辑： 刘磊，刘潇（6<60 的六个影片介绍文字由张宜轩提供）
印务协调： 谢琼枝，孙粹

作品摄影： 陈尧，陈楠，Miljenko Bernfest，陈若愚
翻译： 贺聪，宫林林，罗媛，张静，袁文珊，韩丽
平面设计： wx-design

在整个编辑过程中，难免挂一漏万，还望诸同仁体谅！双年展筹备期间，获得众多朋友与同仁的支持与协助，在此深表谢意，难免差池，还望海涵。

Special Contribution / 特别贡献

Leading Sponsors / 主赞助商

Special Thanks to / 特别鸣谢

Shenzhen Municipal Bureau of Propaganda   深圳市委宣传部
Shenzhen Publicity Ministry for Culture Foundation Office   深圳市宣传文化事业发展专项基金领导小组办公室

Exclusive Telecommunication Service Provider 独家电信服务供应商		**Rest Area Design / 休息区设计** 宜家家居	

Institutional Supporters 文化机构支持	 

Venue Supporters 场地支持		**Designated Tea Break Service** 指定茶歇服务	**Printing Sponsor** 图像输出支持

Chief National Media Partners 首席合作媒体	艺术界

National Strategic Media Partners / 国内战略合作媒体

Exclusive Omnimedia Support 独家媒体支持	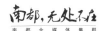

Chief Web Media 首席网络媒体		**Mini-Blog Support** 国内微博支持	**Exclusive LBS Cooperation Partner** 独家真实位置社区

Web Media Partners 网络合作媒体	

National Media Partners 国内合作媒体		**International media Partner** 国际合作媒体	designboom